the SUBTLE KNOT

the SUBTLE KNOT

EARLY MODERN ENGLISH LITERATURE *and the* BIRTH *of* NEUROSCIENCE

LIANNE HABINEK

McGill-Queen's University Press
Montreal and Kingston · London · Chicago

ISBN 978-0-7735-5318-7 (cloth)
ISBN 978-0-7735-5429-0 (ePDF)
ISBN 978-0-7735-5430-6 (ePUB)

Legal deposit second quarter 2018
Bibliothèque nationale du Québec

Printed in Canada on acid-free paper that is 100% ancient forest free (100% post-consumer recycled), processed chlorine free

Funded by the Government of Canada | Financé par le gouvernement du Canada | Canada | Canada Council for the Arts | Conseil des arts du Canada

We acknowledge the support of the Canada Council for the Arts, which last year invested $153 million to bring the arts to Canadians throughout the country.

Nous remercions le Conseil des arts du Canada de son soutien. L'an dernier, le Conseil a investi 153 millions de dollars pour mettre de l'art dans la vie des Canadiennes et des Canadiens de tout le pays.

LIBRARY AND ARCHIVES CANADA CATALOGUING IN PUBLICATION

Habinek, Lianne, 1979–, author
The subtle knot : early modern English literature and the birth of neuroscience / Lianne Habinek.

Includes bibliographical references and index.
Issued in print and electronic formats.
ISBN 978-0-7735-5318-7 (hardcover). – ISBN 978-0-7735-5429-0 (PDF).
– ISBN 978-0-7735-5430-6 (EPUB)

1. Neurosciences and the humanities. 2. Science in literature. 3. Mind and body in literature. 4. Medicine in literature. 5. English literature – Early modern, 1500–1700 – History and criticism. I. Title.

PR408.S34H33 2018 820.9'356109031 C2018-900995-0
C2018-900996-9

This book was typeset by Sandra Friesen in 10.5/13 Minion.

For Mom and Dad

Contents

FIGURES ix
ACKNOWLEDGMENTS xiii
INTRODUCTION 3

ONE
Untying the Subtle Knot: Anatomical Metaphor and the Case of the *Rete Mirabile* 35

TWO
Altered States: *Hamlet* and Early Modern Head Trauma 68

THREE
Labour Pains: William Harvey and the Travails of Conception 91

FOUR
The Mechanics of Reproduction in the Art of Cavendish 119

FIVE
The Bookish Brain: Moxon, Willis, and the Transformation of Flap Anatomy 153

CODA
The Brain of Dr Deijman 205

NOTES 215
BIBLIOGRAPHY 251
INDEX 273

Figures

I.1 "O wretched man that I am!" Hugo and Arwaker, *Pia Disideria*, 1686. 11

1.1 Head with skull and cranium removed, *dura mater* peeled back. Vesalius, *De humani corporis fabrica*, 1543. 46

1.2 Cranial dissection after Vesalius. Crooke, *Mikrokosmographia*, 1615. 47

1.3 "[T]he wonderful Net as Galen describeth it." Crooke, *Mikrokosmographia*, 1615. 60

1.4 The *rete mirabile* (B). "Arteria magna," Calcarensis in Vesalius, *Tabulae anatomicae sex*, 1538. 61

1.5 The circle of Willis (centre, surrounding E-E and Y-Y). Illustration by Wren. Willis, *Cerebri anatome*, 1664. 62

2.1 The cerebellum of a "foolish youth." Willis, *Five Treatises*, 1681. 76

3.1 Avicenna's ventricular localization. *De generatione embryonis*, dated 1347. 101

3.2 Top: The womb opened to show the inner membranes. Rüff, *The Expert Midwife*, 1637. Bottom: The cranium opened to show the cerebral cortex. Crooke, *Mikrokosmographia*, 1615. 107

3.3 Top: The womb at the time of the growth of the brain. Rüff, *The Expert Midwife*, 1637. Bottom: The brain dissected to show the blood vessels supplying it. Crooke, *Mikrokosmographia*, 1615. 108

3.4 "Athena born in a shower of gold." Maier, *Atalanta fugiens, hoc est…*, 1617. 110

5.1 Remmelin's "Adam," 1670. 154

5.2 Andreas Vesalius performing an anatomy. Vesalius, *De humani corporis fabrica*, 1543. 156

5.3 "Eve," whose abdomen may be opened. Vogtherr, *Anathomia oder abconterfettung*, 1544. 158
5.4 *Adam and Eve*, Albrecht Dürer, 1504. 161
5.5 A Vesalian Adam and Eve. Vesalius, *De humani corporis fabrica, librorum epitome,* 1560. 162
5.6 Adam and Eve following Vesalius. Geminus, *Compendiosa*, 1545. 163
5.7 Remmelin's Adam and Eve of 1613. 165
5.8 Dürer's serpent, from *Adam and Eve*. Dürer, 1504. 167
5.9 Remmelin's Eve of 1613. 168
5.10 Detail of Remmelin's Adam, 1670, skull flap lifted. 170
5.11 Detail of Remmelin's Adam, 1670, *dura mater* flap lifted. 171
5.12 Detail of Remmelin's Adam, 1670, "yard" flap lifted. 172
5.13 Left: Detail of Remmelin's Adam, 1702, skull flap lifted. Right: Original engraving from Willis, *De anima brutorum…*, 1672. 174
5.14 Left: Detail of Remmelin's Adam, 1702, pasted on verso of first flap. Right: Original engraving from Willis, *Cerebri anatome*, 1664. 175
5.15 Left: Detail of Remmelin's Eve, 1702, both cranial flaps lifted. Right: Original engraving from Willis, *Cerebri anatome*, 1664. 176
5.16 Left: Detail of Remmelin's Eve, 1702, between legs. Right: Original engraving from Willis, *Cerebri anatome*, 1664. 177
5.17 A progressive cranial anatomy. Valverde, *Historia de la composicion del cuerpo humano*, 1556. 178
5.18 Brain dissection and anatomy. von Gerstorff, *Feldbuch der Wundartzney*, 1528. 179
5.19 First cranial section. Estienne, *De dissectione partium corporis*, 1545. 180
5.20 A later cranial section, with the *dura* and *pia mater* removed. Estienne, *De dissectione partium corporis*, 1545. 181
5.21 A cranial anatomy with epidermis unfolded. I. Dumoulin, after Blanchin, 1679. 182
5.22 The fourth cranial section. Estienne, *De dissectione partium corporis*, 1545. 183
5.23 Joseph Moxon, with selection of his published works. Moxon, *Mathematicks Made Easie…*, 1692. 185
5.24 The 1704 catalogue of Daniel Midwinter and Thomas Leigh. 186
5.25 Title page of Remmelin's 1670 *Survey*. 192
5.26 Title page of Remmelin's 1695 *Survey*. 195
5.27 Top: Detail from Remmelin's 1702 *Survey*, showing a dissection of cardiac muscles. Bottom: Plate showing a cardiac

muscle dissection by Richard Lower. Lower, *Tractatus de corde*, 1669. 197
C.1 *The Anatomy Lesson of Dr. Deijman*. Rembrandt, 1656. 206
C.2 Rembrandt's original sketch (1656) for *The Anatomy Lesson of Dr. Deijman*. 207

Acknowledgments

As with every academic endeavour of this sort, the list of people to whom I extend my heartfelt gratitude is a long one. This book had its first life as my doctoral dissertation at Columbia University, but it drew even then on my work towards an MPhil at King's College, Cambridge University, and even more directly on my undergraduate studies in literature as well as in brain and cognitive sciences at the Massachusetts Institute of Technology. Indeed, an erstwhile friend from my undergraduate years unwittingly set this whole project into motion when she opined, when we were both studying at MIT, that literature touches the heart whereas her field of mechanical engineering (and by extension the hard sciences in general) touches the head; therefore the one was a mere pastime, the other life-saving and essential. My friend did not intend her characterization with any malice, and if you speak with her today she is likely to recant or soften her youthful theory. But the fact remains that my disquiet with the assumption that literature is somehow "lesser" because it doesn't engage the head has driven my desire to seek out means by which this field and neuroscience could converse amicably. I hope this book accomplishes at least that.

Anne Prescott, Pamela Smith, Julie Crawford, and Jenny Davidson read and offered invaluable commentary upon early drafts of this project – though Jean E. Howard was the driving critic of my initial work, and were it not for her, this book would not exist in any form. My debt to her for her unstinting support cannot be overstated.

I wish to thank as well the editorial team at McGill-Queen's University Press, in particular my editor Mark Abley, as well as the two external readers of this manuscript, whose insights and direction shaped this book very helpfully in its later stages. Thank you, too, to Ryan Van Huijstee, Kathleen Fraser, and especially Grace Seybold for seeing the manuscript through to its material realization; after their painstaking ministrations, all remaining errors here are entirely my own.

My colleagues at Columbia, Bard College, and elsewhere helped me develop individual chapters, offering new and exciting directions for my research. I would like to thank in particular Allison Deutermann, András Kiséry, Sharon Fulton, Brynhildur Ómarsdóttir, and Joanna Scutts; Elizabeth Holt, Hoyt Long, and Benjamin Stevens; Alex Benson, Celia Bland, Maria Cecire, Lauren Curtis, and Erica Kaufman; and Christian Crouch, Adhaar Desai, Tabetha Ewing, Collin Jennings, and Alice Stroup. The creativity and passion of my students at Bard College further contributed to how I framed this project. I am also grateful to Mary Thomas Crane, Laurie Johnson, Gail Kern Paster, John Sutton, and Evelyn Tribble; the advice and support received from them in conferences, seminars, and other academic venues was invaluable.

This book would not have been possible without the generous support of various fellowships and research stipends – and I certainly would not have received these without the keen practical sensibility of Susan Elvin Cooper, who guided me through a bevy of fellowship applications. In particular, my tenure at Wellesley College's Newhouse Center for the Humanities was essential for the time, space, and resources necessary to complete my book. I shall cherish forever the generosity, discernment, and general good will of my cohort there: thank you immensely to Arthur Bahr (in particular), Cristelle Baskins, Joel Burges, Brigid Cohen, Pauline de Tholozany, Brenna Greer, Jane Kamensky, Lesley Curtis, Wang Ao, and Kristen Williams. A special thanks goes as well to Carol Dougherty and Jane Jackson, who oversaw our time at the Center.

I would also like to thank the National Endowment for the Humanities and the Institute for Historical Research, for summer research grants. The sustained engagement with primary resources that drives *The Subtle Knot* I owe to the opportunity to work in a range of different libraries. I remain deeply obliged therefore to the Huntington Library in California, the Folger Shakespeare Library in Washington, DC, and Yale University's Cushing/Whitney Library for short-term grants, and in particular to their respective librarians for their expertise and wisdom. I would like to thank especially Melissa Grafe and her colleagues at Yale for their inestimable help with the images herein. This book also benefited from the knowledge and benevolence of librarians at the British Library and the Wellcome Library in London.

Chapters 1 and 2 had very early lives as essays on the literary review website *Open Letters Monthly*, and a version of chapter 1 appeared in that entity's first print collection as "Finely Woven Webs: On John Donne and the Seat of the Soul," in *Open Letters Monthly: An Anthology, 2007–2010*, edited by John Carter (Boston: OLM, 2010). These initial forays benefited from the editing of John Carter, Steve Donoghue, and Sam Sacks.

Expanded earlier versions of chapters 1 and 2 also appeared in the journal *Configurations* (20, no. 3, 2012: 239–77) and in *Embodied Cognition and Shakespeare's Theatre: The Early Modern Body-Mind*, edited by Laurie Johnson, John Sutton, and Evelyn Tribble (New York: Routledge, 2014), 195–215, respectively. I would like to extend my gratitude to the editors and anonymous readers at *Configurations* as well as to the editors of *Embodied Cognition and Shakespeare's Theatre* for their criticism and advice.

Finally, though last, not least, I thank my family: James and Darlene Habinek; Bethany and Andy Blair; and Shankar, Lila, and Mira Raman. Their guidance, support, and love will forever exceed my ability to justly pay homage. They are the stars to my wandering bark. To Shankar especially I owe my deepest gratitude, not only for reading and re-reading every iteration of this manuscript, but for salvaging my sanity in the process. I love you all, and it is my sincerest wish that this book will make you proud.

the SUBTLE KNOT

Introduction

> [E]nter I beseech thee, into the Sacred Tower of *Pallas*, I meane the braine of Man, and behold and admire the pillars and arched Cloysters of that princely pallace, the huge greatnesse of that stately building, the Pedistals or Bases, the Porches & goodly frontispice, the 4. arched Chambers, the bright and cleare Mirrour, the Labyrinthaean Mazes and web of the small arteries, the admirable trainings of the Veines, the draining furrowes and watercourses, the living ebullitions and the springings up of the sinnews, and the wonderfull foecundity of that white marrow of the back[.][1]

> The brayne in the syghte of man is of a wonderfull and marveylous substaunce to be consydred, and it is also very straunge.[2]

> Then forasmuch as they [poets] were the first observers of all naturall causes & effects in the things generable and corruptible, and from thence mounted up to search after the celestiall courses and influences, & yet penetrated further to know the divine essences and substances separate … they were the first Astronomers and Philosophists and Metaphysicks.[3]

This is the story of how history came to view the brain not simply as grey matter, but as a wealth of other wondrous possibilities: a book in which to read the soul's writing, a black box violently to be unlocked, a womb to nourish intellectual conception, a creative engine, and a subtle knot that traps the soul – and thereby makes us human. As these identifications suggest, the route traversed in this book winds not only through the scientific and the medical, but through the literary as well. Early neuroscience manifested itself, after all, not on a remote island of scientific thought, but in a cultural milieu for which the modern boundaries between disciplines were either porous or nonexistent. Consider, for example, my first epigraph, taken from the beginning of an immense early-seventeenth-century anatomical treatise by Helkiah Crooke. Crooke's prose is often heady, but this passage is particularly arresting for its array of allusions; he describes the brain in elegant metaphorical language as a grand sacred palace, that of Pallas Athena, goddess of wisdom. Architecture overwhelms biology, with the effect of rendering the brain as something we could walk through: by considering the arteries as a series of "Labyrinthaean Mazes" and the cortical folds as "furrowes and water-courses," we are presented with a vivid picture of an entity we may never have seen but can now visualize thoroughly. Such an understanding is the gift of the architectural metaphor, which joins and connects the immaterial to the material.

Yet for all of Crooke's metaphorical acrobatics, the brain was still, at that point, a mysterious entity: perhaps "wonderfull and marveylous," as another early anatomist, Thomas Geminus, puts it, but certainly "also very straunge." Prior to the discovery of forensic fixatives and preservatives, it is indeed wonderful and marvelous that anatomists were able to discern anything at all from what would have presented itself as a fast-decomposing pile of mush.[4] But the early modern period obsessively sought to dispel the organ's strangeness; it testifies to an impetus to study and comprehend the brain, stimulated by the desire to map the soul onto the body in order the better to understand both – and, despite its transformations, this desire is arguably recognizable in modern neuroscience as well. Part of the work of this book will be to do as Crooke does: to make the birth of neuroscience readable by teasing out lines of continuity between the early modern period and our own time. As with the brain itself, much will seem foreign, but other, often unexpected, concordances will emerge to cast modern thought in a new light. For instance, the seventeenth century evinces a kind of anxiety about the location or the locatability of the soul that sometimes appears remarkably similar to our modern attempts to localize, say, semantic memory in the brain's hippocampus or emotion in the limbic system. At the same time, the brain did not belong to science and medicine alone. As an increasing number of early modern thinkers argued, the location of the soul was the human brain, making it not only the site where body and soul connected, but also where various domains of knowledge – including, of course, literature and theology – productively converged.

Geminus's words suggest the ambivalence with which early modern thinkers approached the brain, admiration contending with incomprehension. Both "wonder" and "marvel" were loaded terms in the early modern period, expressing fascination as well as a particular pull towards empirical investigation. As Lorraine Daston and Katharine Park note, natural philosophers in the early modern period (such as Robert Boyle, who confesses to wonder upon examining a glowing joint of meat) "saw wonder as a goad to inquiry, and wonders as prime objects of investigations"; indeed, such a focus "on wonder and wonders in the study of nature marked a unique moment in the history of European natural philosophy, unprecedented and unrepeated."[5] The brain becomes – for Geminus, for Crooke, and, as we shall see, for so many others – an object of intellectual curiosity, distinct from and superior to the rest of the body, rather than a passive participant in the body's systems (or, as earlier theories had it, subordinate to the heart or even the liver). It will be marked out variously as the handmaiden to the soul; or the house of the soul; or the prison of the soul; or the foundation of the soul's hydraulic control centre. That the brain must interact with the soul will mark out the organ, too, as the crucial point where the material and the immaterial touch; the brain thus takes on the role of metaphysical testing-ground, its architecture crucial for explaining and exploring problems of consciousness, identity, and intellect.

But Geminus also remarks on the brain's strangeness, asserting that there is something alien about this entity that lies at the very core of our sense of self. Connoting foreignness, "strange" in the early modern period implies an external presence, as if the brain were something that had come from elsewhere, from a place distant from one's self. This "substaunce," then, is at once a marvel that provokes scientific inquiry (because of its fundamental position in the construction of the self), as well as something whose presence within us demands explanation and analysis. Of course, Geminus is speaking of the brain of the cadaver he is (ostensibly) examining and not his own brain. The fantasy persists, however, then and now, that one might be able to witness one's own brain in action, to see the soul's spark and thereby understand exactly how it interfaces with the substance of one's own body.

My third epigraph presents the words of the early modern literary critic George Puttenham, and may seem at first not to follow from the other two, which aim to describe the brain. What place have poets in relation to scientific discourse? One answer lies in the distinctive ways in which the early modern period approached the creation of knowledge: not just as the product of scientific inquiry but, perhaps less obviously, as an object of poetic inquiry. Elizabeth Spiller succinctly explains why Puttenham's assertion might seem odd to a modern eye: "The interval that separates us from early modern understandings of how knowledge is made," she writes, "can be charted in the shifting history of terms such as 'fact,' 'experience,' and 'experiment,' as well as 'natural philosophy' and 'science.' For us, facts typically point towards questions about what is, but for early modern readers they instead mark that which is made."[6] Poetry is inherently something crafted; so, as such, it has particular value for the way knowledge can be constructed and conveyed. Poetic devices enable that knowledge to be retained because, as Puttenham writes elsewhere, "it is briefer & more compendious [than ordinary speech], and easier to beare away and be retained in memorie."[7] Poetry can condense and consolidate, the better to mark and advance scientific fact.

Puttenham also credits poets as "observers" who "moun[t] up" and "penetrat[e] further," uncovering and creating the facts that help construct scientific disciplines. The knowledge for which the poets search ascends as they do, so that they might map out the whole of the universe – which, after all, is related to the human being as macrocosm to microcosm. In this, Puttenham draws closer to Crooke, who points out that "what the Heaven is in the worlde, the same in man is the Braine … Heaven is the habitation of the supreame *Inteligence*, that is of God; and the Braine the seate of the *Soule*, that is the *demi-God* of this *Little-world*." The analogy uncovers the drive in the early modern period to map out the brain and its functions: anatomizing this "demi-God" raises us up to a better understanding of God.[8] Moreover, mapping the soul onto the body satisfied early modern anxieties about and desires for a specific location in which the soul could be found. If only you could pinpoint the place where immaterial soul interacted with material flesh (the reasoning went), you could then have a chance of examining the soul empirically – and, if not that, you could at least anatomize the structure and function of the body part that did interact with the soul.

Prior to the early modern period, debates raged about the soul's house in the body: was it the brain? The heart? The liver? Was it something you could not actually localize, infused instead throughout the body? Though they had not faded away entirely, these controversies had largely been resolved by the beginning of the seventeenth century, as the brain had been broadly accepted as the "palace" of the soul. Yet with the theory of a neurocentric soul came a new set of problems. Did the soul reside in the cortical matter, or did it swim about in the ventricles? Perhaps it pushed the pineal gland this way and that, to control the flow of not-quite-immaterial spirits through the nerves. And if the soul used ethereal spirits to interface with the body, where were these spirits manufactured? How did the faculties of the soul – commonsense, cogitation, imagination, and memory – manifest themselves in the brain? How did thoughts form, and what were the accompanying physiological changes that occurred in the brain?

As currently taught and practised, modern neuroscience deploys a wide variety of metaphorical relationships to describe the workings of the cellular processes that create the mind itself: *neural plasticity* and its associated changes may be evident in a *neural network*, while the activity of *mirror neurons* suggests the importance of imitation for learning, and *coding* describes a connection between sensation and perception. Yet the study of science nonetheless sets itself apart from the study of metaphor (and by extension the literary). Little attention is paid to these descriptions as metaphors or to the fact that metaphoric language remains necessary for us to understand complex and abstract systems. The seventeenth century, on the other hand, made explicit the need for metaphor: its success depended upon the intertwined labours of literature and science to understand the relationship between the soul and its ostensible home in the body, the brain.

The Subtle Knot offers a careful consideration of how such metaphors worked as figures through which abstract ideas about the soul and mind could be made to cohere with the physical form of the brain – and, fascinatingly, vice versa, since the early modern period aimed to chart the soul with map-like precision but often found itself faltering when describing the strange grey matter that made up the physical organ. The five metaphors covered in this book – the net/knot, the cut/lesion, the womb, the writing machine, and the book – present ideas that appear simple, but upon fuller consideration open up multiple connections between a vibrant anatomical discourse and an equally animated literary one. Emerging from a cultural milieu that passionately longed for some clear link between body and soul, they describe a brain alive and lively, extended into its environment, and fully connected both to an abstract concept of the mind or soul and to the physical body itself. The metaphors trace, too, a gradual, uneven shift away from an identification of the soul with the brain; the mind becomes the marker of the individual, setting up or anticipating what is now the standard assumption in neuroscience.

After sketching the physiology involved in the construction of the early modern soul and body/brain and examining the role metaphor played for writers in the period, this Introduction turns briefly to modern theories of grounded or extended cognition. These theories share the claim that psychological and physiological states are interdependent, to the extent that mental representations are seen as depending crucially on bodily experience. Metaphor becomes a crucial means of traversing this mind/body connection; its study reveals, as George Lakoff and Mark Johnson famously proposed, that much of our abstract thought depends at its core upon lived physical states. At stake in *The Subtle Knot* is the notion that a distinctive set of metaphors enabled the interdisciplinary communication necessary to localize the soul in the brain – and indeed paved the way for later attempts to derive the mind from the brain's workings.

What Is a Brain?

One vivid answer to this question surfaces in a literary episode that considers the consequences of the organ's absence. Ludovico Ariosto's epic poem *Orlando Furioso* contains a scene in which unrequited love has driven its eponymous hero *furioso*, to the extent that he sighs, "I am not who my face proclaims me; the man who was Orlando is dead and buried … I am his spirit sundered from him."[9] Orlando's lines oddly reverse a more typical scenario, in which, upon death (even if metaphorical), the body rather than the spirit would be the earthly remainder. Clearly Orlando himself is still embodied, his melodramatic argument notwithstanding. But why should he identify himself with his spirit, which has been sundered from – what, exactly? Because of his love-madness, Orlando's identity has been split across his body and his spirit; his body remains "buried" under his emotional pain, while his spirit roams the forest apparently on its own. In order to unite body and spirit, a crucial third entity must be brought back into the picture.

This, as it turns out, is Orlando's brain, which has (through various turns of magic) ended up on the moon. Orlando's compatriot Astolfo ventures up there in an attempt to cure his friend's lunacy, only to discover a world full of various half-realized metaphors. "[E]verything that was lost on earth … collects up there," an apostle assures him, and indeed he encounters in the lunar realm "a lofty pile of tumid bladders from which seemed to emanate a hubbub of cries," "a heap of gold and silver hooks," "[v]erses written in praise of patrons [in] the guise of exploded crickets." This farrago seems to be the connective tissue tying one concept to another, but in the absence of the concepts themselves, all connections to referents and meanings languish, lost.[10] Eventually, Astolfo encounters "the substance which, it seems, is so innate in us that never were prayers offered to God for its possession: I mean brains,"[11] and in particular Orlando's brains, the means by which the hero's body and spirit might be reunited.[12]

Curiously, John Harrington's translation of this episode, which circulated throughout England in the late sixteenth century, deviates from the Italian source at one point. While his rendition into the English vernacular tends to cleave more closely to the original Italian than its modern counterpart, it nevertheless teases the reader with a vivid transposition. What Astolfo finds in Sir John Harrington's *Orlando* is "[T]he thing which no man thinks he needs, / Yet each man needeth most, to him [Astolfo] was showne, / By name man's wit [*senno* in Italian]."[13] So far, so reasonable: a modern translation renders the *senno* the *brain*, but Harrington's *wit* is probably closer to the original meaning. The real difference lies in how the material organ is described: where the modern English translation refers to the brain as "a soft, tenuous liquid, apt to vaporize if not kept tightly sealed" – and the Italian concurs: *un liquor suttile e molle* – Harrington refers to it as "[a] body moyst and soft,"[14] thereby solidifying a liquid/liquor into an actual body.[15] While "body" has a variety of connotations, the term, then and now, generally evokes the idea of solidity rather than fluidity, so it is noteworthy that Harrington should depart from the original on this particular point.[16] For though he does not mention the brain directly, his description of "wits" in these terms points to the particular physical and metaphorical properties of the brain: its softness and moistness (whether in reference to organ or to wit) imply that it is malleable enough to receive and integrate sensory impressions. Wit, then, and by extension Orlando's very identity, must for the early modern English reader of the epic depend upon a *body* that could unify the corporeal information received through the senses with the intangible sense of self that emerges thereby.[17]

One of the intriguing connections between early modern conceptions of the brain and our own sense of cognitive processes may be found in the desire to see the brain as this site of a crossing, between the immaterial and the material, between culture and individual body. For many contemporary critics who engage with neuroscience, culture is understood as shaping personal identity and thus affecting how the brain acts and what it is capable of doing; conversely, the brain's integration of sensory experiences fashions that very culture. As Alan Richardson and Francis F. Steen note in a piece describing cognitive literary criticism, "To construct culture, human beings intimately rely on immensely complex bodies, nervous systems, and sensory systems; these structures have a history that is neither identical to nor separate from the culture that they make possible."[18] The through-lines connecting the early modern brain to the modern brain will help expand this milieu beyond the seventeenth century, with the emerging picture being one of continuity rather than – as a two-culture version of history would have it – a series of breaks. Modern cognitive science is well poised at this point in time for such a broader historical outlook, given that, in Mary Crane's words, as a discipline it "provides increasingly convincing evidence that the body does shape thought and language, that the early experiences of living in the body are the armature on which consciousness and thought are formed."[19] In this book, we will consider how such construction works and what the nature of its history is.

To do so, we ought first to ask what a brain actually is. Today, the answer may seem obvious (if not necessarily definitive): the brain is the centre of the nervous system, an organ comprising billions of neurons in complex interconnections, the organ responsible for bodily functions and mental operations whether conscious or unconscious, the organizing principle of mind and identity.[20] Yet early modernity delighted in describing the brain in terms other than the purely physical or functional; we have just seen one example in Crooke's traversing of the soul's palace. For this period what the brain *does* and *is* remains intimately bound up with *how it can be described*.[21] The brain must also ascend to its role as the seat of the soul, and eventually the mind, against a number of other organs contending for this title. Describing the historical divisions in thinking about the brain versus the heart as potential locations for the soul, Scott Manning Stevens points out that insofar as the heart takes prominence in Christian iconography it does not work metaphorically; rather, "Christ's heart operates as the reification of his human, fleshly, and mortal body."[22] The question about what he terms the "locus of self" in (and prior to) the early modern period lies in why, say, Christ's selfhood could be identified with his heart "precisely at a juncture in history when medical science is coming to recognize the brain as the center of the human self."[23] One way of approaching an answer lies in reconsidering the brain's openness to metaphor precisely on the basis of the difficulty of grasping what exactly it did and how it functioned; while the heart becomes tied to a specific locus and a specific function, the brain in the period seems to expand to take over the myriad operations that comprise the mind. Each operation, to be comprehended, needs in turn to be metaphorized, so that the brain takes on increasingly complex and bizarre identities in a way that the heart perhaps could not.

What the brain can do is considered, in the early modern period, as a function of its interaction with the soul. The soul itself can be best

I.1 "O wretched man that I am!" Herman Hugo and Edmund Arwaker, *Pia Disideria, or, Divine Addresses in Three Books* (London, 1686), 218. The Huntington Library, San Marino, CA. RB 111626.

construed, for early modern natural philosophy, as a collapse of multiple terms: it combines theological considerations of immortality with philosophical renderings (as of the rational or intellectual soul), as well as medical notions, as of the sensitive (i.e., sensing and perceiving) and organic (i.e., driving the force of life) soul.[24] Its offices could include operations of the lower order, such as digestion, as well as higher-order functions, such as memory. The relationship between the soul and the body was often a fraught one, as the eponymous speakers in Andrew Marvell's poem "A Dialogue between the Soul and the Body" make clear. "Oh who shall, from this dungeon, raise / A soul enslav'd so many ways?" cries the Soul, while the Body poses a similar and similarly histrionic question: "Oh who shall me deliver whole / From bonds of this tyrannic soul?"[25] These were entities ever striving with one another, often portrayed in popular literature as foes joined cruelly (but necessarily) to create the human being. Illustrations of the soul inhabiting the body drove home forcefully the sense of imprisonment. Figure I.1 presents a view more sympathetic to the soul: cast as a pious petitioner, the soul grasps the ribs of the skeleton, sign of an always-already-decaying body, as one would the bars of a prison cell. The verse, taken from Romans 7:24, depicts man's spiritual striving to overcome the sins of the flesh; the soul, tainted by the original sin, can only pray for deliverance from its jail. The body, on the other hand, could be just as tormented by its inmate, as earlier debate poetry, the antecedent for Marvell's "Debate," makes clear.[26] Imprisonment was, then, one key, if draconian, metaphor for understanding the relation of soul to body.

Yet there were more positive ways of depicting even this particular metaphor. As Crooke suggests:

> For seeing the soule of man being cast into this prison of the body, cannot discharge her offices and functions without a corporeall Organ or instrument of the body; whosoever will attain unto the knowledge of the soul, it is necessarie that hee know the frame and composition of the body.[27]

Crooke acknowledges the problems each entity poses for the other but creates a tense truce: the body is necessary for the soul to exist in the world. Casting the body/soul problem in this light allows him to authorize and clarify the study of anatomy: we study the body so that we may know the soul. This impulse leads to the metaphor cited earlier, likening the brain as "seate of the Soule" to God's habitation in relation to the earthly world. For Crooke and others, the brain, being the house of the soul in the body, is metaphorically related to heaven, and the soul to God. Here, too, a variation on the metaphor of the body as the soul's jailor renders intelligible deeper and more complex scientific, philosophical, or theological truths.

But if we wish to make the body the focus of our study, with the aim of better understanding the soul, we need first to grasp how the body was put together in the late sixteenth and early seventeenth centuries. Humoural theory – the notion that the body comprised four humours, blood, black bile, yellow bile, and phlegm, and that these interacted in various ways to produce both character and disease – had largely lost its hold over the anatomical community, but remained an important residual presence nonetheless, especially useful for physicians and surgeons who used it to diagnose illnesses and prepare medicines. Often ascribed to Hippocrates, humoural theory operated via balances, such that if any one of the humours was found either in abundance or in scarcity, the person's attitude, inclinations, and physical health would be affected. Different organs were responsible for the production and regulation of each of the humours, and each humour was also constituted as the combination of two points from the hot/cold and wet/dry binaries: the heart was associated with blood (hot and wet), the spleen with black bile (cold and dry), the liver with yellow bile (hot and dry), and the brain with phlegm (cold and wet). Perfect health required that all four humours exist in equal proportions, and to that end, doctors prescribed diets and behavioural regimens to help their patients attain this balance.[28]

At the same time, however, anatomists offered another model. Though it coexisted with humoural theory, their description of the body offered an alternative organizational structure that eventually came to be the dominant means of representing the interaction between soul and body. This theory, derived from Platonic and Aristotelian tradition and refined by Galen, posited the existence of three distinct sections (venters) of the human body, each presided over by a central organ. The liver ruled the first venter, the heart the middle venter, and the brain the highest venter. These organs were responsible for manufacturing *spirits*, which in turn governed the activities each venter could perform. These spirits were of a mean between the body and the soul; they were physical entities that, upon their production, were refined as they passed through each venter. In one of his sermons, John Donne describes the spirits as "the thin and active part of the blood … of a kind of middle nature, between soul and body," and these "do the office, to unite and apply the faculties of the soul to the organs of the body."[29]

Anatomists relentlessly theorized just what the mechanism for this union could be. Central to the process was *pneuma*, which in this context is that portion of the air drawn into the body as breath. In broad strokes: food ingested by a person would be digested in his stomach, with the useful elements sent on to the liver to be turned into blood and mingled with *pneuma* to create the vegetable spirit. The movement of this spirit allowed the person's body to carry out the basic operations of life – growing, digestion, reproduction – in short, those biological activities common to humans and to plants (vegetables). A portion of the vegetable spirit would ascend to the second venter, where, in the heart, it mixed with more *pneuma* and blood to be refined into the vital spirit. This spirit, thinner and more active than the vegetable, moved the body, allowed for aspiration, enabled rudimentary perception, and helped pump blood around the body, activities humans and animals shared. The vital spirit then rose to the brain, where it was further purified to become the most crucial of the spirits, the animal spirit. *Animal* here refers to the *anima*, or the soul, for it was this spirit that enabled higher-order operations including cognition, imagination, and memory.

This theory, though widespread, was not without its controversies, the most important of which I will treat in chapter 1, "Untying the Subtle Knot: Anatomical Metaphor and the Case of the *Rete Mirabile*." Where and how, exactly, did the transformation of the vital spirit into the animal spirit take place? Potentially, answering this question would enable the soul to be localized, since that portion of the brain where the transition took place would also be the point of contact between body and soul. Here, somewhere in the brain, would the faculties of the soul be united and applied to the organs of the body. Chapter 1 focuses on the metaphor of the *knot* or *net* – central in the period to describe the physical relationship between body and soul – which found its physical expression in a minuscule structure posited as existing in the brain: the *rete mirabile*. This "wonderful net/knot" was held by seventeenth-century anatomists to be the actual location where soul and body converged. Thus the *rete mirabile* became useful both as a physical marker of the desire for a locatable soul and as a metaphor for the complex interaction between material and immaterial. With this dual role, anatomists at once charged the network of vessels with a sense of mysticism (*mirabile*) and marked it as an elegant (non-)solution to the problem of material interaction (with a web, the soul could be everywhere and nowhere at once).

While the early medical authority Galen (second century AD), drawing on prior Classical and Arabic texts, insisted that the *rete* explained the physical interaction between soul and body in man – and his disciples in the following centuries followed suit – the revolutionary anatomist of the mid-sixteenth century Andreas Vesalius (rightly) submitted proof that the *rete* did not in fact exist in humans at all. Vesalius's magisterial *De humani corporis fabrica* quickly spread throughout Europe in a host of pirated editions. But English publishers made one crucial change (though they copied wholesale virtually everything else): whereas Vesalius rejected the *rete*, English writers kept the structure where it was, insisting on its centrality to their model of the human body.

Why this substitution took place is the question at the heart of the first chapter. An answer arises in the form of the powerful imagery evoked by the idea of the net or knot. Especially in metaphysical poetry, the net could signal anything from intricate bindings to a tool for capturing and jailing the soul within the confines of the body. The chapter's most extensive reading is of John Donne's poem "The Ecstasy," in which two lovers discover that, though their souls mingle beautifully, their love is nevertheless enabled by the connection between their bodies – "because," says the speaker of the body's blood, "such fingers need to knit / That subtle knot, that makes us man." The *subtle knot* thus represents the mysterious but necessary bond between soul and body, which found its metaphorical incarnation – and made necessary the belief – in the *rete mirabile*.

Un-culturation

While histories of neuroscience have become widely available in recent years, they tend to focus on medicine and anatomy, often not bothering to consider the broader cultural context in which important discoveries about the brain emerged.[30] Too frequently, these works emphasize more recent developments in the study of the brain – and, indeed, with good reason, since it was only with advances in microscopy, histology, and staining techniques in the late nineteenth century that individual neurons could be isolated and examined, a feat impossible in earlier periods. Yet many contemporary theories about how the mind functions in its physical instantiation are nonetheless underpinned by major "non-scientific" theories – for example, distinctions between perception and memory, which originated in philosophical texts and reveal a lineage that runs all the way back to Plato and Aristotle. Having been proposed well before the most studied technological innovations, these contributions from outside the domain of what we call science today are usually ignored by works that focus narrowly on neuroscientific history. What *The Subtle*

Knot does *not* do is project a history of neuroscience solely in terms of the scientific. One of this book's main tenets is that we need to take seriously the fact that for early modern writers there was no such thing as an isolated scientific discourse. Indeed, the work of Bruno Latour and others in contemporary science studies shows that it is doubtful, despite all appearances to the contrary, that such isolation can be insisted upon even today.[31] At any rate, this book focuses on a prior cultural milieu that has shaped our modern understanding of the brain, one in which multiple discourses operated, influencing and interacting with one another in ways that are difficult to conceive of now, when even inter- and cross-disciplinary work assumes that relatively stable distinctions among disciplines do in fact exist. But for the early moderns, to isolate any one thread of the interwoven discourses is to ignore the rich background in relation to which that thread becomes visible, and it is the job of *The Subtle Knot* to offer a glimpse of the whole tapestry.

Thus we will begin here by assuming that there were no functional barriers between literature and science in the early modern period – or that, to the extent these were emerging, the gaps were easily traversed by such writers as Crooke, John Donne, Shakespeare, William Harvey, and Margaret Cavendish. This approach allows us to bypass contemporary disciplinary boundaries (because they did not yet exist and have been imposed upon the period from the long view of history), and thereby also to short-circuit, as it were, the modern problem of the "two cultures."

For in the background of this book lies another debate – now almost six decades old – that has become relevant again in our time: C.P. Snow's famous articulation in the 1950s of the difference between the so-called "two cultures" of the humanities and the sciences. More recently, that characterization has evolved and been further developed through "third-culture" writers, whose work puts literary and philosophical texts into play in part to allow scientists to communicate with a lay public. Psychologist-turned-grammarian Steven Pinker has often been singled out, and rightly so, as a (problematic) standard-bearer of this new culture; along with him we might name V.S. Ramachandran, the late Oliver Sacks, and Antonio Damasio. These writers explore literary and artistic texts through a scientific lens in the name of a novel form of interpretation. But even as third-culture writing claims to bridge the humanities and the sciences, its main representatives thus far have been scientists who speak about the humanities as though its texts were transparent historical documents rather than creative artifacts in their own right – disregarding, that is, what is artistic about the arts and what is human about the humanities.

Casting back to a time when the arts and sciences were intimately intertwined, *The Subtle Knot* abandons as flawed the notion that third-culture thinking can paint an accurate portrait of the inception of neuroscience. It argues instead that we must retrace the common ground from which our modern understanding of the brain first grew.

In a 1959 Rede Lecture at the University of Cambridge, C.P. Snow (in)famously identified two distinct cultures into which his academic acquaintances seemed to fall. While ostensibly equivalent in intelligence and work ethic, the two cultures nonetheless refused to communicate and often regarded one another with open hostility. Snow illustrates what he terms this "mutual incomprehension" by an anecdote: at a dinner party populated by humanities scholars, he asks his friends to describe the Second Law of Thermodynamics:[32]

> The response was cold: it was also negative. Yet I was asking something which is the scientific equivalent of *Have you read a work of Shakespeare's*?
>
> I now believe that if I had asked an even simpler question – such as, What do you mean by mass, or acceleration, which is the scientific equivalent of saying, *Can you read*? – not more than one in ten of the highly educated would have felt that I was speaking the same language.[33]

Now, more than half a century later, it is likely that scientists can read, and can read Shakespeare, just as it is likely that more than one humanities scholar has worked out what the Second Law of Thermodynamics is. Nevertheless, each domain possesses its own internal cultures, insofar as we take *culture* to imply particular shared modes of thought, uses of language, sets of references, standard practices, and the like. What Snow fails to recognize, though, is that being able to read, and being able to read Shakespeare, are fundamentally different from the actual work of literary analysis; had he posed the question as "Can you perform a close-reading?" or "Can you rationalize Hamlet's 'To be, or not to be' speech as a celebration of commonplace book culture rather than as a meditation on suicide?" he might have hit closer to an equivalency of terms. However, even modified in this way, his query still leaves us with the problem of an apparently impassable barricade separating the two cultures. Can science and literature ever reach out and touch one another?

Theorists of the so-called third culture imagine this to be possible, but in a manner limited by the same oversimplification evinced in Snow's example. These thinkers examine the humanities from the standpoint of the sciences, and generally treat their literary examples as transparent artifacts to either prove or disprove a scientific point.[34] They have made crucial inroads into our modern understanding of the brain and how it grapples with artistic problems, and have disseminated these ideas to an eager, broader readership. But it nevertheless remains that, because these writers position themselves as scientists crossing over and laying claim to the humanities, their work relies on the very cultural divide they claim to bridge. Indeed, it is not unfair to characterize their work as taking what Steven Shapin and Simon Schaffer memorably term the "stranger's viewpoint"; they are outsiders who know they are outsiders, and while this perspective lends their readings a freshness, it also bars them from the sort of understanding that comes from prolonged and serious engagement with the tools and texts central to the humanities.[35]

To establish a genuine conversation between the sciences and the humanities therefore requires not just that humanists truly engage with science but that scientists give up the assumption that the humanities are a simply a window or a transparent plane revealing a view of a point in history, and nothing more. The particular relationship between literature and neuroscience has been especially fraught in this regard, often succumbing to the temptation that understanding how the brain works will help us to "solve" various problems of literature. If we only unpack the mechanisms underlying the translation of olfactory and visual data into sense-memory through its processing in their respective cortical centres (the reasoning goes), we could appropriately analyze the symbolism of the iconic madeleine that so consumes Proust's narrator. It is no mistake that the opening of *Swann's Way* is a favourite among cognitive neuroscientists, for it seems to offer up Marcel's experience of involuntary memory as just such a (delicious) artifact ripe for analysis.[36] No doubt, that often-cited episode yields fruit as a useful documentation of subjective human experience, and scientific fascination with it has sparked an entire realm of memory studies. But to reduce its significance thus is equally to strip *Swann's Way* of its status as literary text by rendering it equivalent to a similar statement one might evoke in a clinical memory test.

The "fourth" culture emerging in response to such reservations is less well delineated. One idealistic suggestion is that research and writing in the fourth culture may already transcend the apparent rift between the two cultures; popular science purports to demystify the intricacies of the discipline far enough to allow it to usefully "read" the humanities. Setting aside the evident condescension in such a move (why must science be made legible, but not the humanities?), the key problem with this formulation is that it replicates the means of studying the humanities favoured by third-culture writers. That is, it continues to treat poetry, artwork, music, and so on as unambiguous exemplars that science can somehow explain (or, vice versa, that can somehow be used to explain science).

A better and more useful rendering of the fourth culture comes from an emerging intersection of literary study and cognitive theory, as is found in the work of scholars deploying extended or embodied cognition to unpack concepts of language use and selfhood. The work of such writers as Crane, Richardson, Gail Kern Paster, John Sutton, Laurie Johnson, and Evelyn Tribble (on the literary side), and Lawrence Barsalou, Anna M. Borghi, R.W. Gibbs, Arthur M. Glenberg, Jessica K. Witt, and Janet Metcalfe (on the neuroscience side) promises a connection and a new culture where prior members of prior divisions are placed on increasingly equal footing. Starting from an idea of a mind that, as Crane and Richardson put it, "because of its biological and embodied features, is inextricably implicated with thought processes that seem emotional [and] metaphorical," such critics emphasize that "[t]hought and its representation in language ... are shaped not solely or even primarily by cultural forces but also by the innate and universal physical parameters of our bodies and brains, as we attempt to make sense of and successfully negotiate what surrounds us, in nature and culture."[37] At the same time, work in studies of brain activity patterns suggests that "language comprehension is a simulation process ... [W]ords and phrases are transformed into a simulation of the situation described ... in neural systems normally used for action, sensation, and emotion."[38] Now, this description should not lead us to such tasks as monitoring neuronal activity using an fMRI while a subject reads Jane Austen: the value of such experiments remains doubtful, as there are far too many variables to account for, even in the simplest iteration of such an experiment, and one of the consequences of theories of grounded cognition is the recognition that structure and function are ever more deeply entangled.[39] Such reductionism aside, it remains the case, though, that such studies of literature have drawn productively on modern knowledge about the brain's functions to offer new insights into the cognitive dimensions underpinning literature and interpretive practices.

But the focus on applying the results of modern neuroscience to literatures of the past has also led to an insufficient recognition of contributions arising from the rich and bizarre history of neuroscience, as well as of the connections that still bind the modern discipline to its pre-modern pasts.[40] It is here that this book hopes to offer a corrective. *The Subtle Knot* engages both neuroscientific history and literary studies on the common ground of the early modern period to argue that they are necessarily linked in a narrative about how the mind came to be identified with the brain. It suggests that the centrality of shared and connected metaphors in both literary and scientific contexts enables a bridging of the modern gap between the two disciplines – and reveals how intertwined they were in the early modern period. To witness the birth of neuroscience, we must first recognize that the emergence of the study of the brain involved understanding drawn from numerous fields, including theology, natural philosophy (the precursor to empirical science), philosophy, literature, art history, and medicine. These fields cited, borrowed from, overlapped, stole from, and understood one another on a fundamental level that would be perhaps inaccessible to us today. To be certain, they were relatively discrete areas of study, but their boundaries were blurred more often than not: Donne, remembered primarily as a poet, became dean of St Paul's Cathedral and wrote some of the most elegant and moving sermons in the English language, and reveals in all his writings a keen interest in the scientific thought of his day. Margaret Cavendish, the subject of chapter 4, published her bizarre fictional utopia *The Description of a New World, Called the Blazing World* bound together with her *Observations upon Experimental Philosophy*; she uses both modes of thought (the poetic and the philosophical) to advance a critique of the overenthusiastic empiricism practised by the Royal Society. Both Donne and Cavendish proposed ideas about the relationship between the soul and the body, and the soul's specific location in the body, their writings spurred by and drawing upon contemporaneous studies of the brain.[41]

Thus I want to suggest the need for a kind of culturelessness. Instead of worrying about (and away at) the distinctions between disciplines, we will take their always-already-being-connected as a given in the early modern period. The ways in which the individual chapters in this book function will hopefully make clear the need for such culturelessness in understanding how this formative period of neuroscientific history engaged with the brain. For example, chapter 2, "Altered States: *Hamlet* and Early Modern Head Trauma," considers the metaphor of the *lesion* or *cut*, as a figure that marks trauma to the brain but also enables a form of neuroforensics, the art of understanding usual or normal functions by investigating the damaged or impaired brain. As I argue, *Hamlet*'s preoccupation with the brain can be connected to a technique pioneered by Galen, whose work with head-wounded gladiators revealed the first connections between the brain's structure and its mnemonic functions. At the same time, the metaphor looks ahead to the most famous modern instantiation of this method, under the auspices of what is now called the lesion study: the case of Patient H.M. (Henry Molaison), whose bilateral lobectomy in the 1950s led to the current conventions for mapping the transfer of short-term to long-term memory, and laid the basis for various other memory categorizations.

Connecting Galen to contemporary neuroforensics, the guiding metaphor of the *cut* or *lesion* allows us in turn to make sense of the literary bases of memory. Hamlet, for example, infamously pledges to wipe from his memory "all trivial fond records … That youth and observation copied there" so that the Ghost's commandment (to revenge and/or to remember) will live alone within the "book and volume" of his brain. Hamlet's erasure constitutes an act of violence, a self-inflicted brain-wound that subsequently shapes his memory and his madness throughout the play: every mention of Hamlet's brain has to do with that organ being physically or metaphorically problematized in some way. This chapter proposes that the roots of Hamlet's violently willed oblivion – a sense of the damaged brain shared by other Shakespearean tragedies, notably *Macbeth* – may be traced back to the history of treatment for head trauma, and that the play opens a productive link between that medical past and present-day brain lesion studies. (As Hamlet's pledge suggests, the early modern period also makes the crucial step of likening brains to books and memory to the act of writing, thus anticipating the topics of the last two chapters of this book – but, as this chapter also shows, neither writing nor remembering was an innocent act.)

A broadly accepted account of neuroscience today reads as a conventional narrative of scientific progress; it sketches a trajectory from soul to mind, showing how the brain was progressively delinked from the soul in order to be merged instead with the mind. From *pneuma* to electrical impulse, we see the slow crawl from faith-tinted belief in historical authority to scientific empirical demonstrability and reproducibility. The story involves casting off the shackles of apparent authority as precondition for the emergence of the independent spirit of the newly created investigator. The form of this emancipatory narrative is itself familiar: after all, a similar story was told and retold several times over in the early modern period itself, most obviously in the clash between Vesalian and Galenic models of anatomy, but also more insidiously in Descartes's supposedly definitive severing of soul from body. Yet such descriptions overlook the myriad minute compromises, skirmishes, and backsteps involved in the history underlying the story. In tracing these, *The Subtle Knot* thus offers a needed corrective to accounts of the development of neuroscience already extant. As suggested above, while the traditional narrative often excavates valuable historical data and draws useful connections between early and modern thought, it remains nonetheless plagued by a backwards-looking historicity – the sense, that is, that science as such has evolved to a better and more accurate point, and that the job of the researcher is to demonstrate the deviations expressed in the past from current accepted medical doctrine. Whatever its technical validity, this approach discounts the valuable contextualization that a broader and more open-minded consideration can provide; here we will argue instead that only an immersive history can properly revivify the materials of the past.

Tracing a picture of the brain *in situ* historically, as it were, involves at least a temporary disengagement with our modern understanding about the way that organ operates. Here instead we will rebuild the early modern brain, itself an uneasy conglomeration of received historical authority and contemporary experimental discovery. But though it presages the idea of the brain we have today, the early modern brain is an entity fundamentally distinct from the modern – not necessarily physiologically, of course, but in terms of how it was understood to operate with both body and soul. Thus the distinction I make here is less one of *correctness* versus *incorrectness* than between one form of thought and another. Only when we have considered the early modern brain in context does it become useful (and necessary) to examine the threads of continuity between that brain and the one we believe we know more of today. As we will see, anterior ways of understanding the brain do not simply vanish, as shadows to be dispelled by the bright lights of science, but often re-emerge, translated, even though the centre of interest shifts from brain/soul to brain/mind. The connections we will encounter between past and present conceptions of the brain, as well as between the scientific, broadly conceived, and the literary are not always direct. Nor do they always operate through the direct citation of one domain in and by another. More important for us are the metaphors that allow the movement of thought between areas and provide as well the grounds upon which connections were made and perceived.

Metaphor on the Brain

For it is through the key notion of the metaphor that this book reads the early modern period and indicates its continuations in the present. Drawing on Greek etymology, Puttenham's 1589 *Arte of English Poesie* called metaphor "the figure of transport," and saw its operations as crucial to the poetic responsibility of rendering God's works intelligible. Through translation – that is, the shifting across places – metaphor relates aspects of a more readily accessible object to another, less intellectually accessible one so that "every man can easilie conceive the meaning" of the latter.[42] In this sense, metaphor accomplishes for meaning or signification what the spirits do for the relationship between body and soul, joining seemingly disparate domains together, transporting sense and meaning across their boundaries, and showing how these mutually act upon one another. Puttenham's description thus opens out onto the primary terrain covered in *The Subtle Knot*: the particular metaphors used by writers and thinkers in the sixteenth and seventeenth centuries to describe the very *act* of the soul's operating upon the body.

As Denis Donoghue reminds us, "metaphor is a cognitive act, or at least it may be, a device to make further experience possible."[43] Metaphor underpins our processes of making sense of the universe, and while true even now, this was perhaps especially obvious during the early modern period, where analogies play a crucial role in the production of knowledge. In so doing, metaphors drive not just the descriptions of pre-existing phenomena but also open the way to not-yet-conceived possibilities. The metaphor is at once fundamentally *literary*, as its status as a "made" or "crafted" linguistic entity suggests, and at the same time essential to our broader understanding of and communication about the world around us. Indeed, metaphor is so much a standard mode of thought that we easily gloss over its appearance in our everyday lives. As Lakoff and Johnson's *Metaphors We Live By* explains,

> [o]ntological metaphors … are so natural and so pervasive in our thought that they are usually taken as self-evident, direct descriptions of natural phenomena. The fact that they are metaphorical never occurs to most of us … Explanations of this sort seem perfectly natural to us. The reason is that metaphors like THE MIND IS A BRITTLE OBJECT are an integral part of the model of mind that we have in this culture; it is the model most of us think and operate in terms of.[44]

As we shall see, the sciences of the brain in the early modern period deployed key metaphors, drawn from an array of different fields, to undertake ontological work, and in particular to solve the basic problem of how to link the theoretical to the physical, the abstract to the concrete, the immaterial to the material. How the brain came to be viewed as the organ in which the link between the material and the immaterial was forged depended fundamentally upon putting literary metaphors to work in scientific and anatomical texts in order to describe the connection between soul and body. In their capacity for bridging the gap between spiritual substance and matter, such metaphors rendered the soul an object of empirical inquiry by tying it to the brain.

In fact it is important to recognize, as N. Katherine Hayles emphasizes, that *all* scientific language is already in a sense metaphoric, even if we do not immediately register that metaphoricity:

> When a carpenter says that a room is 7 yards long, he is comparing the length of the room with the length of an Anglo-Saxon girdle. When a scientist says that a molecule has a diameter of 2.5 angstroms [one hundred-millionth of a centimeter, the preferred unit of measuring wavelengths or interatomic distances], the standard has changed but the principle is the same; the object is still understood in terms of its relation to something else. A completely unique object ... could not be described. Lacking metaphoric connections, it would remain inexpressible.[45]

Taking seriously the idea that scientific understanding depends on metaphors allows, as Hayles points out, for a "field notion" of culture that is not constrained by the need to demonstrate an explicit transfer of knowledge between scientific and extra-scientific contexts. She suggests, for instance, that Ferdinand de Saussure's relational linguistics shares common ground with Albert Einstein's theories of relativity without the one having explicit knowledge of the other's work – in both cases, "meaning" (or its equivalent within relativity theory) arises through relations rather than pre-existing as a definable, definite quantity that has only to be uncovered.[46]

The case for such connections is even stronger for the early modern period, since for the cultural milieu upon which this book focuses, the "field" is already porous and fluid, metaphor already a vital means of relating the abstract to the concrete and everyday. Michel Foucault's influential account in *The Order of Things* of what he calls the classical episteme brings into view the central place of resemblance and representation in the construction of knowledge in the sixteenth and seventeenth centuries – which would yield towards the end of the period to what is still the dominant understanding of scientific language in particular, as a transparent window upon the natural world. For resemblance to enable knowledge, the Foucauldian *signature* was important, for it helped to reify and ground ideas of analogy and metaphor. Signatures were signs that established a relation amongst things, especially with respect to microcosmic-macrocosmic relationships; thus, Foucault notes, "[a] knowledge of similitudes is founded upon the unearthing and decipherment of these signatures," a claim with which many in the early modern period might have agreed.[47] The world-as-a-book analogy is, after all, among the most salient of Foucault's examples:

> the face of the world is covered with blazons, with characters, with ciphers and obscure words … And the space inhabited by immediate resemblances becomes like a vast open book; it bristles with written signs; every page is seen to be filled with strange figures that intertwine … And all that remains is to decipher them.[48]

This description echoes the very impulse voiced by Galileo, "to comprehend the language and read the letters in which it [the book] is composed." Analogy and metaphor work to make these written signs visible and understandable, the latter in particular working to "translate" one (perhaps inscrutable) point into another (more legible) one.

Foucault situates his discussion of analogy amongst what he identifies as the four fundamental similitudes governing thought in the period being examined here: *convenientia* (convenience or juxtaposition), *aemulatio* (likeness at a distance), *analogy*, and *sympathy*. In the case of analogy, he writes,

> *convenientia* and *aemulatio* are superimposed. Like the latter, it makes possible the marvelous confrontation of resemblances across space; but it also speaks, like the former, of adjacencies, of bonds and joints. Its power is immense, for the similitudes of which it treats are not the visible, substantial ones between things themselves … [analogy] can extend, from a single given point, to an endless number of relationships.[49]

Analogy derives its power from its ability both to span and to compress distances – for example, the distance between the physical and the immaterial, or between body and soul – through likeness and similitude. But there does exist one intimate yardstick that stabilizes what would otherwise be an infinite displacement from signature to signature, and through which we can map the world and cosmos: the human body stands as "one particularly privileged point … [Man] stands in proportion to the heavens, just as he does to animals and plants, and as he does also to the earth, to metals, to stalactites or storms."[50] Thus Crooke's paean to the human body as a microcosmos, a sentiment the seventeenth century engages with – and ultimately explodes.

But, more particularly, one single given point may be even more privileged: the brain. The resemblances it gives rise to are legion: it may resemble coral, clouds, a womb, or a walnut. The last of these comparisons is in fact cited by Foucault as an example of a sign being legible via analogical sympathy: the walnut and the human head are alike in that "what cures 'wounds of the pericranium' is the thick green rind covering the bones – the shell – of the fruit; but internal head ailments may be prevented by use of the nut itself 'which is exactly like the brain in appearance.'"[51] That the brain is here represented by referring to the wounds that would potentially damage it will become important in chapter 2, but the general likening of brain to walnut because of the visual resemblance between the two entities is exemplary for construing how the early modern period depicted the brain. As in Crooke's labyrinthine example above, and as we shall see throughout this book, the brain had to be stabilized, concretized, by securing it to another more tangible or relatable object via some form of metaphor. Thus the crucial anatomical connection between body and soul resides in the brain as a *net* or *knot*; a *cut* suggests how the brain may malfunction; the *womb* expresses the brain's generative capacity (and vice versa); and, ultimately, a *book* allows one to leaf through this most reverend, though simultaneously fraught, of organs.

Given Foucault's emphasis on the disjunction between epistemes, however, issues of continuity across historically successive knowledge formations are underplayed in his account. And in this regard Hans Blumenberg's theory of metaphorology opens a different, supplementary line of thought that will allow us to bring out the continuities across historical periods, even as we acknowledge their radical differences. In Blumenberg's rich historical excavations, transformations in the very status of metaphor itself are brought to the fore, in ways that resonate with both the early modern period and *The Subtle Knot*. Considering philosophy and natural philosophy in the mid-sixteenth century, Blumenberg argues that metaphor had a prior significance largely as a rhetorical device, but through the work of mechanistic philosophers such as Descartes and Hobbes, it became a philosophical tool. In the period preceding them, however, "[w]hether the rhetorical artifice of *translatio* [a form of and itself a translation of the notion of metaphor] could do anything more than arouse 'pleasure' in the truth to be communicated remained undiscussed."[52] We ought not, in any case, to discount the pleasure in truth which metaphor could convey – not for nothing is there a playfulness in the references surrounding the brain, in which this occluded organ was likened to clouds, a walnut, a labyrinth, to name just a few possibilities.

But the importance of Blumenberg's analysis for us lies in his seeing that metaphors are not only "*leftover elements*, rudiments on the path from *mythos* to *logos*" – that is to say, metaphors are more than superseded formulations that, retaining "the inauthenticity of figurative speech," remain useful only as expressions of a supplementary pleasure that accompanied the uncovering of truth. More importantly, he identifies a range of metaphors that function as "*foundational elements* … [as] 'translations' that resist being converted back into authenticity and logicality."[53] As foundational elements, metaphors are tasked with producing more than pleasure – rather, they provide a stable "absolute translation," bearing with them an aspiration to truth that will render them crucial to concept formation rather than being merely decorative. For metaphors also constitute something like a pre-conceptual ground, to which we repeatedly return in order to build new concepts and new ways of understanding the world. What Blumenberg calls absolute metaphors function as reservoirs upon which the production of knowledge depends.

Such metaphors must "therefore have a *history*," and their study (metaphorology) "endeavors to stake out the terrain within which absolute metaphors may be supposed to lie."[54] On a broad scale, then, Blumenberg's project parallels the narrower undertaking of *The Subtle Knot*, which is in part an historical study of how a particular set of foundational metaphors were constructed and persisted, and of how they then shaped knowledge – in this case, knowledge about the brain and how the soul/mind and body interact within it. Such an imperative connection between science and language is peculiarly essential to the early modern period, when, as Claire Preston observes, "[h]ow to *do* science was not just an empirical question but also a rhetorical one: it was a question of how to *say* it."[55]

This aspect of my book may perhaps be best illustrated through chapters 3 and 4, which grow in their different ways out of a metaphor that has never really fallen out of use, even if its status as metaphor is far more muted for us: conception. Likening the generation of thoughts to childbirth, the word remains in widespread use, in contexts much broader than the scientific. But the early modern use of the metaphor was far more concrete, making the posited connection between the mental and the physical strikingly evident. "Labour Pains: William Harvey and the

Travails of Conception" examines a moment in the seventeenth century when "conception" forged a peculiar link between intellectual and biological processes, and between the very structures in which those processes took place – that is, an unexpected but highly productive link between the brain and the womb. In a short and generally overlooked essay titled "On conception" (appended to his better-known work *Experiments concerning animal generation*), the physician Harvey casts about for some means of explaining exactly how biological generation takes place in the absence of a clear motive force connecting sperm and egg in the womb. He settles on the metaphorical concept of conception, which enables him to liken thought in the brain to the child in the womb, thereby associating the artistic act with the biological act, and the brain with the womb.

Harvey's use of the metaphor taps into the pulse of a broader association between brain and womb that flourished in a wide variety of contemporaneous discourses. The womb-as-brain metaphor abounds in midwifery manuals, which worry constantly about the effect the mother's imagination could have on the developing fetus: on a fairly benign level, if the mother craves strawberries, a strawberry-shaped birthmark could appear, but, more troublingly, maternal discontents might manifest monstrosities that exposed not only the mother's disordered thinking but disruptions in her surrounding environment. Conversely, the brain-as-womb metaphor served to elaborate upon the creative process, as writers (with much false modesty) described the difficult gestation of their ideas and the laborious birth onto paper that resulted in the child-book being delivered into the hands of the reader.

Co-opting the metaphor likening brains to wombs would prove especially productive for one of the seventeenth century's most intriguing writers, Margaret Cavendish, Duchess of Newcastle, as chapter 4 argues. "The Mechanics of Reproduction in the Art of Cavendish" turns to the dizzying array of metaphors the early modern period produced to describe how thoughts occurred in the brain, and no writer was more prolific in spinning these than she: thoughts for her were like stars, citizens, forests, and even pancakes. These proliferations lead back to Cavendish's distinctive celebration of the generative capacity revealed even in instances of the maternal imagination gone awry, resulting in monstrosities of various stripes. Hence pregnant women were warned away from things that could cause them to be afraid or that could provoke excessive desire in them, for the mother's imagination held curious sway over the developing fetus. In Cavendish's hands, the sense of such admonitions is reversed in that she sees them as bringing into focus the very power of the maternal imagination, allowing her thereby to shape or represent her self-creation as authoress.

Building on contemporaneous associations of brains with wombs, Cavendish exploits the notion that thinking could have material and even automatic effects, in order to connect thought production to the mechanics of writing and print. Given her brand of vitalist philosophy, these connections would prove especially fruitful for exploring and critiquing the Cartesian divide between body and mind. In both her philosophy and her fiction, Cavendish would espouse instead the quasi-monist idea of "thinking matter," seeing the brain as self-animated matter upon whose surface the senses "patterned out" information. The organic nature of such "writing" points in turn to a different metaphor to describe her constantly productive brain: that of the organ as a *writing machine*, not entirely under her conscious control, through which the role of fancy, imagination, and wit ("thought" most broadly) in fictive production could be circumscribed. The danger, as with the case of maternal imagination, lay in the possibility that such an uncontrollable machine might produce aberrances, be they monstrous births or poorly received texts.

The inseparability of mind from brain, and the notion of writing as the best means of describing thought, point this chapter towards the claim that metaphors for understanding the brain necessarily had to do with literary production. Cavendish's considerations of the brain and mind, both in prose and verse, are particularly instructive, not only as a way of reading the mind-body problem in opposition to Cartesian duality, but for how they look ahead to the notion of extended or embodied cognition, a conceptual turn in contemporary neuroscience familiar to our eyes.

For while I have already said that *The Subtle Knot* will not seek to apply concepts drawn from contemporary neuroscience retroactively upon the early modern period, it is nonetheless to be hoped that the reader will see powerful continuities with the present, continuities borne precisely through the persistence of metaphors, however translated or transformed. The metaphors used in the early modern period to describe the brain point to distinct functions and operations that needed to be accounted for, and they helped produce "solutions" to problems that needed to be addressed. Indeed, the echoes between the metaphors underlying contemporary neuroscientific concepts and those deployed in the early modern period speak to the persistence of these problems across historical periods, to continuities between what exactly each period felt had to be explained about what the brain did. Thus, the metaphor of the *rete mirabile* arguably finds its transformed equivalent in the modern metaphor of the neural net (chapter 1); the cut – and its attendant sense that we learn about the "normal" brain by examining it in its perturbed state – opens up connections to modern methods of the lesion study (chapter 2); likening the brain to a womb (chapter 3) reveals the long history underlying the double sense of conception; we hear the echoes of Cavendish's vitalist understanding of the brain as an engine for thought (chapter 4) resound in contemporary discussions of grounded cognition; and, finally, the bookish brain of chapter 5 anticipates contemporary techniques of slicing and scanning the brain in order to "read" its operations.

In 1813, the German Romantic Jean Paul evocatively noted the significance of metaphor to spoken language, observing that

> [o]riginally … metaphors were forcefully derived synonyms of body and mind. Just as in writing pictography preceded letters, similarly in speaking the metaphor, in so far as it signified relations and not objects, was the earlier word, which had to grow pale in order to turn into the term proper … Thus every language, as far as mental relations are concerned, is a dictionary of faded metaphors.[56]

"A dictionary of faded metaphors" is quite a lovely way of thinking about how metaphors serve to connect and transport meaning, and of how they underpin language both spoken and written.

From this perspective, *The Subtle Knot* is concerned with giving back to such metaphors their force, seeking to recover the living and vibrant metaphors that served to advance a variety of descriptions of the brain and its relations, in particular as a means of exploring the necessarily imagined (because not empirically measurable) link between body and soul. Paul's suggestion that metaphors began life as "forcefully derived synonyms of body and mind" is useful for navigating the catalogue of images paraded in the early modern period. If we keep in mind that the substitution so fervently desired by knowledge-producers was of tangible (body) for intangible (soul/mind), we see the centrality of metaphors in traversing that gap, both then and now.

From the Book of Nature to the Book of the Brain

Among the most cited of foundational metaphors for the early modern period is Galileo's evocation of the "book of nature," which appears in *The Assayer* (*Il Saggiatore*, 1623), a key text for describing and elaborating on what would come to be known as the scientific method. Naturally, significant critical attention has been devoted to Galileo's metaphor, [57] but the entire passage in which the idea appears is worth consideration and quotation at length. For Galileo introduces his famous metaphor on the back of another comparison to books, through which he criticizes reliance on authoritative opinions inherited from the past:

> In Sarsi[58] I seem to discern the firm belief that in philosophizing one must support oneself upon the opinion of some celebrated author, as if our minds ought to remain completely sterile and barren unless wedded to the reasoning of some other person. Possibly he thinks that philosophy[59] is a book of fiction written by some writer, like the *Iliad* or *Orlando Furioso*, productions in which the least important thing is whether what is written there is true. Well, Sarsi, that is not how matters stand. Philosophy is written in this grand book, the universe, which stands continually open to our gaze. But the book cannot be understood unless one first learns to comprehend the language and read the letters in which it is composed. It is written in the language of mathematics, and its characters are triangles, circles, and other geometric figures without which it is humanly impossible to understand a single word of it; without these, one wanders about in a dark labyrinth.[60]

For Galileo's opponent Sarsi, the discipline of philosophy – and concomitantly scientific reasoning – legitimates its claims by citing prior authorities, resting upon the shoulders of giants who came before. The call to free oneself from the "opinion" of "some celebrated author" is not in itself unusual, but Galileo's attack on this obsequious relation to the past takes an unusual tack nonetheless in that in rejecting such authorities he also changes their status: from philosophy to fiction. So, Aristotle and Plato become equivalent, in the brave new world of scientific modernity, to Virgil and Ariosto. Before the book of nature can be read, prior philosophies of nature must be treated as if they were books of fiction. Galileo thus seems less concerned with rejecting authoritative opinion as false than with preserving it in a different way, assimilating it to those epic "productions in which the least important thing is whether what is written is true."

There is substantial precedent, which chapters 4 and 5 will examine in more detail, for placing fiction beyond the true/false opposition – in England, for instance, it stretches at least from Sir Philip Sidney through Cavendish. And central to this defense of poetry was a celebration of the power of the imagination, and of its capacity to break out of constraints imposed by facts in order to communicate deeper truths through pleasure. It is perhaps no accident that the *Iliad* and *Orlando Furioso* are singled out by Galileo as representative of the fictional foundations upon which Sarsi and thinkers like him have built their epistemologies. These books, these artificial and composed constructions, exemplify fiction's power to captivate, to sway its readers in a way quite different from how the "language of mathematics" operates. For Galileo and others, the latter is intuitive rather than discursive: "triangles, circles, and other geometric figures" constitute, rather than describe, nature. As he points out, these figures make up the foundation upon which philosophical language is built, rather than being entities which language can be used to describe. Without such an understanding, one could never properly "read" nature's book.

It is therefore less the falsity of their descriptions that Galileo finds disturbing in earlier philosophies of nature than the nature of their hold upon the mind, and in this they work in the same way that explicitly fictional and imaginative works do. This has consequences for how Galileo approaches his own texts. As Lorraine Daston puts it, "Galileo's vision of natural philosophy as the mathematical redescription of phenomena, coupled with his suspicion of the imagination on traditional Aristotelian grounds, narrowed the scope for the use of explanatory analogies in his work."[61] And yet, as much as Galileo here distrusts the fictive and the imaginary, he must in his broader work, Daston continues, "appeal constantly to the imagination to unfold his argument by forging analogies between the macroscopic and the microscopic, between the physical and the mathematical, and between the finite and the infinite."[62] Just so, in the quotation from *Il Saggiatore* we have been examining, does Galileo maintain and exploit the metaphor of the book, even as he seeks to draw a sharp line between his book and those of others. Indeed, only through this metaphor is he able to bring together the immaterial and the material, the intelligible and ideal figures of geometry with the visible phenomena of the natural world.

From macro- to microcosm: given the power and resonance of Galileo's metaphor, it is appropriate then that *The Subtle Knot* closes with the brain as book. My concluding chapter examines this metaphor by studying the complicated textual history connecting two books: Johann Remmelin's *Catoptrum microcosmicum*, first published in Augsburg in 1613,[63] and Thomas Willis's 1664 *Cerebri anatome*. The former testifies to the proliferation of a new kind of anatomical illustration in the sixteenth century: the flap anatomy. These generally consisted of one or two loose pages (also called fugitive sheets) comprising male, female, or paired figures with torsos that could be lifted to display viscera, bones, and arterial and venous systems. They presented three-dimensional pictures of the human body to be manipulated by the reader. By opening up the printed illustration, the reader opened up the human body, reading turning into dissecting. Flap anatomies were a valuable and innovative means of disseminating knowledge to a broader public, beyond those who could afford to buy such monumental volumes as Andreas Vesalius's *Fabrica*. But these early modern texts and images also compensated for fragmentation by entreating the reader to consider the body as a wondrous creation of God. Often framed by or containing a moral message or a poem, they reconstituted the body in parts, showing how the pieces fit together, materially and spiritually, to make a whole. Indeed, this reintegration was arguably as important as the dissection, since the actual anatomical information in these anatomies was often out of date. Directed at literate lay audiences eager to know and understand themselves, the flap anatomies were thus aimed in a sense at fostering a particular mode of self-knowledge, whereby information about the body produced by anatomists could be led back to the place of that body in a divinely regulated universe.

Yet strangely enough, virtually no flap anatomy in the early modern period showcased the human brain, despite its centrality to anatomical discourse as well as to popular consciousness surrounding the organ and its spiritual functions. The one exception belongs to the text at the heart of this final chapter, Remmelin's *Catoptrum*: it contains a weird and extremely complicated set of flap prints – and includes, in particular, a flap anatomy of the brain, concealed beneath a large skull at the foot of the upright male figure in one of its *visios*. The flap prints from early-seventeenth-century European editions were taken up later in the century in England, where pirated copies were sold in various editions ranging from the plain unbound to the sumptuously leather-bound. The full translation of the *Catoptrum*, along with images produced from Dutch strikes, would be published beginning in 1670 by Joseph Moxon and his son, under the title *An exact survey of the Microcosmus, or Little World*. The choice of an anatomical text was in any case an unusual one for the Moxon printing house, which mainly focused on globes, charts, and more mathematically oriented scientific texts. By and large, the early Moxon editions faithfully follow the original, reproducing its flap representations of body and brain, even though these reflected anatomical knowledge from about a century prior.

But the 1695 Moxon edition, published by his son after Joseph Moxon's death, proves to be exceptional even with regard to Remmelin's exceptional text. It is the only edition that replaces the *Catoptrum*'s original flap images of the brain with impressively miniaturized versions of the most up-to-date renderings then available, which had been created by Christopher Wren for Thomas Willis's seminal book on the brain. Willis's magnum opus, the *Cerebri anatome*, has the distinction of being the first text dedicated to neuroanatomy, and has for this reason has been held, retrospectively, to mark the beginning of a new discipline. No other British edition of Remmelin's work bears these careful emendations based upon ongoing work at the very frontiers of brain science in the seventeenth century – and these changes have also hitherto escaped critical attention from contemporary scholars. I conclude *The Subtle Knot* by reconstructing the complex path that led from Remmelin, to the Moxon publishing house, to this particular edition, to argue that its appearance not only sheds light on the relationship between the early modern printing press and the scientific labours of the Royal Society, but also signals a transformation in the kind of knowledge that anatomies were expected to provide. With the updating of brain images in the 1695 edition, the English *Catoptrum* moves from being a traditional anatomy, in the sense of a passive repository of inherited knowledge, to asserting its place as a scientific and artistic instrument in its own right. It becomes a body-book that displays to its reader the most current knowledge of this miraculous organ. And by so doing, it projects a future that would be realized only centuries later in the transparent-layered anatomical textbooks of today. The resemblance is in fact uncanny: we "read" the "pages" of Remmelin's and Willis's brains in much the same way as modern fMRI studies read data and images from virtual slices of the brain. That the brain can be read as a book, with successive leaves, and a language inherent to it, constitutes the final metaphorical exploration of *The Subtle Knot*.

CHAPTER ONE

Untying the Subtle Knot: Anatomical Metaphor and the Case of the *Rete Mirabile*

[O]ur blood labors to beget
Spirits, as like souls as it can,
Because such fingers need, to knit
That subtle knot, which makes us man.
– John Donne, "The Ecstasy" (61–4)[1]

"This ecstasy doth unperplex"

To unperplex the soul's relationship to the body, we must first read the body's book. The lovers in John Donne's poem "The Ecstasy" knew as much, as did Donne's sixteenth- and seventeenth-century contemporaries in English medicine and philosophy. For them, the body, as the most readily accessible instrument for measuring the universe around man, became necessary for understanding the soul and its workings. Let us, for the moment, place ourselves in the position of these seventeenth-century thinkers. Were we searching for the soul in a living body – imagine perhaps a virtual vivisection, the only means by which a glimpse of the soul might become even a remote possibility – what questions would we need to answer? Presumably, we would want to know precisely *what* we were looking for, and *where* we would be most likely to find it.

This chapter addresses the second of these questions. The answer to the first lies in part outside our grasp for the uncomfortable reason that seventeenth-century thinkers tended to describe the soul in paradoxical terms. Consider, for example, the contorted definition Edward Popham offers the reader in his 1619 religious treatise, *A Looking-glasse for the Soule*:

> First, it [the soul] is an immatierall substance: While it doth revive the body, it is the *Soule*; when it willeth or chooseth any thing, it may (though improperly) be called the *Minde*: While it knoweth any thing, it may be called (though improperly againe) the *Understanding*: While it judgeth, some have tearmed it *Reason*: While it doth breathe or contemplate, a *Spirit*: While it calls any thing to minde, the *Memorie*: While it thinketh any thing (though more grosely) the *Sense*. But to speake of the *Soule* as it is, it is an immateriall substance, and Reason, Memory, Sence, &c. are the severall faculties and divers opperations thereof.[2]

Popham's description – by no means an unusual one – is characterized by the unresolved tension between what the soul is and what the soul does: the soul is almost, but not quite, the sum of its "severall faculties and divers opperations." Any simple identification of the soul with its capacities clearly discomfits Popham – witness his doubts about what he himself has proposed, viz., "improperly," "improperly againe," "some have tearmed it," "though more grosely." Ultimately the only description to which Popham fully assents is the one he repeats: that of "an immateriall substance." While the soul acts in various ways and possesses diverse faculties, its essence nevertheless lies beyond these, and is most accurately characterized by its encompassing both the physical ("substance") and the spiritual ("immateriall").

There were many in the seventeenth century who would have agreed with Popham, for while "immaterial substance" was not the only way to think about the soul, it was certainly one of the most prominent. In a mid-century tract on souls and procreation, Henry Woolnor proposes that, though "the soul is far from such a grosse & visible substance as the body," the body nevertheless must possess "some spiritual kind of substance" in order for it to join with the soul at all; the soul must therefore be "made of that matter [which] is immateriall."[3] Poets, too, conveyed a similar sense of the soul's self-contradictory and inmixed nature. Joseph Fletcher likens the soul to the Holy Trinity, noting that it "[c]ontaineth in't a threefold facultie" of mind (God the Father), will (God the Son), and the power of moving the body (God the Holy Ghost).[4]

As these definitions suggest, while early modern writers could not quite say what the soul was, they had a clear sense of what it *needed to be* for the body both to be alive and to be connected to a divinity whose nature was fundamentally incommensurate with that of corporeal existence. This impulse becomes evident in Cambridge Platonist Henry

More's acknowledgment of the difficulty we have in contemplating the nature of the soul: "the notion of an *Incorporeall Substance* is so subtile and refined, that it leaving little or no impression on the Fancy, it's [*sic*] representation is meerly supported by the free power of Reason."[5] Here, the *quid* of the soul becomes a necessary assumption, something that reason must posit despite the inconvenience of its leaving virtually no trace of itself behind.

The manifest difficulties of definition impel the need for a more intellectually apprehensible approach to the problem of the soul, and in particular to uncovering the mechanisms whereby it interacts with the body. These may be found by turning to the question of its location – or, more precisely, by considering its locatability, the possibility that it could be located at all. As I shall suggest, there was a concerted effort in seventeenth-century England to ascribe a potential *physical* site for the soul, as British anatomists began asking and answering questions with which Continental anatomists had grappled in previous centuries. The possibility of fixing and naming its corporeal place worked to soothe early modern anxieties about the soul and how it worked in (and upon) the body, anxieties evinced in the paradoxical insistence on the soul as an immaterial substance. As a consequence, what the soul needed to be turns out to be closely connected to where it was to be found: in the *rete mirabile* or the "wonderful net" (sometimes "wonderful knot"), a network of minuscule arteries ostensibly located at the base of the human brain. Here was one possible location for the immortal, rational (the object of study for philosophers and theologians) soul and its faculties (in large part the object of study for anatomies) to join with the body most fully.[6]

As we shall see, though, the existence of the *rete* was by no means a given. Often it was proposed that the *rete* either was not present in every human or disappeared as soon as a person died. In fact, the *rete* escaped observation for the simple reason that it did not actually exist in human beings at all. And yet, English anatomists were determined to make it a part of the human body. For even if it could not generally be glimpsed in action, the *rete*'s "presence" at least corroborated the contemporary anatomical understanding of the body and its means of communication with the soul. In an important sense, the persistence of the *rete mirabile* in the English anatomical imagination is one example of what Katherine Eggert terms "disknowledge," wherein the early modern period exhibited an affinity for holding onto ideas that had been empirically proven incorrect: Eggert describes disknowledge as "the conscious and deliberate setting aside of one compelling mode of understanding the world" – here, the *rete*'s nonexistence – "in favor of another," not out of "pure ignorance," but rather a conscious choice "to be wrong rather than right."[7] The net's structure was particularly convenient in the English neuroanatomical scheme since it allowed the soul to be at once in a particular location and at the same time not immediately localizable there; the soul existed in the net insofar as it used the net to communicate with the body, but the soul need not be pinpointed at any one particular node of the net. Here and nowhere, immaterial matter got stretched fine across the wonderful fabric of the *rete*.

Crucially, the web testifying to the soul's locatability was itself located in the organ most nearly associated with the soul and the soul's functions: the brain. The seventeenth century arguably witnessed the birth of modern neuroscience,[8] with the brain becoming an object of empirical inquiry alongside an increasing anatomical and philosophical concern with what Popham calls the soul's "opperations" – thinking, willing, memorizing, and so on, all the operations which we today associate with the mind. By 1664, Thomas Willis and Christopher Wren had published the first dedicated neuroanatomy, the *Cerebri anatome*, in which Willis laid out theories of (for example) sympathetic and parasympathetic nerve systems, the cerebellum's functions, and the connection between damage to the brain and damage to memory, character, or linguistic ability. To be able to tie the soul to one particular area of the brain represented the first step in identifying mind and thought with the little grey cells – but only by first identifying (or at least joining) soul with body.

However, what ultimately allowed these two entities to be bound together thus was less a physical reality than a rhetorical one: the *metaphor* of the net or knot. I want to suggest that the burgeoning field of early modern neuroscience unfolded in the seventeenth century by drawing upon a field of metaphors shared by other intellectual domains, and, as importantly, that these metaphors were used in those other domains to describe the soul.[9] This chapter focuses on one particularly compelling intersection: between the emerging discourse of neural anatomy and a contemporaneous poetic discourse invested in examining the soul's ties to the body.[10] As we shall see, the contemporary poetic investment in the metaphor of a knot or net to describe the soul-body link led anatomists to cling to the idea of a net- or knot-like physical structure in the body – even in the face of mounting doubts as to its actual existence.

The lines from Donne's "The Ecstasy" that provide my title and epigraph point to the complex translation between anatomy and poetry which the metaphor enables. His image of the blood generating soul-like spirits responsible for "knitting" the "subtle knot, which makes us man" echoes the anatomical image of the *rete mirabile* as the necessary corporeal structure that makes the human body something more than either vegetative or animal matter. But the issue is less whether Donne poeticizes (putative) anatomical discoveries or whether, conversely, the anatomists corporealize (imaginative) poetic discoveries. Rather, it is the metaphor of the net or knot itself, as a response to deeper concerns about how bodies and souls are joined together, about how human beings are made human, that underwrites both poetic and anatomical discourse, and enables the translation between them. As Richard Sugg points out, Donne in his poetry is concerned with "an ingeniously contrived transitional zone, blurring the boundaries between matter and spirit."[11] I want in this reading to press upon this transitional zone and to situate it within a broader context of anatomical investigation into its nature and specific location.

Metaphor was one important means whereby art and science communicated with one another in the seventeenth century, at a time when disciplinary boundaries were still relatively undefined and hence far more fluid and porous than they are today. Though today associated primarily with the literary, the utility of the metaphor was (and is) much wider. It lay in its opening up traffic between diverse ideas, and in making such translations accessible to "every man." As intimated in my introduction, for George Puttenham, because poets were "the first that entended to the observation of nature and her works," and also "the first observers of all naturall causes & effects in the things generable and corruptible," they were likewise "the first Astronomers and Philosophists and Metaphisicks."[12] While Puttenham perhaps enthusiastically overstates the case, he nevertheless makes it clear poetry shared with other intellectual domains at the time the responsibility to render intelligible God's works. One means of accomplishing this end was the metaphor – in Puttenham's words, "the figure of *transport*" – through which aspects of a more readily accessible object could be transferred onto another, less readily grasped object so that "every man can easilie conceive the meaning" of the latter.[13]

The concept of metaphor underpins, in fact, the scientific object in the seventeenth century. Even today, the standard for measuring the world is the human body, that most familiar and instantly accessible of forms. To calculate using hands, feet, or cubits is to cast the calculation in human terms, in essence to use parts of the body as metaphors for space or distance. Indeed, for the early modern natural philosopher, the reverse could equally hold true.[14] The common conceit that man's body presented a miniature version of the heavens – that is, a microcosm wherein each individual part bore a relationship to an entity in the outside world – offers the clearest evidence of a scientific investment in a literary figure.

Richard Sugg speculates that the relationship between scientific and rhetorical anatomy in early modern England was more complex than that of simple cause and effect:

> It may even be that literature spurred medicine to some extent … Often, indeed, uses of dissective rhetoric appear not merely fashionable but highly compulsive, sometimes lacking an integral semantic motivation to the extent that the body must be seen as actively invading the English literary imagination. Frequently, anatomy uses writers, as well as vice versa.[15]

Though I too focus on the deployment of a metaphor shared by anatomical and poetic texts, it needs to be emphasized that a wide range of disciplines in the seventeenth century was particularly invested in teasing out the connections between body and soul. Anatomists mapped and catalogued the human body in order that they might better comprehend God's work and elucidate the operations of the soul; they merged their knowledge with that of the physicians, whose task it was to treat the body, prolonging life for the further glorification of God. Mathematicians and natural philosophers were devoted to uncovering the patterns of the world which had been laid out by the First Mover. Artistic work aimed to reveal man's soul to himself. Consequently, in answering questions about the soul and the brain, thinkers could make use of the advancement of knowledge on a variety of fronts – and central to these disciplinary translations and transferrals was the deployment of metaphors. Their use enabled communication and adaptation across different intellectual fields, reminding writers and thinkers in the process that the separate domains of human understanding were all filtered through a common human experience, and that the macrocosm, by design, corresponded to a potentially intelligible microcosm.

Spirits in the Material World

In 1543 the young anatomist Andreas Vesalius published in Basel his immense eight-volume anatomical text, *De humani corporis fabrica* (*On the Fabric of the Human Body*, hereafter *Fabrica*).[16] Historians K.B. Roberts and J.D.W. Tomlinson term its arrival "the great leap forward," and it is certainly difficult to overstate the importance of this opus, especially for English anatomists of subsequent generations. The *Fabrica*'s impact in England from the late 1500s to the mid-1600s is evidenced by the amount of material – verbatim text and woodcuts – that anatomical texts "lift" from it. Vesalius styled himself as a sort of revolutionary: among his goals was not only to provide the most realistic visual and textual depictions of the dissected body ever seen in printed form, but also to revise a tradition that had been in place for well over a millennium. The young anatomist advocated a hands-on empirical approach to teaching and learning anatomy in place of a medieval model heavily invested in lecturing.[17] Likewise, he strove to produce images of dissection that would set the standard of anatomical representation for at least two centuries to come. Nor was Vesalius the only revolutionary on the scene. Another was Jacopo Berengario, who composed, as R.L. Lind points out, "the first full-scale description of the entire human body between that of Mundinus, who wrote in 1326, and that of Vesalius, in 1543."[18] But for our purposes here, the critical difference Vesalius introduced involves what we have identified as being at issue in Donne's metaphor of the "subtle knot": the *Fabrica* was to challenge existing theories (and desires) about a structure in the brain capable of linking body and soul.

The broader tradition Vesalius set out to overturn amounted to an uneasy amalgam of Aristotelian and Platonic theories about the body, filtered through Greek and Arabian physicians, as well as through medieval Catholicism.[19] The most influential of these theories came from the second-century-AD Greek physician Galen. The Galenic body worked by means of spirits, purifications of the blood responsible for everything from the most basic tasks of existence to the complex machinations of the human mind. According to Galen's popular model, spirit-production occurred at three different levels in the body. The lowest of these levels produced the vegetative spirit, which governed basic functions of life (such as growing, reproducing, taking in nutrients, and so on). These were essentially the actions plants could carry out, hence the descriptive name "vegetative." The liver produced this spirit, using food, blood, and air, then passed it on to the heart, where a second spirit – the vital spirit – was manufactured. The vital spirit was responsible for more complex life functions such as the preservation of the life flame, and all the basic operations animals could perform. The third or animal spirit was produced when the brain sublimed the vital spirits from the heart. This was the most refined of the three spirits, belonging to humans alone, and as such it was charged with the most important duty: it communicated directly with the soul and controlled the soul's psychological operations, namely intentional moving, perceiving, and thinking. After the transformation in the brain, the soul could then send this finest of spirits through the nerves to instigate in an hydraulic manner movement and sensation, which in turn allowed the soul to gather information from and interact with its environment.[20]

The conundrum facing anatomists concerned where, precisely, the final transformation of vital spirits into animal spirits took place. Pinpointing the location would answer at least the mechanical aspect of the question of how soul and body interacted. It is difficult to overstate the importance of the anatomical desire for such locatability. As George Rousseau notes, the broader history of neuroanatomy "has indeed been, to some extent, the record of [the] search for [such] fire, fluid or ether [as the animal spirit that] mediates between mind and body in the eternal search for consciousness, emotion, and memory."[21]

The reigning theory until the mid-1500s was Galen's, which made a densely woven net of fine arteries at the base of the brain responsible for the sublimation. This intricate set of intertwining vessels was known as the *retiform plexus* until Bartholomaeus Anglicus renamed it in a thirteenth-century anatomical text.[22] Galen depicts a sublimely subtle structure: the composition of the net is such that it exceeds by its nature anything manmade – or at the very least it represents a significant complication of human craftsmanship. Galen describes the retiform plexus – "the most marvelous of structures" – as

> not merely a net-like structure but ... as if you had taken several fisherman's nets and stretched one out over the other. But it is characteristic of Nature's net that the meshwork of one layer is always attached to the other, thereby making it impossible to remove any one net by itself ... Naturally, because of the fineness of the members composing this network, and their intimate conjunction, you could neither compare this plexus to any manmade net, nor ascribe its formation to chance.[23]

The finely wrought filaments are bound to each other, but not in a disordered or tangled fashion; rather, the "network" extends neatly in three dimensions. Its order is so precise that it could only have been designed by Nature's hand. In fact, Galen's point was precisely that Nature had been so exacting in her crafting that no human hand could hope to undo its skein. His overlaid fisherman's nets proved an apt early expression of the metaphor: the physician had painted the *rete* so vividly as to be easily imaginable, but the image nonetheless preserved a mystery in that it went beyond the limits of human workmanship. The *rete*'s importance had much to do with its place in the system of spirits, but, as Galen's description demonstrates, the notion of the subtle knot had important metaphorical connotations as well – a point I shall return to in the subsequent section.

Galen's describing the retiform plexus as "the most marvelous" of the brain's structures doubtless spurred Anglicus's eventual coining of the name *rete mirabile* as well as what Rocca calls "the almost mystical devotion" of later anatomists to the structure. Variously translated in the early modern period as the wonderful or miraculous net, knot, or network, the *rete* proved a puzzling entity, but one that nevertheless made sense: knots and nets suggested the perfect physical forms to bind together soul and body, and the brain was the best – and perhaps the only – place for this joining to occur. Galen's anatomical theories were accepted, almost without hesitation, for centuries. Rocca reads in this fact "the historical development of *Galenism*, and the authority of his *ipse dixit*," and indeed the response to the *rete* in particular is representative of the scope and tenacity of Galen's influence.[24] Given such devotion and the holding pattern of the anatomical scene throughout the Middle Ages, it is perhaps not surprising that the *rete*'s reign lasted as long as it did.[25] The *rete*, with its elegant ability to simplify in metaphoric terms the most complex and glorious process of which the flesh was capable, easily infused itself into the cultural consciousness – so much so that it was preserved as an explanatory paradigm even after Vesalius proved that it did not exist.

No doubt, there had been others who, like Vesalius, had questioned the existence of the *rete* in order to propose a different structure in which the soul might inhere. As a 1521 commentary by the anatomist Berengario da Carpi concluded, "[s]o I believe that Galen has imagined the *rete mirabile* and he never saw it and I believe that other men after Galen believe in the *rete mirabile* more because of the opinion of Galen than because of fact."[26] In place of the *rete*, da Carpi maintained the prominence of what, both then and now, was known as the *choroid plexus*, networks of blood vessels which lined the ventricles of the human brain. Da Carpi himself argues thus:

> Yet I have never seen this net [the *rete mirabile*], and I believe that nature does not accomplish by many means that which she can accomplish by few means. But nature can subtilize these spirits in the smallest branches of the arteries descending … upon the dura

> mater which adheres to the basilar bone and ascending through the pia mater up to the center of the brain; therefore this net is not given in that place between the dura mater and the basilar bone. I have given many other occasions [for the nonexistence of this net] in my *Commentary on Mundinus*, to which I refer my readers for the sake of brevity; among these reasons the experience of the sense is my chariot-driver. (And if this net is given, it is above the dura mater immediately below the medulla of the brain from whose smallest arteries therein is made a certain membrane continuous and similar to the pia mater [i.e., the choroid plexus] in whose branches the vital spirit is subtilized and made into animal spirit. Thus subtilized, it passes to the ventricles by means in part of the worms of the brain mentioned before.)[27]

Note that da Carpi does not categorically dismiss the possibility of the *rete*'s existence; he simply points out that he himself has not observed it. As he prefers to rely on his "chariot-driver," he appoints the choroid plexus the seat of the final transformation of vital to animal spirit.[28]

By contrast, Vesalius's self-appointed task in the *Fabrica* would be to undo all that the wonderful net had tied together, overturning in the process his own earlier descriptions of the *rete* in the 1538 *Tabulae anatomicae sex* as well. In fact, Baldasar Heseler, one of Vesalius's students, describes an anatomical demonstration the master anatomist had given in 1540, during which the *rete* was ostensibly unveiled in its full glory:

> Eventually, he showed us the network of winding arteries around the rete mirabile [*ipsas revolutiones arteriarum circa rete mirabile*] in which the spiritus animales are produced, transmitted there from the heart as spiritus vitales. And I saw particularly those which run in coils around the base of the cranium, which are called rete mirabile [*Et notabiliter eas vidi involutas et revolute circa os basis cranei, quod rete mirabile vocant*] … And it was a reddish, fine, netlike web of arteries lying above the bones, which I afterwards touched with my hands, as I did with the whole head [*Et erat rubra subtilis contextura supra ossibus ex arteriis ipsis, quam postea manibus meis sicut totum caput tetigi*].[29]

In the light of later anatomical work, however, Vesalius would reverse this earlier stance, throwing into jeopardy the Galenic account of the *rete* that had assured the soul's operation upon the body.

The net's construction had allowed the soul to be both everywhere and nowhere, hovering delicately between embodiment and disembodiment. To untie this set of knots was thus tantamount to displacing the soul itself. Importantly for our purposes, Vesalius argued for some crucial alterations to Galen's theory of the brain and its functions. The workings of the brain were, perhaps more than any other body part, problematic for all anatomists. Even Vesalius confesses that while he could "in some degree follow the brain's functions in dissections of living animals," he could not understand "how the [human] brain can perform its office of imagining, meditating, thinking, and remembering."[30] Yet the impossibility of observing the brain in action did not dissuade Vesalius from expressing his admiration for this *admirabili machinae*:

> For who … can fail to be astonished at the host of contemporary philosophers and even theologians who detract ridiculously from the divine and most wonderful contrivance [*admirabili machinae*] of man's brain. For they fabricate, like a Prometheus, out of their own dreams … some image of the brain, while they refuse to see that structure which the Maker of Nature has wrought, with incredible foresight, to accommodate it to the actions of the body.[31]

Vesalius's enthusiastic description suggests that the key term in the title of his magnum opus – *Fabrica* – reflects a hidden fascination with the brain. As Roberts and Tomlinson put it,

> The word *fabrica* may be translated as *structure* or framework, but, as used by Vesalius, it has greater functional overtones. It could also mean a place where a thing is made, as well as the thing made. Obviously the word *fabric*, most commonly referring to material or cloth, can also be used as a metaphor for the intricacies of the tissues and organs of human anatomy.[32]

The brain thus epitomizes in its layered and textured complexity the very enterprise that engages Vesalius: to describe a machine so admirable that it exceeds even the imaginary "fabrications" of those anatomists blind to its reality.

Moreover, Vesalius's expression of wonder virtually displaces the brain from the body, in that he describes its role as merely accommodating "the activities" of the body, even as it is instead assigned the offices of "imagining, meditating, thinking, and remembering" – offices which, most

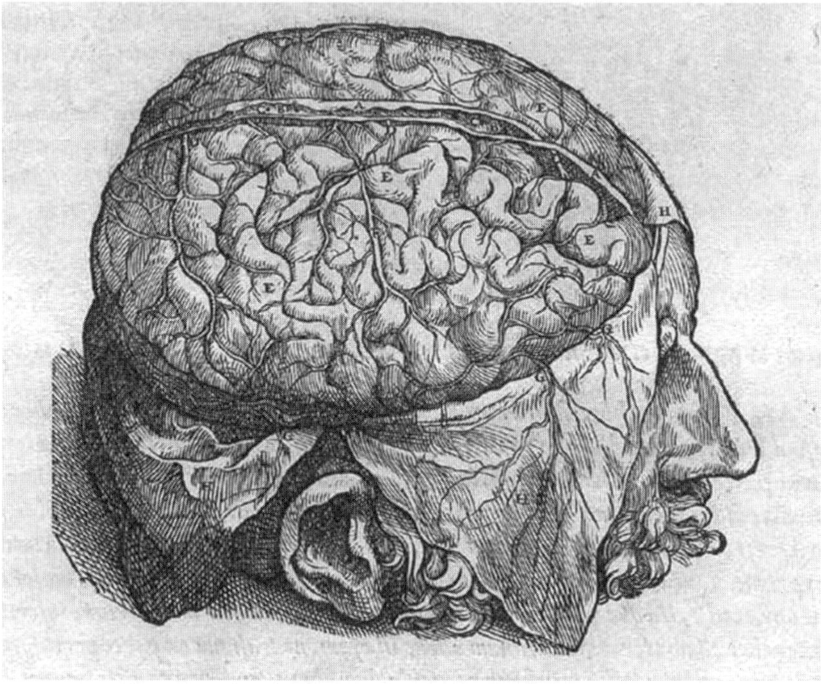

1.1 Head with skull and cranium removed, *dura mater* peeled back, to reveal the brain. Andreas Vesalius, *De humani corporis fabrica* (Basel, 1543), 606. The National Library of Medicine, National Institutes of Health, Bethesda, MD.

anatomists agreed, arose from the workings of the soul upon the brain. In arguing in this way for the preeminence of the brain, Vesalius effectively elides the distinction between it and the soul it housed. In so doing he anticipates the sense of the soul as "immateriall substance," as something both real yet incorporeal. His awe at the *admirabili machinae* suggests his willingness to place the seat of the soul somewhere *within* the brain, and not stretched across the *rete* as it generally had been in the Galenic tradition. Thus, though Vesalius had included the *rete* in the earlier fugitive anatomy entitled the *Tabulae sex*, he had realized by the time he wrote the *Fabrica* that the wonderful net did not exist in humans; rather, that arterial structure was only to be found in hoofed ungulates, a fact which Vesalius used as further evidence to debunk Galenic tenets of anatomy (since it implied that Galen had never even dissected a human body).[33]

For the most part, Vesalius's work spread quickly and was deeply absorbed into the contemporary English anatomical scene. English

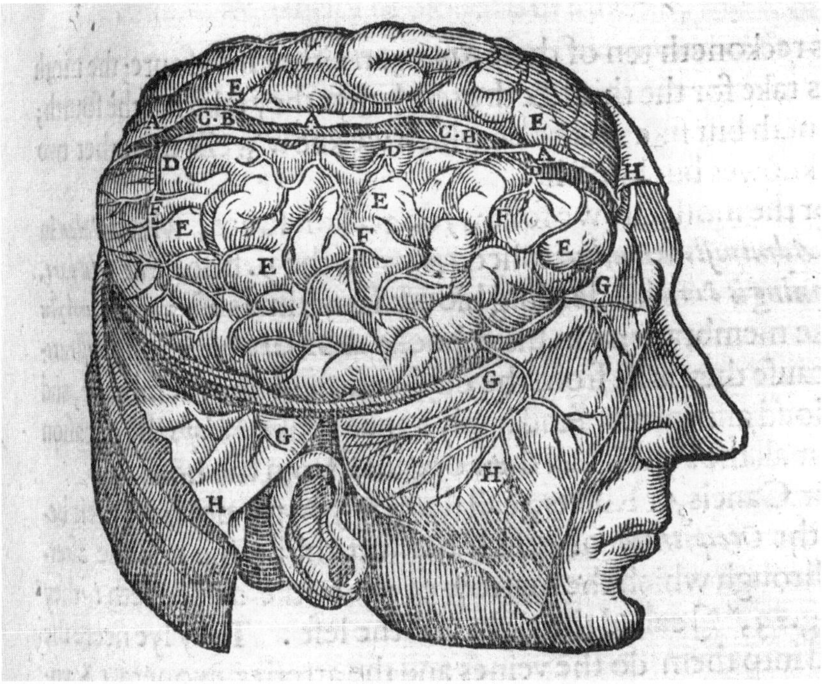

1.2 Cranial dissection after Vesalius. Helkiah Crooke, *Mikrokosmographia* (London, 1615), sig. Rr_1^r. The Huntington Library, San Marino, CA. RB 53894.

anatomists unabashedly copied Vesalian text and pictures for their own treatises. An early example is the 1552 English quasi-translation of Thomas Geminus's *Compendiosa totius anatomie delineatio.*[34] The Latin *Compendiosa* had been published in England in 1545, a mere two years after the publication of Vesalius's *Fabrica*, from which it plundered most of the illustrations along with large sections of text.[35] Such pilfering demonstrates clearly the immediate cultural impact of the *Fabrica.*[36] And Geminus's case was not an isolated incident: in 1615, one of the most famous and important anatomists in early modern England, Helkiah Crooke, too, happily purloined images from the *Fabrica.*

There was, however, one vital difference between Vesalius and his English successors. While they were willing to incorporate nearly everything else the *Fabrica* had to offer, they refused to give up the little, so-very-nearly-invisible *rete mirabile*. This curious inclusion does not seem to have happened by accident, and belies critics' claims that the

Fabrica triggered a total scientific revolution or "leap forward" to modern anatomical representation. Despite Vesalius's proposing another structure in the brain to assure the soul's control over the body – the interior lining of the ventricles, where the refinement of the spirits could occur, thereby still assuring the soul's control over the body – his new theory did not lend itself to as tenacious a metaphor as the net. Helkiah Crooke later insisted, for example, that "wee knowe and are able to demonstrate that no cavity of the bodye is the seate either of the sensative [*sic*] soule or of any of her Faculties, but rather the solid substance of the parts" ($Ss_1{}^r$). The nonexistence of the *rete* was the crucial difference that Vesalius introduced – and that English anatomists stubbornly denied.

Significantly, the denial concerned what was arguably the most important part of the whole body: even though it was so minuscule as to be nonexistent, it was charged with the all-important task of producing the transition from animal to human, as can be seen in Donne's "subtle knot which makes us man." This persistence of the wonderful net thus marks the integration of a key metaphor into anatomical parlance. It offered a simple and vivid image that anyone could comprehend and believe in. Such belief or faith proved necessary for these English anatomical heirs to Vesalius. Even as evidence mounted against the existence of the *rete* – in no small measure as a result of their own dissections – the metaphor persisted in their texts as if it were established reality.[37] Importantly, the seventeenth century witnessed a key transformation of the *rete*, from its original, Galenic instantiation as a crucial part of the arterial system, to a structure more thoroughly linked to the brain, marking, even, the gateway between earthly body and superior cortex. And yet the net, despite its nonexistence in humans, continued to intrigue anatomists because of its versatility as a metaphor.[38] This double response, as I wish to argue, testifies to the crucial importance of the metaphor's functions in the surrounding culture, functions that were extended and deepened by the otherwise very different "anatomies" of contemporaneous English poets.

"But yet the body is his book"

Charles Mitchell's oft-cited discussion of "The Ecstasy" notes of Donne's argument that "[i]n order for the soul to unite with the body, it must first separate itself from the body. The knot must first be untied in order to be tied: 'This Extasie doth unperplex' in the sense of 'untie,' since *perplex* derives from the Latin *plectere*, 'to plait' or 'interweave.'"[39] What the ecstasy marks is thus a form of unbinding, unfastening, and untying, in order to form a different kind of connection – a process that echoes the complex interweavings so central to the anatomies. Anatomical readings of Donne's poetry abound, most recently in the work of Sugg.[40] Most of these interpretations assume, however, a one-way flow of information: namely, that Donne wrote under the sway of medical knowledge, having read and understood a range of contemporary work about the body. That Donne made use of available medical theories is evident in particular in his sermons. Consider, for example, the direct parallels between the Galenic model of the passage of spirits (described above) and the following lines from one of Donne's 1619 sermons:

> [T]he spirits in a man which are the thin and active part of the blood, and so are of a kind of middle nature, between soul and body, those spirits are able to doe, and they doe the office, to unite and apply the faculties of the soul to the organs of the body, and so there is a man.[41]

Given how thoroughly some of its concepts infuse his writings, then, it is both valid and useful to insist on Donne's active engagement with contemporary anatomy.

Yet my concern here is less with the manifest influences of science upon poetry (or even vice versa) – and specifically upon Donne's poetry – than with the shared ideas that allowed the different disciplines to communicate with one another. That is, I am not arguing for a specific, causal-historical link between particular anatomical texts and a certain poetic example, but for a vehicle of transportation that connects the two.[42] Thus far we have seen how the *rete mirabile*, the wonderful net or miraculous knot, persisted as an explanatory metaphor, treated as if it existed even though its existence had been disproven by one of the foremost anatomists of the early modern era. Now I want to turn briefly to the corresponding idea in Donne's rich and complex poem to show how his use of the metaphor discloses some of the cultural investments shared by the anatomists.

The overarching "plot" of "The Ecstasy" seems easy enough to make initial sense of, despite the extraordinary difficulty posed by its details. Two lovers strive for a more perfect union. First they attempt to bring together their bodies, but fail to do so:

> So to intergraft our hands, as yet
> Was all our means to make us one,
> And pictures on our eyes to get
> Was all our propagation. (9–12)

Donne's lovers crave a thorough union, and the initial quandary lies in the fact that their attempts at physical connection fall short of their desires. Simple "intergrafting" of fingers and imagined pictures in one another's eyes only go so far, particularly since, as the narrator coyly reminds us, "it was not sex" that would ultimately conjoin the lovers (31).

Thus when physical means prove unsuccessful, the lovers coax their souls heavenwards for another try at commingling:

As, 'twixt two equal armies, Fate
Suspends uncertain victory,
Our souls (which to advance their state
Were gone out) hung 'twixt her, and me.

And whilst our souls negotiate there,
We like sepulchral statues lay[.] (13–18)

This upward flight, which separates souls from bodies, represents the ecstasy in question, and seems likelier to achieve the "intergrafting" by gaining the lovers that for which they long. Indeed, the moment of the ecstasy transforms the bodies from living owners of the souls to (literally) ex-animated near-corpses who are silent and whose postures are the same all day. The ecstasy allows the lovers' souls to cohere and for the lovers themselves to transform into "all mind," making of the two "a new concoction" mixed by love, which "makes both one, each this and that," thereby creating a subtle balance whereby the two are at once a whole entity and distinct parts (23, 27, 36). Each soul informs and repairs the "defects of loneliness" (46) in the other soul yet nevertheless retains its own identity:

But as all several souls contain
Mixture of things, they know not what,
Love these mix'd souls doth mix again,
And makes both one, each this and that. (33–6)

Ramie Targoff notes that the poet's previously invented term "inanimation" describes "the process by which spirit gets infused into a person"; "interinanimation," then, "conveys a sense of motion, a forward thrusting of soul into body in a manner that the ordinary term 'animation' lacks … [B]oth parties are giving life to each other, enabling them to form from their purely spiritual exchange a new creature."[43] Yet even this kind of creation proves inadequate, keeping separate the "two souls" and making a distinction between "that abler soul" and its "poor, and scant" counterpart (42, 43, 39).

It is at this point in "The Ecstasy" that the metaphor of the subtle knot enters the poem. For the inmixed souls, while separated from the bodies, nevertheless remain incapable of achieving the necessary kind of connection for which they seem to strive. Instead, the body must reappear to offer an alterative model to the interinanimation of the souls. Thus, the souls are now recalled to the bodies: the lovers

> owe them [the bodies] thanks, because they thus
> Did us, to us, at first convey,
> Yielded their forces, sense, to us,
> Nor are dross to us, but allay. (53–6)

The very fact of the ecstasy *depends* upon the bodies: without a physical compulsion there would be no need for the lovers to strive for a higher, more perfect form of marriage, and no means by which they could eventually join completely. Not only are the lovers' physical selves that which draws them to one another in the first place, but contemplating the limitations of physicality leads to the ultimate admission (by the "we" of the lovers' ecstasied souls) that *both* spirit and matter are required in a quintessential union. In fact, the bodies point the lovers to the fact that there exists in physical form the very entity responsible for "mak[ing] us man."

But, while critics have hitherto recognized the necessity of bodies in Donne's poem, the *form* of their coupling has not been as yet adequately analyzed: it takes as its paradigm the metaphor used to express the different kind of joining or knitting that brings together body and soul. In other words, Donne seeks to depict the desired connection *between* the two lovers in terms of the metaphor that expresses how each *individual* lover's soul is connected to his or her body.[44] Just as each person is "made man" by the operation of the subtle knot, so too must the souls of the lovers descend into their bodies again, and be subtly knotted together if they are to achieve the true ecstasy they seek.

Targoff's observations that Donne's *Songs and Sonnets* overall evince "a profound distaste for separation," and that the poems "are marked by his desire to ensure the possibility of future reunion in the face of impending division," are pertinent here.[45] For whereas anatomical discovery assumes a prior cutting, the *rete* and the "subtle knot" counter such divisions to guarantee the sort of union the lovers themselves enter into in

"The Ecstasy." Love has already rendered the pair nearly indistinguishable from one another at the poem's opening: witness the insistence on their mutuality in phrases such as "we two," "one another's best," "our hands," "our eye-beams," "our eyes," "our means," and so forth. The only time that pronouns separate the lovers is at the exact moment of the poem's recounting of the ecstasy's inception, at which point the two are in any case (equally) "two equal armies," when "our souls ... hung 'twixt her, and me" (15–16). That the distinction between genders should be observed here – rather than, overtly at least, anywhere else in the poem – serves as a reminder of the heterogeneous starting-points that love and its subsequent ecstasy will dissolve.

Donne is no monist, though we might be pardoned for thinking he aspires to something like monism in "The Ecstasy."[46] In a 1622 sermon he laments, "When I look into the larders, and cellars, and vaults, into the vessels of our body for drink, for blood, for urine, they are pottles, and gallons; when I look into the furnaces of our spirits, the ventricles of the heart and of the brain, they are not thimbles; for spiritual things, the things of the next world, we have no room; for temporal things, the things of this world, we have no bounds."[47] His frustration points to what is for him the irreconcilable difference of body and soul: the body, with its vast and innumerable store-houses, has no difficulty mapping and expressing the physical, while the soul is at best an uneasy tenant in these larders, cellars, and vaults.

While we must begin somewhere (hence the initial distinction of gendered pronouns), we end up with what Sugg terms a "mystic union of the lovers ... achieved through a number of quite precise materializing conceits, culminating not only in a broad joining of souls but in a psychological fusion of bodily spirits" and what Targoff describes as "[t]he ultimate indistinguishability of spirit from flesh once the soul is reincarnated."[48] This union is documented via the transformation of collective pronouns across the poem. The *we* adheres first to the purely physical in the first five stanzas, where "her, and me" are the owners of the "our souls" of line 15. By the time the "Dialogue of One" begins in line 29, the *we* has become indeterminate; true, "souls' language" is being spoken, but the nature of the identity of the speaker remains uncertain:

> "This Ecstasy doth unperplex,"
> We said, "and tell us what we love;
> We see by this it was not sex;
> We see we saw not what did move[."] (29–32)

The next three stanzas, which describe the nature of the union between the ecstasied souls themselves, are "spoken" in the abstract. When *we* returns in line 45, the shift from body to soul has become apparently permanent, for the speaker identifies itself as "this new soul" and remains in the realm of the spiritual for the rest of the poem – yet the problem of soul-from-soul distinction (explained above) remains, given that the "new soul" speaks of itself in the plural and describes the "atomies of which we grow" as being, themselves, "souls" (45–8).

It is only via the coupling of the coupled souls with their bodies that the true union comes about. The ecstasied souls gently berate themselves for forbearing their bodies – great thing of them forgot – and, prickly possessive, declare, "They are ours, though they are not we" (51). This seems a peculiar turn from the opening of the poem, but it marks the extent to which the lovers have become conjoined. The union, in turn, effaces the gendered distinction of line 16 and even strips the bodies of articles ("it to body" in line 60, and "we are to bodies gone" of the final line). The souls further point out that the physical must precede the spiritual union: "So soul into the soul may flow, / Though it to body first repair" (59–60). Bodies reveal love and ecstasy for "weak men" to read, as in a book, and such revelation requires that the souls "descend" to reanimate their physical selves (70, 65). The subtle knot thus not only "makes us man" by uniting immaterial and material substances – it also renders the union of the lovers in broader metaphorical terms. Just as these spirits knit the individual knot binding body and soul, so too love conjoins the bodies and souls of the lovers.[49]

The metaphors of knitting and chaining that surround the image of the subtle knot were not unique to Donne. John Davies, for example, uses the same verbs several times in his poem *Nosce teipsum* to remind his readers that, although souls themselves are "*bodilesse* and free," they must be "to grosse materiall bodies knit" in order to exercise their powers.[50] Likewise, the embattled soul in Andrew Marvell's "Dialogue between the Soul and Body" complains that it has been bound to the body "in chains / Of nerves, and arteries, and veins," enveloping constraints crafted by the very vessels meant to aid in the union between body and soul.[51]

Despite the fluidity and utility of such metaphors, however, there were also those of Donne's contemporaries who resisted the figure. Thomas Traherne, for example – whose scholarly study included works by Galen, Hippocrates, Galileo, and Plato, among others (thus intersecting eerily with Donne's own studies), and whose works were often in nineteenth-century scholarship confused with those of other metaphysical poets – often evinces what Barbara Lewalski describes as a "distrust of language and figures" to the extent that "[h]is poetics disparages metaphor as obscuring naked truth."[52] In his "Thanksgivings for the Soul" (in *Thanksgivings*), Traherne describes his soul as "A living Centre, wider than the Heavens," which nonetheless is bounded and constrained by the highly literal, if "endless," body:

> A Body endless, though endued with Sense,
> Can see
> Only visible things,
> Taste
> The Qualities in Meat and Drink,
> Feel
> Gross or tangible Bodies,
> Hear
> The harshness or melody of Sounds,
> Smell
> The things that have Odours in them.
> But those things which neither Sight, nor Smell, nor Taste, can
> discern, nor Feeling try, nor Ear apprehend ... Come not within
> the sphere of Sense:
> All are Nullities to such a Creature.[53]

Traherne here celebrates the subservience of material body to immaterial soul with pleasure quite similar to that of Donne and Herbert (and certainly none of the pessimism infusing Marvell's account). Yet his emphasis is on the strict physicality of the senses, thus his language is insistently corporeal; as Lewalski observes of Traherne's lyricism, "[t]here is much naming of beauties or joys or glories" – as in the example above – "as if such naming will call forth the essence of the thing."[54] Traherne is more invested, then, in the concrete roles of body and spirits insofar as they serve the soul – the ultimate relationship is still necessarily metaphorical, but the terms of that relationship are for Traherne thoroughly grounded. Donne – and by extension here, Davies and Marvell – presses instead upon other, analogical senses to render the soul-body connection legible.

Beyond the direct echoes of Galen's hydraulic model of bodily spirits, the shared investment of the Galenic anatomical texts and Donne's "Ecstasy" in the *rete*'s metaphoric labours and achievements is conveyed most powerfully in Donne's use of the word *subtle* to describe the "knot, which makes us man." *Subtle* expresses the workings of something keen, deft, or nearly imperceptible.[55] Yet the word bears another, older meaning, of which Donne would certainly have been aware: it derives from the (inferred) Latin **subtelis*, **subtexlis*, meaning "finely woven," words themselves formed from the conjunction of *sub-* and *tela*, *texla*, meaning "woven stuff" or "web."[56] The "spirits" or "fingers" doing the weaving represent the most sublimed substances in the human body, as they had been identified in the Galenic anatomical texts upon which Donne draws. But extracts from the blood, passed through and purified by the body's organs and arterial systems, are not in themselves sufficient: finally, they have to create the web-like structure which binds the all-too-material body to the immaterial soul, the "knot" that crowns the glory of God's creation. The word *subtle* hints at this process even as it marks the fact that such a structure would have to be so refined as to be virtually invisible.

Davies offers perhaps the most developed defense of the peculiar aptness of the metaphor of the net to make manifest the relationship between soul and body. Written around the turn of the century, *Nosce teipsum* presents a sort of historical overview of the different metaphors through which the soul's union with the body had hitherto been represented in order to ask the question: "*But* how shall we this *union* well expresse?" Davies rejects a number of metaphors for the union, complaining that they are too clumsy. The soul does not sit in the body "as in a tent" or the pilot of a ship; she does not leave her print in it, as in wax; she is not like liquid in a vessel, like heat in fire, or a voice spreading "throughout the aire."[57] The only comparison he can finally back is that of the knot. In language reminiscent of Donne, he settles on the idea of diffusion, of sunlight spreading her fingers through the ether, weaving the delicate connection. The delicacy of the connection is sublime: "Nought ties the *Soule*, her subtiltie is such, / She moves the body, which she doth possesse, / Yet no part toucheth, but by *virtues* tuch."[58] Davies's description, and in particular the homophonic pun *nought/knot*, points to what was evocative about the idea of the net or knot: it allowed the soul not to be tied to one specific location, but to be instead spread out over the web, or to exist in the various fibres of the knot. After wondering whether Donne could actually "confidently say what or where the subtle knot was," Sugg postulates that "The Ecstasy" "uses love to respiritualise a human body whose innermost core has been radically dislocated and problematised by the anatomists' scalpels."[59] Given the anatomical and linguistic evidence, I do not think it a stretch to identify directly the "subtle knot" with the *rete* and thence to read Donne's poem as expressing both a desire for and an uncertainty about the ultimate existence of such an anatomical structure.

The theological overtones of the metaphor evident in both Donne's and Davies's poems consistently accompany as well the anatomical texts which seek to shed light on the operations of the human brain. For example, the prominent English physician John Banister waxes eloquent in describing the arteries passing through the pineal gland into the brain (where they ostensibly formed the *rete*), telling us that they

> bring[] to passe a large and notable plicature, or weavyng: which … may wortheyly, for the marveilous workemanshyp thereof, be called the marveilous nette … Therefore of foure Arteries this marveilous nette is made, most notable to eche studious Anathomist, both for the dignitie … of the thyng, which before … was never taught or noted.[60]

Banister's admiration for both the craft of this "marveilous nette" and its "dignitie" is manifest in these lines. The "notable plicature, or weavying" recalls again Galen's image of the *rete*: the net is yet again the object of wonder and veneration, for of course the workman in question is God.[61] In a similar vein, Geminus – who along with Banister helped translate Vesalius's work for England – writes that "the brayne in the syghte of man is of a wonderfull and marveylous substaunce to be consydred, and it is also very straunge."[62] Its strangeness derives in no small part, though, from the very structure that Vesalius had rejected. Interjecting the opinion of a Persian physician into an otherwise faithfully Vesalian text, Geminus continues, "And herein I note greately the saynge of holye Abbas,[63] where as he speaketh of these small Arteries, of whome he affirmeth to be made a marvelous nette or caule, in the which the brayne is most necessarelye infolded and bewrapped."[64] It is telling that Geminus should transmute the usual rendering of the Arabic name of Haly Abbas to "holye Abbas," thereby turning an anatomical observation into a "saynge," something reverential and oft-repeated. Abbas would seem to bear a (theological) authority to surpass even that of Vesalius. While Geminus spins out the minute *rete* into a "nette or caule" long enough to wrap about the brain, his deployment of the metaphor remains very close to Galen's.

As with Banister, the association of holiness and wonder with the net expresses the attribution of the ultimate credit for this most marvelous of constructions to God (rather than to the anatomical authority who discovered it), suggesting perhaps that the connection between spirit and flesh would forever remain a mystery outside of anatomy's grasp. Sugg aptly summarizes the situation: "The wonderful net was not only a physiological zone of transition but – more subtly – a psychological one. Rapturously lost in the contemplation of this organic labyrinth, the viewer experienced a wonder sufficient to carry them, as well as the rarefied spirits of blood, across the intriguing space."[65] It is this intangible divide between soul and body that the metaphor of the net sought to bridge, through an intermediary entity which offered a richly viable image of connection even as it left room for further mystery and wonder.

"Labyrinthes and Meanders"

But in another sense, the emerging anatomical emphasis on the brain prepares the way for the *rete*'s dissolution in the second half of the seventeenth century. While Vesalius and Galen may have disagreed over the existence of the net, they both conceived the site of the final purification of bodily spirits to be part of the *arterial* system. Over the course of the seventeenth century, however, the *rete* is first re-located in the *brain*, as the brain itself comes into prominence, preliminary to being discarded altogether. Geminus's description of the rete as a "caule" and Banister's location of this organ in the pineal gland already indicate this growing fascination with the brain, signalling the end of the long and rich career of the metaphor of the net.

The transition can be glimpsed with some clarity in Helkiah Crooke's 1615 *Mikrokosmographia*, which argues (as its title suggests) that the human body is a readable version of the vast universe. On the one hand, Crooke does not yet relinquish the *rete*, for it not only continues to fulfill the Galenic dictum of translating vital into animal spirits, but it also marks the importance of the process. Even as the *rete*'s hold on anatomical thought gradually slackened, the earlier emphasis on the peculiarity of the knot and how essential it was to creating the human being infused the seventeenth-century treatment of neuroscience. For the goal was not simply to map the brain, but to probe its possibilities and to consider the philosophical implications of uniting flesh with spirit. Like his predecessors, Crooke, too, expresses awe and wonder over the *rete*, noting that it seems

> as if you should lay many fishers Nets one above another; wherein this is admirable, that the replications of one are tyed to the replications of another so that you cannot separate the Nets asunder, but they are all of them so wrought into one another as if it were a bodye

> of Net meshed together not into breadth onely, but even into thicknesse also. (Ss_1^v)[66]

Crooke describes a complex geometry accessible only through the familiar metaphor of the net. The "replications" of the net are not simply balled up ("meshed together"); rather, they are connected in an orderly fashion in three dimensions, as Galen says they should be. The fact of these connections is "admirable" and the minute and mysterious arteries seem apt for the refinement of the animal spirit, as thinner passages make for more ethereal vapours. Because this "Wonderfull Net" is "so wrought" as not to be any complete "bodye" but a porous set of "replications," it serves as a useful receptacle for the soul itself.[67]

At the same time, beginning with the *Mikrokosmographia* the *rete* as such would be depicted pictorially not simply as a part of the arterial system (as in earlier anatomies) but as a being in its own right, associated strongly with brain anatomy. Thus, net-like metaphors of complication persist, but are increasingly transferred on to the brain itself, and linked in particular to an emergent reconceptualization of the nervous system. Consequently, Crooke speaks equally reverently of the brain in a way that recalls his celebration of the *rete*:

> the most noble part of the whole body and framed with such curiositie, so many Labyrinthes and Meanders are therein, that even a good wit may easily bee at losse when it is trained away with so divers sents in an argument so boundlesse and vaste. (Oo_6^r)

Crooke sees the brain as a complex and potentially unknowable structure (and in so doing echoes the *admirabili machinae* of Vesalius).[68] He sets up the brain here as something stranger than – and perhaps even foreign to – any other part of the body, for "curiositie" implied not just novelty or strangeness but exquisite craftsmanship and scientific investigation.[69]

The seventeenth century emphasized the place of the *rete* in contexts beyond what was purely necessary. For example, in a 1610 sermon, Alexander Chapman develops the metaphor to describe the dangers facing the sinner:

> Anatomists do write, that in the brain of man, there is a *rete mirabile*, an admirable net, that is, a heap and clod of arteries, that for the many windings, and turnings, and intricate infoldings, can not be anatomized; and so indeed, as if *this* of the body were for to signify

> *that* of the mind; in the brain, and the wit, and the wisdom of man, there is a *rete mirabile*, an admirable net, a heap and clod of manifold infolded subtilties; that for the windings and turnings, and intricate devices, can not be anatomized; with which admirable net, we do catch the poor fish and fowl that we do deal withal: of these it is, I exhort you to repent.[70]

Chapman's language here recalls Galen's description of the *rete*, but, even though he translates *mirabile* as "admirable" (and repeats that word three times in this passage), he explains the ways in which these "windings, and turnings, and intricate infoldings" signify as negative; Galen's fisherman's net is transformed into a trap for the "poor fish and fowl" of good intentions. Chapman does also make a point to deploy the word *subtle*, which in turn amplifies the potentially devilish nature of the *rete*, both anatomical structure and spiritual extension.[71]

Then, too, we may recall Donne's fingers which knit the subtle knot; the final transformation of the soul in "The Progress of the Soul" finds it lodged at last in a human body. The soul nevertheless remains in motion, reluctant to pin itself down to any one place. Once the liver and heart have developed in the body, the brain is forged from the stew and form and alchemists' fires of the womb:

> Another part became the well of sense,
> The tender well-armed feeling brain, from whence,
> Those sinewy strings which do our bodies tie,
> Are raveled out, and fast there by one end,
> Did this soul limbs, these limbs a soul attend,
> And now they joined.[72]

The play on "well-armed" and "well of sense" to describe the brain suggests both that the brain is the fount of sensation and rationality and that arms spread from the brain throughout the body. We can imagine the brain spouting nerves like arms – "sinewy strings which do our bodies tie" – which then "ravel out," snaking forth through the channels of the body. These nerves are what the soul would use to perform its will in the body, as, lodged in the *rete mirabile*, the soul had access to the nerves through which it sent the rarified animal spirit. Donne mentions nets directly elsewhere in the poem, too: women weave nets to trap men in sin, nets are cast to catch birds, fish find themselves entangled in nets. In these other examples the net functions more as a prison for capturing the body to

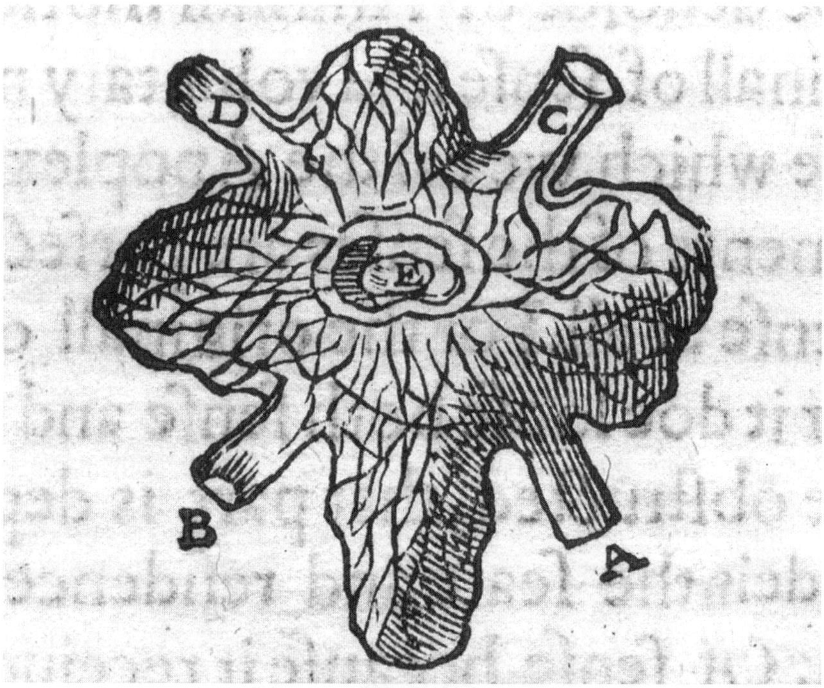

1.3 "[T]he wonderful Net as Galen describeth it." Helkiah Crooke, *Mikrokosmographia* (London, 1615), sig. Ss$_2^{r}$. The Huntington Library, San Marino, CA. RB 53894.

corrupt the soul, like the "nerves, and arteries, and veins" binding Marvell's soul to its body. Yet Donne's description of the soul threading its strings through the body, unravelling itself in order to join with its physical counterpart, is elegant, even beautiful.

Crooke obligingly provides us a picture of the subtle knot. Divorced from its earlier arterial context, Crooke's picture of the *rete* places it instead among the anatomical depictions of the brain and nerves. Unsurprisingly, Crooke's image owes a debt to Vesalius's *Tabulae sex* of 1538, in which the Italian anatomist had included the *rete* before recanting its existence in the *Fabrica*.[73] Intriguingly, Vesalius did include a representation of the *rete* in the *Fabrica* (which also aided Crooke's reproduction) – but, as John Forrester points out, that was because "tradition was still strong in his mind."[74] Indeed, Vesalius very nearly taunts his readers when he explains the inclusion: "[W]e have fabricated the *rete* the way it ought to be to suit Galen's descriptions."[75] The structure is, in both cases,

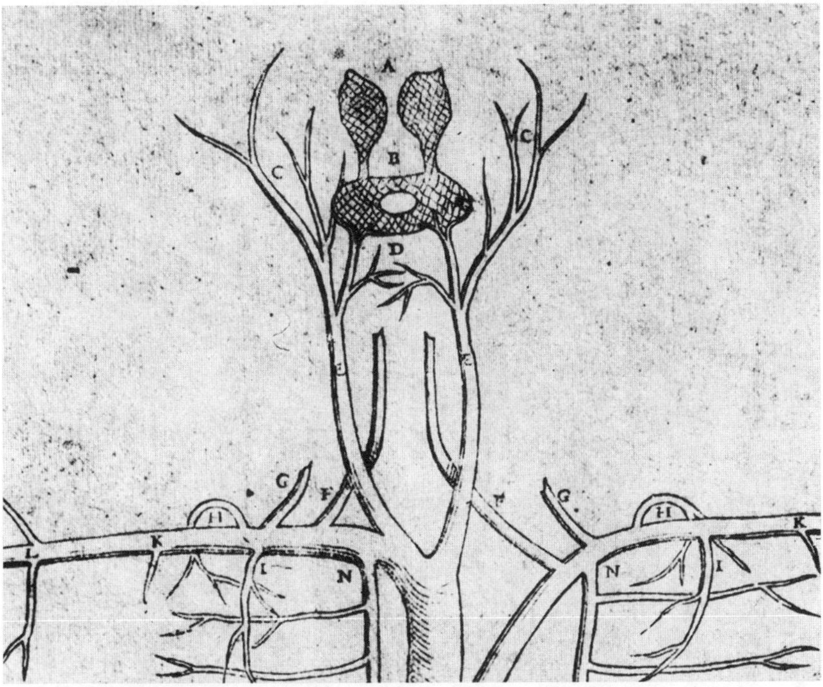

1.4 The *rete mirabile* (B). "Arteria magna," Venetiis, B. Vitalis Venetus sumptibus Ioannis Stephani Calcarensis, in Andreas Vesalius, *Tabulae anatomicae sex* (Venice, 1538), plate 3. The Wellcome Library, London.

represented by fine crosshatching, though Crooke's rendition appears decidedly more filament-like.

But there is a crucial distinction between these two images: Crooke's *rete* is completely disembodied, whereas Vesalius's *rete* (of the existence of which he was at that point confident) retains its connections to a rudimentary arterial system. Vesalius's *rete* links directly to the other arteries in the head, whereas Crooke's *rete* is isolated from its contextual surroundings but does have (at A, B, C, and D) dissected arteries stemming from it. Not only is the latter image one of the stranger pieces of anatomy within the brain, it is not even obvious from the figure that the structure sits at the very base of the brain. Indeed, this ambiguity expresses a more general instability in the early seventeenth century. Though many pre-Crooke English anatomists grouped images of the *rete* with the rest of the arterial system, their discussion of the structure took place within sections devoted to the brain. This disjunction suggests in

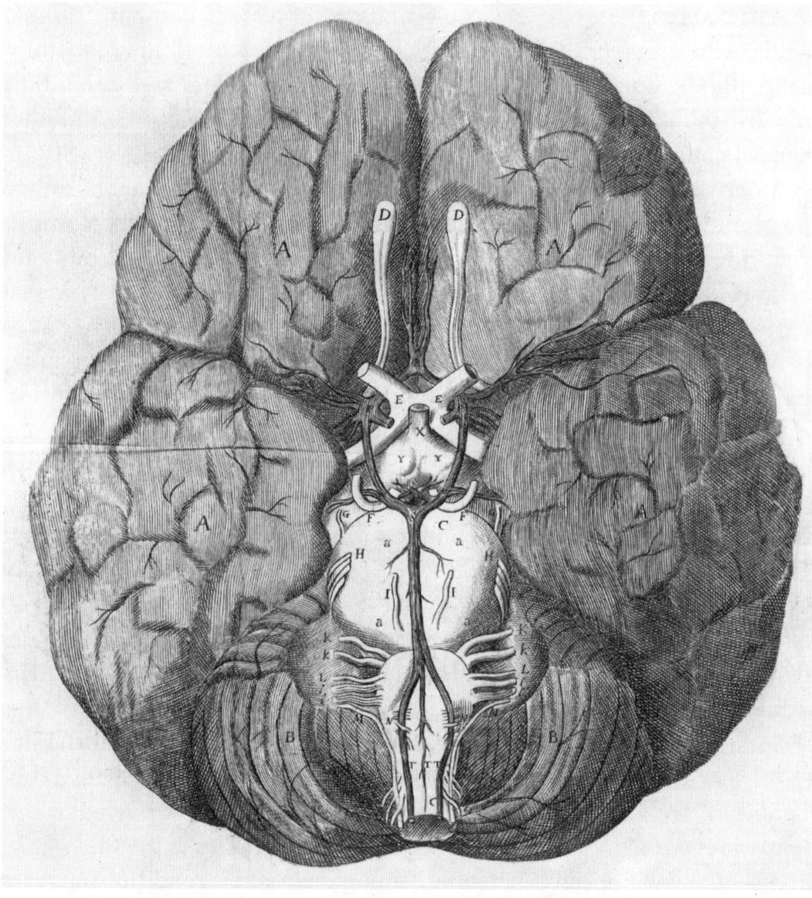

1.5 The circle of Willis (centre, surrounding E-E and Y-Y). Illustration by Christopher Wren. Thomas Willis, *Cerebri anatome* (London, 1664), Figure 1a. The Huntington Library, San Marino, CA. RB 357286.

part the uncertainty anatomists must have felt as they probed the brain for a net that was not there, but it also points to the *rete*'s privileged position as something both inside and outside the brain, something which transcended the functions of the arterial system to join itself with the far superior brain. The fact of its puissance is due, again, less to its empirical reality than to its utility as a metaphorical structure: on one hand, it evoked scientific curiosity and reverential wonder, and on the other, it housed the mystical union between body and soul.[76]

As a coda to the works of Crooke and his successors, compare Thomas Willis, whose *Cerebri anatome* of 1664, the first dedicated neuroanatomical textbook, coined the word *neurology*, and was illustrated splendidly by Christopher Wren. (It was translated into English in 1681.) Willis's "special genius," as Rousseau argues, "lay not in the scientific veracity [in a modern sense] of his theory" of the localization of the soul entirely to the brain, "but in his ability to deflect men," in a Kuhnian sense, towards what Rousseau considers a paradigm shift in the brain's responsibility for the soul: Willis was "the first scientist unassailably to posit that the seat of the soul is strictly limited to the *brain*, and nowhere else."[77] While Willis mentions the "admirable" or "wonderful" net several times, he notes that it is only sometimes found in humans – and drastically attenuates its use and function. He does point out that the *rete* consistently exists in certain four-footed creatures; while most humans do not typically possess it, those who "lacked all vigor of mind" may be found to have it, confirming their status as more brutish than cultured.[78]

Where the *rete* should be, Willis discovers a different anatomical structure, for which he is best known today: the circle of Willis. The "circle" is no net, but rather a set of arteries responsible for supplying blood to the brain; it represents in this sense a transmutation of the earlier Galenic understanding of the *rete* in that it restores its link to the arterial system. This physical function of keeping the brain (and by extension the mind) alive had also been one of the functions attributed by Galen to the net. In a way, the idea of the wonderful net, albeit in altered shape, has survived to this day. However, what the *rete* loses with Willis are its duties to the soul.

"This net thrown / Upon the heavens"

This chapter has narrated the development of a particular idea. Often critics trace in the literary signs of the scientific; that is, the concern is with evidence that an author has read, absorbed, or become familiar with a given scientific theory. But the argument, for example, that Donne's treatment of the human body in his poetry is indebted to the interest in medicine he inherited from his father makes the connection between the two discourses only in one direction. As readers of literature, we have often sought proof of the influence of other, non-literary discourses upon the text. This critical method has served us well for the past century, elucidating as it has our understanding of the way an author's mind works, or how an aspect of culture shapes the creation of the literary text. Yet we might also try to frame the question in different terms altogether: for it will appear to the modern reader/thinker that nigh-insurmountable walls have been erected between literature and science. Where then are the places through which the observer may glimpse the other side?

As we noted in the introduction, C.P. Snow famously addressed this issue by identifying the two distinct "cultures" into which his academic acquaintances seemed to fall; this observation, made in 1959, has arguably had a hand in the wall's construction, even if the foundations for the wall had already been laid centuries before. The denizens of Snow's two cultures possessed equal amounts of intelligence and enthusiasm for their respective fields, yet each culture regarded the other with what Snow saw as open hostility – which foreclosed on the possibility of negotiation or the creation of a common language. Scientists and their foils, the "literary intellectuals" who had been charged with maintaining a "traditional culture," suffered from a "mutual incomprehension."[79]

It is indeed the case that a handful of thinkers and writers since Snow have attempted to forge Snow's third culture, albeit with mixed success. The difficulty is that most modern writers of the "third culture" are scientists (e.g., Steven Pinker, Brian Greene, Richard Dawkins); they necessarily write from the perspective afforded them by their training (i.e., their culture) – not in itself a bad thing, as it represents a movement in the direction Snow suggests. But the critique of these scholars from the side of the humanities is that they often miss important subtleties, and as such they rule out the effects of society, culture, and history on the development of artistic thought. Thus, while these works represent steps toward the third culture, because they impose scientific readings on literary texts (for example), they tend to divide rather than to unite.[80]

The idea pursued here has not been so much to read science through the lens of literature, or vice versa, but to take a step back in an attempt to comprehend *as a whole* the projects of what appear to be two discrete disciplines. This kind of step proves especially achievable in the context of early modern literature, because the sorts of rifts among thinkers that are part and parcel of our modern experience (and that Snow laments) were far less apparent then – and in some cases, it is only due to the force and prevalence of today's assumptions of a sharp division between the arts and the sciences that these rifts seem to us to exist at all in texts from an earlier era.[81]

Intriguingly, the modern intellectual space most ripe for a reintegration of the two, three, or even four cultures appears to be neuroscience. Consider, most recently, a collection of essays celebrating what its curators term the "neuroscientific turn": Melissa Littlefield and Jenell Johnson, in *The Neuroscientific Turn: Transdisciplinarity in the Age of the Brain*, argue that "neuroscience is a *translational discipline*: a set of methods and/or theories that has become transferable – sometimes problematically so – to other disciplines. We contend that the complexity of the brain and its centrality in most human endeavors ought to be reflected in equally complex, multifaceted modes of inquiry, perhaps best facilitated by transdisciplinary conversations."[82] Essays in the collection consider not just neuroscience, but other "neuro" fields, such as neuroeconomics, neuroethics, neurotheology, neuroskepticism, and neuroliterary theory. This proliferation of neuroentities and neurodisciplines is, of course, not without its discontents. In an essay in *The Neuroscientific Turn* titled "'The Paradise of Non-Experts': The Neuroscientific Turn of the 1840s United States," Justine Murison prudently both celebrates and cautions against a modern iteration of the nineteenth-century fascination with mesmerism, which "acted as a conduit between medical circles and the public, helping to make knowledge about the physiology of the nervous system widely available." While such a conduit enabled the rise of psychology as an independent professional discipline, Murison argues, it would eventually descend into "merely a laughable and possibly fraudulent pastime."[83]

This is not to suggest that a similar fate will necessarily befall current inter- and transdisciplinary research – but the current fecundity of the field of inclusive neurostudies highlights the care that must be taken by its practitioners to preserve the novel and familiar links between the sciences and the humanities. Such care involves a heightened sensitivity to the problem of influence. As Rousseau cautions, we ought not to assume that "imaginative literature – and by this I mean poetry, fiction, the drama – is influenced by science *at once*, as we know this is not true of [for example the] eighteenth century merely by noticing that it took Newtonian science at least one generation, and longer, to 'demand the muse.' It is no less dangerous at this point and consequential for the future of [historical literary] studies, for us to confuse *imaginative literature* and *speculative science*."[84] Given what we might term the modern interinanimation of neuroscience and literature, Rousseau's concern seems applicable to most interdisciplinary work of this nature.

Thus it is neither strictly necessary nor even exactly true that the *rete mirabile* survived because it *had* to explain the conjunction between the soul and the body, or because it *had* to manufacture a means of communication between the two entities.[85] That was doubtless its primary function, but Vesalius managed to propose other locations in the brain (such as the ventricles) that could have served the same office – yet those did not even come close to proving as popular for English anatomists as the *rete mirabile* did. Rather, the *rete* survived not just because it *explained* something, but also because its very form *marked* something: rather than being simply the means for transmitting "sensory experience of life to the brain" and conveying "motility, the outward sign of life" to the rest of the body, the wonderful net persisted despite anatomical evidence to the contrary because it captured and expressed the cultural anxiety concerning the *place* of the soul.[86] What literary texts worked on as a metaphor to explore the complex connections between body and soul was transposed in anatomical texts into a concern over the extent to which – and the manner in which – the soul could be said to be embodied.

The *rete mirabile* thus made manifest the desire not only that the soul have a physical presence in the body, but that it be an object whose characteristics could be studied, dissected, and explained. The uncountable ramifications and branchings that this wonderful net offered could in principle be detailed; its physical complexity combined with the difficulty of actually finding it thus offered the appropriate vehicle for a soul that was simultaneously manifest everywhere (i.e., distributed across the threads) and nowhere (i.e., not tangible at all). Such a paradox tightly mirrored the relationship between body and soul as it was rendered in poetic and theological texts. In addition, the figure of the net suggested a balance to the actions of anatomy: whereas the latter worked to unpack, dissect, and pull apart, the former bound, joined, knit, and enclosed.[87] The ability to pull together what death necessarily tore asunder must have held some appeal. Finally, the *rete*'s existence required a strong degree of faith to accept: to acknowledge its presence meant to run counter to the very authority – Vesalius – whose work had made possible the kind of investigation that English anatomists were conducting. That the wonderful net didn't actually exist ultimately mattered less than the fact that a century and a half of anatomists were willing to believe that it did.

The reason for the wonderful net's tenacity in early modern English anatomical work thus derives in large measure from its metaphorical utility – a utility of which poets had already taken note. Consider, for example, the following lines from Donne's *First Anniversary*:

For of meridians, and parallels,
Man hath weav'd out a net, and this net thrown
Upon the heavens, and now they are his own.
Loath to go up the hill, or labor thus
To go to heaven, we make heaven come to us.[88]

Jonathan Sawday suggests that Donne's net here is "woven out of the new philosophy of reason" and "functions in both a positive and negative sense": the net promises knowledge and order even as it creates for its thrower a kind of trap.[89] The measurements of longitude and latitude with which the globe was being described encase not merely the world, but the entire universe. But the poetic investment in these lines extends even beyond the physical universe. We ought to recall here Crooke's analogy that the heavens are to the earth what the brain is to the body: the net Donne describes knits the eternal soul to the temporal body, the metaphysical to the physical, even as it binds macrocosmos to microcosmos.

More dangerously, however, the net also becomes an emblem for imprisonment: recall the "chains / Of nerves, and arteries, and veins" that Marvell's trapped soul had to endure. And this, too, is another function of the *rete mirabile.* The *rete*, like a fishing net with its struggling fish, encloses the soul and subjugates its power to the nerves of the body. By putting the soul under the body's control, it can equally be, as we see with Marvell's soul, the agent for an immense amount of pain – to worry at the connection between body and soul is not a task to be undertaken lightly. It is this conjunction of promise and difficulty that also becomes evident when we review Donne's "Ecstasy" through the lens of contemporary anatomy:

> [O]ur blood labors to beget
> Spirits, as like souls as it can,
> Because such fingers need, to knit
> That subtle knot, which makes us man.

The "subtle knot," woven by the spirits, hidden beneath the brain, the work of a metaphor upon anatomical imagination, so minute as to be invisible but charged with holding our most valuable possession – the thing, in short, "which makes us man" – is nothing either more or less than the wonderful net. And yet it is also the product of "labor" and "need" – not just of the body, but also of anatomists and poets.

CHAPTER TWO

Altered States: *Hamlet* and Early Modern Head Trauma

At all times when a man is about to commit any thing in custody to his *Memory*, first let him study to drown all unnecessary thoughts in oblivion, that he may perfectly intend the things he is to learn.

– John Willis, *Mnemonica; Or, The Art of Memory*[1]

"About, my brains!"

John Willis's 1618 memory treatise advises its student to take a course of action strikingly similar to Hamlet's infamous pledge to wipe from the table of his memory "all trivial fond records, / All saws of books, all forms, all pressures past / That youth and observation copied there" (1.5.99–101).[2] The underlying idea – which is not without its discontents – is that a blank slate allows one all the better to entrust to the custodian Memory the important things requiring remembrance. Willis continues:

> *Oblivion* being such a principle of *Memory*, as *Privation* is of *Generation*; and a ready remembrance most commonly proceedeth from right understanding the thing in hand; therefore a man must prepare himself diligently, and so untie the force of his imagination, that he may as it were engrave and imprint occurent things in his Memory.

For Willis, the effacement preparatory to imprinting involves unlinking imagination from memory, thereby negating its potentially distorting sensory and intellective influence so that memory can focus on the "thing in hand," allowing present occurrences to "engrave and imprint" themselves upon the mind.

But how should one go about effecting such an erasure, and where, physically, should that erasure occur? We are of course now aware that the seat of memory is the brain, and our current understanding of the functions and specific localizations of memory's component parts derives from decades of study involving lesions. In such studies, scientists make inferences about the state of a healthy brain by observing perturbed behaviours in subjects with specific (known or intentionally created) types of brain damage.[3] Work of this sort has been invaluable for mapping cognitive function onto brain location, albeit tentatively and with numerous caveats.[4] This method of studying the brain has its roots in the broader history of neuroscience. Moreover, in the absence of modern imaging technology, these roots, as I argue below, are necessarily steeped in violence and trauma to the brain. Drowning the active, sensing mind in oblivion, or wiping clean a mental tablet, is, of course, not an innocent action.

Criticism of *Hamlet*'s fascination with the human mind and body has often focused on Yorick's skull as a *memento mori* upon which Hamlet meditates as he attempts to answer his seminal question, "To be or not to be?"[5] Here, I am more interested in what is (or rather, was) *inside* the jester's skull (and inside all the skulls in the play). I want to suggest that in *Hamlet* the violence of a willed oblivion inaugurates investigations that seek to reconstruct perceived damage to memory, to self, to brain, and to state, in order to understand the nature of such damage and how a healthy or normally functioning system would operate. In short, *Hamlet* is distinctive in its concern with the brain and with what transpires should that brain be damaged in some way. And, as we shall see, the particular investigations carried out in the play have close metaphorical connections to those carried out by contemporary surgeons and anatomists.

The wounds involved in *Hamlet* are evident from the play's inception: the opening question, "Who's there?" presumes an absence, a hole in the proper functioning of a system (i.e., if the machine of perception and cogitation worked fluidly, it would be obvious *who was there*). What, then, can be utilized to work through this lesion, to solve the "crime" of the damage, to restore order? Two central mysteries in the play – the murder of Old Hamlet, and young Hamlet's madness – constitute both real and metaphorical wounds which, in turn, require Hamlet in the first place, and Claudius in the second, to attempt to study and understand the extent of the damage and then to reason backwards in order (ostensibly) to restore justice or health.[6] One investigation, then, involves the work Hamlet must do to test the truth of the Ghost's testimony: in order to "catch the conscience of the king," Hamlet stages "The Mousetrap" as a kind of theatrical laboratory which "comes near the circumstances" of Old Hamlet's death, allowing him to recreate in a controlled space and manner the initial scene of trauma (i.e., Claudius's poisoning of Old Hamlet: 2.2.540; 3.2.72).

Hamlet's play-within-a-play has been convincingly read as a precursor to the early modern scientific experiment; so, setting that aside, I examine here a different line of affiliation centred upon the initial traumatic event that triggers the other investigation in the play, namely that of Claudius (and by extension the rest of the Danish court) into Hamlet's mind.[7] Whether real or feigned, the prince's madness occurs after his visitation from the Ghost, and after he pledges to wipe his memory so as to inscribe the Ghost's commandment in the "book and volume" of his brain. I focus on this pledge, for it is, I will suggest, a self-inflicted wound, a violent act which damages Hamlet and renders his memory and brain a locus of forensic investigation for Claudius.

A curious thing about *Hamlet* is that every instance of the word "brain" has to do with the organ being ill or shaken or emptied or mistaken. In other words, the brain is constantly seen to be fallible in the play, an entity forever malfunctioning and which no amount of thought or action seems to be able to set quite right. Horatio, for example, fears that the Ghost may put "toys of desperation" into the young prince's brain (1.4.75–6); Hamlet offers to wipe his brain so that the Ghost's commandment will live there (1.5.98–103); Hamlet rails against his own brain for making him so pigeon-livered as to lack the gall to stand against his uncle (2.2.522); Claudius worries that Hamlet's brain puts him "from fashion of himself" (3.1.173–4); Gertrude, after Hamlet has seen the Ghost which she has not, fears that it was an hallucination, a "coinage" of his brain (3.4.135). The human brain in *Hamlet*, then, is always apt to misfire – and it is crucial that the brain in question is invariably Hamlet's.[8]

To recentre the interpretation of this play on the brain further allows us to see its indebtedness to an increasingly important area of study in early modern medicine. Physicians and anatomists working in the sixteenth and seventeenth centuries evince a kind of anxiety about the location or the locatability of the soul that appears remarkably similar to our modern attempts to localize, for example, semantic memory in the brain's hippocampus or certain types of emotion in the limbic system.[9] As we saw in chapter 1, an increasing number of early modern thinkers argued that the location of the soul was the human brain – and the brain therefore was not the domain of science and medicine alone; rather, it was the site not only where body and soul connected, but also where various disciplines productively converged. Indeed, the work of early modern physicians describes an intriguing tie between experimentation and an understanding of how the mind and body were linked via the brain, a link that *Hamlet* productively exploits. To probe the play in this manner demands, though, a brief contextualization, by considering the kind of investigation that contemporary surgeons and anatomists were conducting when faced with damaged brains. This historical context will provide a framework both for my reading of *Hamlet* and for thinking about the emergent practice of neuroforensic investigation. Finally, a brief foray into *Macbeth* will extend the thread I identify in *Hamlet*, connecting the problems of the Danish Prince to the broader context of Shakespeare's tragedies.

"Private errour"

There are quacks abroad, warns physician Franciscus Arceus in a 1588 treatise. Something is rotten in the state of medicine when "unlearned and unskilfull Chirurgions" are unleashed to practise upon patients: "obscure and unperfect experience, [*sic*] is joyned with most perfect knowledge," laments Arceus, such that these hapless pseudo-doctors "know nothing els then that which they have learned of their Maisters of the same sorte … everie one of them doth follow his owne sense or private errour, having no author at all, although never so unlearned to lead them thereunto."[10] Arceus's key complaint here has to do with the imperfect transmission of knowledge and experience, either via incompetent teachers or through the faults of the students themselves. Indeed, it is the impossibility of distinguishing student from master that causes, in part, such a collapse of authority, for if "every one of them doth follow his owne sense," then where is the original, where the appropriate method of medicine? Transmission errors destroy the proper growth of intelligence; for such errors to be "private" is for them to be like a festering wound or a disease, whose observable symptoms are not merely "covetousnesse," "want of judgement," or "arrogancie of minde," but also "lacke of knowledge."[11] Such serious "errour" erodes the characters of the would-be practitioners – keeping them too content in their work, too dull – but, naturally, it also has serious consequences for the patients unlucky enough to fall into their faltering hands.

It is not by chance that Arceus's admonishment appears at the start of a text on how to treat head wounds. Indeed, the purpose of *A most excellent and Compendious Method of curing woundes in the head* is to staunch this damaging ignorance, by laying bare a method that will curb the tide of "errour" which leads to madness (and worse) in mishandled patients. For the prognosis for head wounds in the seventeenth century does not seem to have been particularly rosy. "Grievous wounds are either *Periculosa*, or *Malefica*, or *Lethalia*, or *Difficilia*," notes the celebrated physician Alexander Read in one of his lectures (1650) at the Barber-Surgeons Hall: "*Periculosa*, or dangerous wounds, are such as sometimes are cured, although for the most part they bring death; such [are] superficiall wounds [that is, wounds to the surface] of the braine."[12] Read cautions that damage to the brain is naturally very dangerous because of the organ's preeminence in the body: "the wounds of the braine ... are deadly, if they penetrate to the basis; because it is a principall part, the well-spring of the animall facultie, in continuall motion, and because inflammation, and a sharpe fever insueth."[13]

Arceus would have been acutely aware of these potential complications. His panacea for what he identifies as "errour" is *method*, the careful accounting of a system, the laying-out of a standard procedure which will not be corrupted by ineptitude, or deviated from by those who think they know better. He rebukes such authorless practitioners and students for their failure to consider the wounds they encounter with anything like order or analysis. The unmethodical physician, according to Arceus, examines only the surface, not caring to dig deeper. Such physicians

> doe neglect to serch out throughlie whether any thing be hurt or perished in the *Rete Mirable*, or any of the other panicles or compactions of the braine, for the partes of the lower or innermost bone ... oftentimes happeneth to be cut in sunder, shivered, dashed, & broken in peeces, and moved out of their places, and that fault is found more oftener in the inner Table, then in the upper. Which thinges first most grievous panges and griefes, and after death it self doth ensue.[14]

The "most grievous panges and griefes," as well as death, follow not merely from injuries to the brain, but from the investigative failures of quack surgeons, their "private errour[s]" turned horribly public.[15]

But what could be done to investigate the "inner Table" of the brain, particularly in the late 1500s, when Arceus was writing? After all, modern scientists have only recently worked out how to peer into the interior of the living, functioning brain. The error of the early modern injured brain would seem to be quite private – so private, in fact, as to be unintelligible even to the person who suffered the wound. The diagnosing physician could only hope to understand the extent of the injury in one of two interrelated ways. Neither of these would grant him access to the original error or fault, but they would at least allow him to investigate obliquely. First, he could attempt to recreate the circumstances of the wound in the hopes of discovering the nature and extent of the damage. Second, he could observe the patient's behaviours – which here would include suffering "grievous panges and griefes" and then perhaps death – and then hypothesize about which parts of the brain had been affected. Such a method represented an early means of scientific systematization: the physician could diagnose the troubled mind by stripping away complications to get at the essential truth of the wound. Anatomist Helkiah Crooke describes a similar methodology, whereby one must reason backwards from effect to cause in explaining how we may use anatomy to discover the nature of God: "all the knowledge of God that can be had," he writes, "must be derived … not from any cause or matter preceding, but from the effects and thinges subsequent."[16] Thus when Crooke argues that the brain, rather than the heart, is the principal member of the body and so is responsible for motion and sensation, he does so by contrasting what happens when something adversely affects each organ. Our own experience, he writes, suggests that "when the ventricles of the Braine, are either compressed, or filled and stuffed up, as in the Apoplexy, Epilepsie, and drowsie *Caros*, then all the faculties are respited and cease from their functions; but when the heart is offended, the life indeede is endangered, but neither motion nor sense intercepted."[17]

Here we can glimpse a rudimentary version of the technique widely employed by modern cognitive scientists in order to understand the brain's functions – namely, the lesion study. In an actual lesion study, cognitive scientists usually recruit patients who have – through disease, head trauma, or necessary operations – suffered lesions or damage to a specific section of their brains. A well-known example of a long-term lesion study with only one subject was the work done with Henry Gustav Molaison (known in standard literature as patient H.M.), who in 1953 had surgery to treat severe epilepsy; doctors removed a significant part of his medial temporal lobe, which houses the amygdala and hippocampus, areas of the brain associated with memory, emotions, and navigation. Knowing the location and extent of the damage to Molaison's brain, those working with him could design cognitive tests to determine the extent of the damage to his memory (Molaison had significantly impaired short-term memory) – and, in so doing, could theorize the roles of those areas in normal cognition.[18] Examining the cognitive abilities of lesion patients could thus potentially allow the investigator to map these abilities onto particular areas of the brain in a normal patient.

While the imaging techniques available today are radically new, the conceptual and methodological principles underlying such studies are much older even than Crooke, and may be traced to the work of the foundational figure in Western anatomy, Galen, whose ideas concerning the brain derived from his work with injured gladiators.[19] Galen was perhaps the first in Western practice to link the idea of the mind or soul and its attributes to the human brain; what is more, he was probably the first Western physician to study the effects of trauma to the brain and nervous system, and thus to begin to draw the lines between specific functions and specific sections of the brain.[20] As surgeon to the gladiators of Pergamum, Galen witnessed first-hand alterations in sensation, movement, and thought that might follow when an unlucky competitor sustained various sorts of head trauma.[21] The deeper and more serious the wound to the brain, the more severely affected the gladiator's behaviour. Through his observations, Galen even managed to discover correlations between the location of the brain damage and the sort of behavioural change which followed: a laceration to one of the brain's lateral ventricles altered motion and sensation, while damage to the third ventricle could garble reason.[22] In one infamous public demonstration, Galen countered his cardiocentric detractors by slicing the laryngeal nerve of a squealing pig, arguing as it fell silent that the brain and nervous system, because of their direct influence from the soul, had far more control over the body than did the heart.[23]

Note that it would have been extremely difficult for Galen to theorize links between brain and mind *without* resorting to damaging the former to consider its effects upon the latter. In the absence of something like functional magnetic resonance imaging, the early physician had to rely on what he could study using his own senses – and naturally there was no way to observe directly the soul acting on the brain, and the brain on the body. In Galen's work, then, are the roots of a very modern conception of cognitive experimentation, namely the idea that the brain must be studied by looking at cases of things that have gone wrong with it.

So, too, the treating of the head wound required a reliable and consistent method. Arceus argues that one should first discern the cause of the wound: was the instrument sharp or dull, did the patient fall or get hit, what were the patient's mental and physical states at the time of the injury, how strong was the person who inflicted the damage, and so on. Next one must examine the skin surrounding the injury; if necessary, the skin should be broken in order to expose the extent of the damage to the skull, and parts of the skull should then be removed to give the brain room to breathe. All of this must be undertaken *before* treatment (which involved applying poultices to the wound), so that the very nature of the injury could be observed and recorded. That is, one must question circumstances; one must make the damage worse before one can make it better; one must remove various obstacles to the access of the original wound; and one must treat only when one understands the nature of the wound.

Arceus holds himself to this standard of medical investigation. He discusses, for example, a case wherein his patient was involved in the construction of a tower adjacent to a church in Valnerde; a stone a cubit in length and width struck the man on the head and fractured his sagittal sulcus. His people bore him home, where "he suffered the passion of the braine three whole daies together, not onely speechles, but also without any mouving, & was moved & turned of the standers by no otherwise then if he had bene dead."[24] Blood issued from his nose and eyes, and, what is more, his neck and head were black and swollen. The physicians worked round the clock and observed that

> the third day he spake, yet unperfectlie, and as men of a troubled minde are wonte to doe, his eyes were open and staring, in the manner of them that have the falling sicknesse, or that lie in a traunce, but he saw nothing at all, nor yet began to see untill the xx. day, but after that hee mended every day in his sight. And after the second moneth, he could see verie well as when he was in health … And he rose presently out of his bed, allthough he could not then well goe, and so by the helpe of God more then by the helpe of man, he was healed. And he escaped his eies looking a squint. One, one way, and another, another way, the which also (as the cure did proceede) at the fourth moneth were restored and looked right, he liveth as yet, and hath married a wife. For he was then a yong man.[25]

Following such a careful method of treatment, Arceus argues, will lead to the recovery, health, and eventual normalcy of your patient. Their young man was treated according to such dictates, and not only did he not lose his sight, he also escaped his "eies looking a squint" and was able to marry – a prognosis considerably more positive than that of a patient not treated according to this method.

Arceus and Read both argue that the aim in such medical investigation is a proper cure; the brain should be allowed to be purged as a last resort. It is much healthier and safer – or so the thinking went, though today we would be skeptical: how safe can it be, for example, to give a ten-year-old small beer while applying a poultice of juniper and thyme to his open

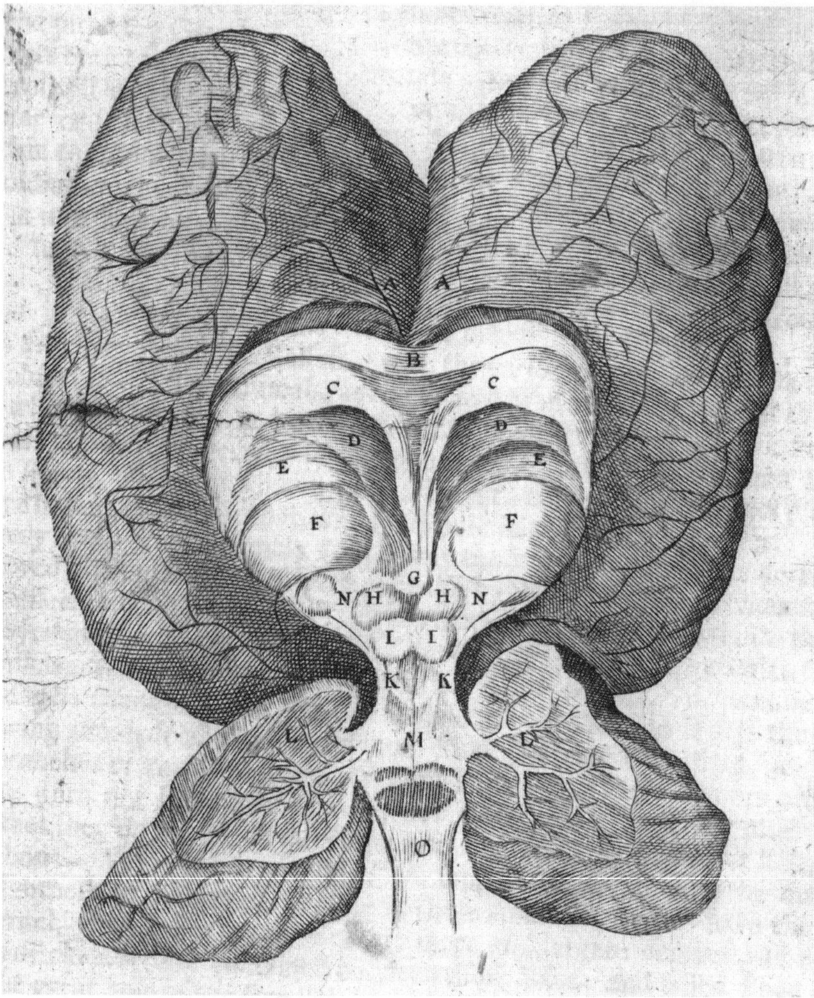

2.1 The cerebellum of a "foolish youth." Thomas Willis, *Five Treatises* (London: 1681), Figure 4 opposite p. 70. The Huntington Library, San Marino, CA. RB 347768.

brain? – to staunch the wound, to clean it, to probe it, but ultimately to try to heal it. Arceus's method is a rudimentary version of the kind of empiricism that would, later in the century, be posited by Bacon and others as a new means of accessing essential natures. England's first dedicated neuroanatomist, Thomas Willis, for example, would note that the difference between the brain of a normal person and that of a "fool" would be apparent in anatomical dissection, as the cerebellum of the former (Figure 1.5) would be large and healthy, while the cerebellum of the latter (Figure 2.1) would be thin and small. Though it is now taken for granted that brains are the receptacles of minds and are responsible for thought, memory, and the like, this was by no means a given in pre-modern times. Galen was among the first to make such a connection during his work with wounded gladiators, and it was upon this simple yet seminal idea that the method of Arceus, modern neuroscience, and – as I am arguing here – *Hamlet* are based.

"My tables"

The method proposed for treating head trauma in the early modern period requires a careful reconstruction of the events surrounding the initial incident of trauma. For a reading of *Hamlet* in this context, the pertinent question is, then, where and how this first traumatic incident occurs. I shall focus my reading here on one such incident, namely Hamlet's pledge to wipe his memory so that he may remember solely the Ghost's "commandment." This pledge then becomes the traumatic event it will be Claudius's self-appointed responsibility to investigate.

> *Ham.* Remember thee?
> Yea, from the table of my memory
> I'll wipe away all trivial fond records
> All saws of books, all forms, all pressures past
> That youth and observation copied there,
> And thy commandment all alone shall live
> Within the book and volume of my brain,
> Unmix'd with baser matter.
> … Now to my word.
> It is "Adieu, adieu, remember me." (1.5.97–111)

Key to this scene is the linking of brain to book, and the depiction of memory via an extended metaphor of writing.

But what is the specific capacity and function of this mercurial object, the brain? From the perspective of textual history, there are a few striking things about how this passage treats the brain. Intriguingly, the brain itself does not appear in the First Quarto (1603):

Yes thou poore Ghost; from the tables
Of my memorie, ile wipe away all sawes of Bookes,
All triviall fond conceites
That ever youth, or else observance noted,
And thy remembrance, all alone shall sit.[26]

In fact, Q1 simplifies both the method and the object of Hamlet's declaration: youth and observance merely "noted" the "sawes" and "triviall fond conceits," so they may be the more readily wiped away, and it is not the Ghost's active "commandment" but simply his more passive "remembrance" which will "sit" in Hamlet's "tables." The action here is decidedly uncomplex – erase the tables and reinscribe a new remembrance. This "remembrance" would allow Hamlet some latitude of recollection; he might have space in his freshly wiped tables to accommodate both the Ghost's description of Old Hamlet's murder and the Ghost's call to revenge.[27] Yet any sense of the location of this wiping-away has itself been effaced from the text.

The elaboration of both the Second Quarto (1605) and the Folio (1623) versions of this speech to include reference to the brain (as well as to "sinnows" and "sinnews," respectively, which refer to the nerves) underscores a related transformation effected by these later texts: a shift in the locus of memory from the heart to the brain. George Williams ingeniously points out that one source for Hamlet's metaphor can be found in the Geneva Bible's Book of Proverbs. Proverbs 7:1–3 urge the wise person to "kepe my wordes, and hide my commandements … Kepe my commandements, & thou shalt live … Binde them upon thy fingers, and write them upon the table of thine heart."[28] Here the memory is kept in the heart, as words and commandments written upon its tables. "The word of the Father, clearly, must supersede all 'trivial fond records, / All saws of books …' etc.," remarks Williams; "that commandment – or the specific part of it Hamlet chooses to record – is 'Remember me'" (141).[29] A similar anatomical metaphor occurs in Heywood's 1607 *A Woman Killed with Kindness*, in which Wendoll, who is charged by his friend Frankford to seduce Frankford's wife Anne, claims her refusal of him is "recorded / Within the red-leav'd table of my heart."[30] Ann and John Thompson note the similarity of this passage to Hamlet's promise: "the permanence of the printed or written, as opposed to the transitory nature of spoken language, is relevant and the 'coveredness' of heart and brain, like the 'coveredness' of books, is seen as helpful."[31] Such a preference for "coveredness" recalls, too, Proverbs' injunction not just to "kepe my wordes" but to "hide my commandements"

– writing on both heart and brain is thus secret, "visible" only to the shared memory of he who commands, and he who is commanded. In Hamlet's case, as we shall see, this "coveredness" will slip dangerously into solipsism as the Prince fails to transcribe his father's commandment properly onto a memory that has been wiped clean of everything else.[32]

What is key about the two examples I have cited above is that both indicate the heart, rather than the brain, as the table-book upon which memories will be transcribed. This point, coupled with Q1's lack of anatomical referent, highlights the uniqueness of the later versions of Hamlet's declaration: the writing will take place in or on the brain. The shift in focus with respect to this passage is part of a broader fascination with the brain. Of course, the brain is not yet malfunctioning here: Hamlet's speech inaugurates the moment of trauma from which his madness will proceed. The question will become, however, whether the trauma can rightly be located in the wiping of the memory or in the memory's inscription with the Ghost's commandment.

Both options involve writing – erasure on the one hand and reinscription on the other – indicating that the metaphor of writing is key in Shakespeare's play to thinking about memory's relationship to the brain and its repercussions should that memory be "wiped." As this instance confirms, the wax metaphor deployed by the Aristotelian tradition was still useful for describing how memory worked in the early modern period; it was particularly attractive to those seeking to link anatomy and humoural psychology.[33] Excessive heat or coldness would damage the brain, and subsequently the memory, as the organ would be rendered more or less impressionable, much the same way that wax would be affected by similar temperature gradients. Laurentius's discourse about vision explains:

> Experience also giveth us to understand, that if the braine have his temperature altered: as for example, if it be too hot, as it falleth out in such as are franticke: or over cold, as it falleth out in melancholick men; it corrupteth presently the imaginative facultie, troubleth the judgement, weakeneth the memorie … [The soul] requireth a good temperature.[34]

A "good temperature" will enable the soul to operate as it is meant to do; coldness in particular will weaken the memory, though we are not here given the precise mechanism. Melancholy itself affects memory, as Laurentius points out, whether the cold creates the melancholy or the melancholy the cold. Locating a suggestive resonance between the graveyard scene and metaphors used by clinical patients nowadays to describe their depression, Perry Guevara usefully draws attention to the earthiness of Hamlet's melancholy: "[A] hole in the head is not so different from a hole in the ground … where melancholy takes hold. The dirt-filled cranium upon which Hamlet gazes is filled not just with the remnants of a human brain, but with melancholic, earth-based humours."[35] Rhodri Lewis, citing Timothy Bright's humoural description of melancholy, argues that, were melancholy affecting Hamlet, he "would not have been able to think in an orderly fashion; his imagination … would have exercised a dominant influence over his words, thoughts, and behavior."[36] Thus John Willis's suggestion that a man wishing to remember something "untie the force of his imagination" from his memory would have been most prudent for the melancholic.

The wax metaphor was attractive for early modern theorists because, as Stephen Greenblatt points out, it conveyed "subtle variations, gradations, and effacements of impressions, some cut deeply, virtually permanently, into the medium of memory, others only lightly, mostly slowly losing their original distinctness of outline and eventually fading altogether."[37] John Willis more elaborately considers the mechanism by which temperature affects the memory:

> Imbecillity of Natural Memory proceedeth from too much heat, coldness; moisture, or dryness of the brain. Overmuch heat wasteth and consumeth animal spirits; too much coldness obstructeth motion of the Spirits lodged in the Cells of the Brain; superfluous dryness causeth such callocity, that the species cannot be imprinted: Redundant moisture doth erase and obliterate forms of things, as soon as they are imprinted.[38]

The animal spirits – manufactured in the brain from the vital spirits in the heart, and key tools of the soul working in the human body[39] – may have their motion obstructed by changes in heat and moisture and, as such, are unable to carry out what John Sutton terms "wriggle-work," a sort of engraving motion of the animal spirits upon the human brain.[40]

Yet there seems to be something more than wax on the Prince of Denmark's brain. As Peter Stallybrass, Roger Chartier, John Mowery, and Heather Wolfe have shown, wax tables have their limits both as a metaphor and as a stage device: they would not be capable of retaining permanently the Ghost's commandment, nor would they be a feasible prop for an actor to lug around onstage. A book whose pages could be written upon in ink is problematic in that it could not easily be erased or wiped clean. Thus some technology was necessary as a halfway point between these two devices: an instrument which could bear both permanent and temporary traces, preserving the one while allowing for the near-infinite removal and reinscription of the other. In a note to the "tables" speech in the Third Variorum *Hamlet*, James Boswell the younger (son of Dr Johnson's biographer) remarks that "I am in possession of three of these *table-books*: one printed in 1604, the date of the first edition of Hamlet: 'Writing Tables, with a Kalendar for xxiiii Yeeres, &c.' The Tables made by Robert Triplet. London. Imprinted for the Companie of Stationers, 1604."[41] Boswell's note anticipates Stallybrass et al.'s claim (though their essay does not in fact mention this prior source) that early modern table-books underpin the actual metaphor at stake here. These generally consisted of printed almanacs bound with pages that had been specially treated so as to allow for inscription and wiping (e.g., with a damp cloth or fingertip).[42] Such tables were easily portable and allowed the carrier to make temporary notes which could later be transferred into, for example, the more permanent record of a commonplace book.

Stallybrass et al. explain how this particular technology is meant to operate in Hamlet's case:

> When Hamlet uses writing tables as a model of memory, they also suggest forgetfulness: Hamlet wants to forget all that "youth and obseruation" have "coppied" in his mind. His mind's tables must first be wiped clean of all this earlier copying so that a single command can be permanently inscribed there – the Ghost's "Remember me." It is surely because there could be no less suitable technology than erasable tables for a permanent remembrance that Hamlet metamorphoses the "Tables … of Memory" into the quite different "Booke and Volume" of his brain, which he imagines as a place of indelible writing.[43]

The key here is that the tables offer Hamlet a space which can be wiped clean, while the "book and volume" of his brain can contain the "indelible writing" of the Ghost's injunction. Thus Hamlet's memory involves a dual system, a marriage of wax and stylus, on the one hand, and paper and pen, on the other.[44]

Some difficulty occurs, however, when we consider that these two writing instruments must coexist within the narrow region of Hamlet's brain.

Table-book and volume would also need to share access to sensory and intellective information, so that temporary notes from the one might be more permanently transferred into the other. Wiping the contents of the table-book to free up space in the "book and volume" suggests that damage to the table-book results in parallel damage to the "book and volume." Because it externalizes the act of memory through the metaphor of writing (even if that writing takes place in the interior space of the brain), the table-book becomes, to borrow a Derridean term, a *supplement* to memory: a helpful but potentially dangerous tool in that it offers the fantasy, but not the actual means, of permanence. Problematically, it seems that in the act of transcribing the Ghost's "commandment" into his memory, Hamlet ensures he will forget it. As Douglas Brooks argues, "Within seconds of hearing the father's last words … the forgetful son suffers a memory lapse the moment he attempts to take dictation"; having sworn his memory to the Ghost, "Hamlet recalls the central analogical tradition for describing the workings of memory in terms of the very exteriorized supplements that memory has come to rely on."[45] A number of modern studies corroborate what Plato demonstrated with the warning fable of the *pharmakon* and what Hamlet proves: as soon as one ceases to actively hold information in one's memory, that information will be lost.[46] Whether one writes down the information, presses a button, speaks, or otherwise communicates the information beyond one's own memory, the neural circuits responsible for maintaining the information in working memory will cease to fire.[47]

In writing his "word," then, Hamlet facilitates his forgetting of it.[48] An additional irony lies in his forgetting the word "remember," for it is precisely in writing down the command to memorialize that forgetting is initiated. Moreover, further forgotten in the act of noting down the need to remember is the Ghost's primary charge:

> *Ghost*. Pity me not, but lend thy serious hearing
> To what I shall unfold.
> *Ham*. Speak, I am bound to hear.
> *Ghost*. So art thou to revenge when thou shalt hear. (1.5.5–7)

The Ghost would have Hamlet seek revenge, but what the prince uses to overwrite his memory is the injunction *remember me*.[49] The act of remembering, particularly when one has suffered damage to the organ of memory, proves difficult – and this difficulty is compounded when the act of remembering is itself what causes the damage. For what else is there to remember when the Ghost's commandment is the sole living entity in Hamlet's brain?

I point this problem out to highlight what I see as the violence inherent in the act of wiping away memory and inscribing a particular commandment. Garrett Sullivan comes close to articulating the effects of such violent forgetting:

> the Ghost's words initiate in Hamlet a fantasy of the annihilation and reformation of the self through forgetting and the subsequent inscription of a memory trace … Interestingly, the forgetting that is to precede Hamlet's inscription would destroy the contents of his memory. Hamlet understands the contents of his memoria and the Ghost's demand that Hamlet remember him as mutually exclusive.[50]

If forgetting is an act of "annihilation," then so too must remembering be. William Proctor Williams, for example, notes at least two performances of *Hamlet* in which the actors whip out daggers to assist in the inscription of the Ghost's commandment.[51] The use of daggers here points, too, to what Jonathan Goldberg has described as the violence of the early modern act of writing; knives were often used in the period to scratch out writing on paper.[52] As Juliet Fleming explains,

> writing, even in ink, can be "erased" ("scraped out") from paper (for that matter, any writing that was not finally erasable would not be writing at all) … But such erasures are neither readily effected nor infinitely repeatable: they are possible, but not practical … [T]o cross or "expunge" ("prick out") writing is to mark it, more or less memorably, for erasure: it is, in effect, to *write it again*. Clearly, then, paper does not preserve writing that has been effaced from it.[53]

Even erasing writing, then, is a sort of writing-again, or writing-over. Inscription of any sort, whether writing or erasing, thereby constitutes a kind of violence. One cannot write, remember, or forget without causing damage – to the physical medium of writing or to the psychic medium of the self.

To literalize the play's metaphor by thinking of such inscription as taking place on or in the brain recalls the physician's inquiry into a case of head trauma. Memory was thought to be housed in the fourth or hindmost ventricle of the brain.[54] Crooke observes that "it is a perpetuall truth in the body of a man, that by how much the cavity is greater, by so much it is the baser. The fourth ventricle is of all the rest the least and the narrowest, and containeth the Animall spirit sincere defœcated and exquisitely purged: the others do onely prepare the spirit, and therefore the hindmost ventricle is the most noble."[55] This fourth ventricle, smallest and housing the "exquisitely purged" animal spirit, deploys this spirit in service of memory itself. How else can we be certain that the fourth ventricle is the highest and best? Crooke explains: "*Galen* teacheth that the upper ventricles are the basest … by the example of a young man of *Smyrna* a Citty of *Ionia*, who being wounded into one of the upper ventricles yet escaped with life … A wound in the hindmost ventricle doth most of all offend the creature."[56] It is through this sort of backwards reasoning, the logic and violence of the lesion study, that we are able to comprehend what we do about the properly functioning brain.

Hamlet's pledge to remember the Ghost's commandment constitutes an act of self-damage which it will be the task of Claudius to probe. The effect of Hamlet's lesion will be madness – real or feigned, we know not which – that will seem, as Claudius will observe, to be "something in his soul / O'er which his melancholy sits on brood" (3.1.166–7). Resolving this something, as we know, incurs further violence – a purgation. Just as Galen's work with gladiators informs his understanding of the effects of "offense" upon the brain, so too does Claudius – and, as we have seen, the lesion study more broadly – utilize damage to the brain to read its overwritten secrets.

At the end of 1.5, the ventricle containing Hamlet's memory has been wiped clean. Upon its surface the injunction to *remember* is inscribed – but the referent has been lost. The prince's mind and memory become an echo-chamber of remembrance without an entity which can properly be said to be remembered. Thus Hamlet delays, and it is not until he encounters Yorick's skull that he is reminded strikingly of the mortality of all things. Crucial, too, is what Guevara construes as the earthiness of Hamlet's melancholy: "As Hamlet reinforces the humiliating trajectory from human to humus, Shakespeare's brain dust [i.e., the dirt filling the skulls in the graveyard] … explicitly materializes the link between dirt and cognition."[57] Throughout the play, but particularly in the graveyard, Hamlet encounters numerous forms of death, to which his living brain responds with wit[58] – but he has forgotten the very "canoniz'd bones, hearsed in death" of his father, and "the sepulchre" which seemed to have "op'd his ponderous and marble jaws" to cast forth not Old Hamlet's body, but merely the ghostly semblance of that body (1.4.47–50). In remembering Yorick, Hamlet remembers that he has forgotten the king.

Since we find ourselves in the graveyard, let us turn to what it contains: the crypt. In his essay "Fors," a preface to Nicholas Abraham and Maria Torok's *The Wolf Man's Magic Word*, Jacques Derrida explores the architecture of the crypt, a constituent part of the self created through, and essential for, the mourning process. The crypt, he writes, "is not a natural place" but a place "*comprehended* within another but rigorously separate from it, isolated from general space by partitions, an enclosure, an enclave. So as to purloin *the thing* from the rest." Inside itself, the crypt constructs a safe, "sealed, and thus internal to itself, a secret interior" which is "the condition, and the stratagem, of the cryptic enclave's ability to isolate, to protect, to shelter from any penetration, anything which can filter in from outside." Yet "the crypt can only constitute its secret by means of its division, its fracture. 'I' can only *save* an inner safe by putting it inside 'myself.'"[59] Derrida's description returns us to the epigraph with which I began this chapter, for it underscores the internal division evoked by Willis's description of the effacement necessary to memory. Following John Willis's rules, a man could force himself to remember by sending to oblivion the rest of his memory and carving out a space untied from other sensory or imaginative inputs. In this sense, the fourth ventricle becomes itself a crypt – in order that, as Derrida puts it, "[I may] pretend to keep the dead alive, intact, *safe* (*save*) *inside me*, but it is only in order to refuse, in a necessarily equivocal way, to love the dead as a living part of me, dead *save in me*" (71). Hamlet likewise only loves his dead father as a thing living inside himself – he has, after all, promised that the Ghost's commandment "all alone shall live" inside his brain. Such a promise folds dead into living and renders the prince's skull more crypt than cranium.

It is not, then, simply a "something" in Hamlet's soul that troubles him so and that incites the madness which Claudius will study and of which, eventually, he will purge the prince. The "something" sitting over Hamlet's soul is the violence of loss and of memory, the uncoupling of sensory and intellective input from recollection, which causes him to keep his dead father's "word" alive inside himself. Old Hamlet is dead *save in* Hamlet; he lives insofar as his son lives with the memory of death. Such an attempt will disease his wit (3.2.313) and blunt his purpose (3.4.111), and will eventually allow his dying brain to recognize its mortal wounds. A normal person, facing death, would be able to utter only, "I am dying." Hamlet, carrying all along the violent inscription of the dead upon the living matter of his brain, will be able to speak the unspeakable, "I am dead," for he has been so all along (5.2.338).

The Written Troubles of the Brain

It is worth noting that the case of *Hamlet* opens onto a broader tendency of Shakespeare's tragedies with respect to the brain and its discontents. In a moment of clarity in his mad scene, Lear complains he has been "cut to the brains"; Othello worries that "[s]ome horrible conceit" may be "shut up" in Iago's brain; Marcus threatens that "[t]he poor remainder of the Andronici / Will … on the ragged stones beat forth our brains" if the Romans can prove their wrongdoing; Octavius Caesar complains to Marc Antony, "It's monstrous labour, when I wash my brain, / And it grows fouler"; Juliet imagines awakening, terrified, in her family mausoleum, "And, in this rage, with some great kinsman's bone, / As with a club, dash[ing] out [her] desperate brains" – to cite just a handful of examples.[60]

The overwhelming tendency in Shakespeare's tragedies is a treatment of the brain as a fallible object, vulnerable to damage, and operating in such a way as to bridge the literal and the metaphorical. That is, Shakespeare's *brain* is a literal, physical entity, grey matter located in the head and very much able to be shaken, knocked about, beaten, or otherwise physically maligned. At the same time, however, this *brain* points also to its own metaphorical equivalent, the house of the mind that can also stand in for the mind itself. The examples cited above weave back and forth between these two representations, as Othello and Octavius treat the brain as metaphorical entity, while Marcus and Juliet treat the brain as physical object. As the privileged site of the soul and its functions, the brain is, as we have seen and as we shall continue to see, perhaps the lone member of the body republic to stand in all its instances astride the boundary dividing literal object and metaphorical reference (*pace* Menenius and his fable of the overbearing belly). This boundary, again as we have seen, is peculiarly marked by violence, its threat and its actuality. In *Hamlet*, in order to construe the brain as memory's operative, the organ must be written upon, a violent act that ensures that memory will be encoded, but at the price of the razing out of other forms of subjectivity. Brain and self are coincident insofar as the bond between them is forged by damage to both entities.

From several of the perspectives we have been engaged with here – ghosts and murder, writing and memory, damage to the brain – the Shakespearean play most closely aligned with *Hamlet* is *Macbeth*, and I conclude this chapter by turning briefly to this later play in order to establish the presence of a more pervasive connection amongst these terms in early modernity.

What is at stake in Macbeth's brain? One peculiar scientific study diagnoses the Thane with prion disease, tracing in his various speeches patterns of disorder consistent with Creutzfeldt-Jakob disease, fatal familial insomnia, and, most inexplicably, kuru.[61] We would not here want to fall into this trap whereby a work of fiction is understood merely as a transparent window into a diagnosable reality; as we have already seen in this chapter, an early modern scientific understanding of the brain developed in actuality alongside a fictional portrayal of the organ. Such a reading enabled our exploration of brain damage in *Hamlet*, allowing it to be informed by contemporary theories that pointed to recognizably modern models. A similar connection is exhibited in *Macbeth*, and though the references to the brain in this play are not as extensive as those in *Hamlet*, they nonetheless unfold the same links to violence, trauma, the malfunctioning of memory and imagination, and the problem of writing.

The Scottish play first mentions the brain when Macbeth is speaking to Banquo, Rosse, and Angus after the first prophecy; the soon-to-be Thane of Cawdor asks the men to give him their favour: "[M]y dull brain was wrought / With things forgotten. Kind gentlemen, your pains / Are register'd where every day I turn / The leaf to read them."[62] These lines render the link between writing, in or on the brain, and memory explicit. Macbeth's lie (that he was distracted by "things" he is trying to recall to memory) is couched in terms that take advantage of this link, albeit in a curious way.[63] The "things forgotten" have "wrought" Macbeth's "dull brain"; *wrought* here implies fashioning, shaping, or using art or artifice to render a finished elaborate design upon something crude or, here, "dull."[64] What Macbeth claims he is trying to remember, the "things forgotten," are thus molding his brain, drawing it out from dullness into a more finished form, and the consequent labour of remembering necessarily draws attention away from his conscious interaction with the world. Yet memory is not meant to work this way – it should operate when called upon, it should not occlude the senses. Memory is meant to assist, rather than disturb, lines of thought. But as with Hamlet's destructive relationship to memory, so here, too, Macbeth's lie signals, whether consciously or unconsciously, a disruption of his normal cognitive activities. Furthermore, these "things forgotten" point to the *failure* of Macbeth's memory to function as it is meant to. Were all in order in Macbeth's brain, there would be no "things forgotten" to be wrought upon.

That an object has been wrought suggests labour of a particular sort: fine work, a change from one state to another, but some manner of violence as well. Consider, for example, how metal is wrought: it must be beaten or hammered to an "airy" thinness (to borrow a phrase from Donne) or shape. Perhaps less violent, but still fundamentally transformative, is material that has been wrought in the sense of embroidered, for a needle must be employed to force thread through the base fabric.[65] I am perhaps belabouring the point with respect to the violence inhering in the metaphor – we would not normally consider embroidery to be an aggressive act – but the general idea is that an enforced physical change is involved in metal or fabric being wrought, just as Macbeth's brain was forcefully wrought by memory's failure. The semantic leap from *being written* to *being wrought* is not a great one, but the connection is significant: in an act we might casually find benign, violence lurks.

Yet what is the nature of the labour memory has wrought upon Macbeth's brain? The following lines embellish the metaphor in a way that should by now be familiar from our consideration of *Hamlet* – namely, with explicit and infamous reference to a specific kind of brain damage. When she is goading her husband to the murder of Duncan, Lady Macbeth recalls:

> I have given suck, and know
> How tender 'tis to love the babe that milks me:
> I would, while it was smiling in my face,
> Have pluck'd my nipple from his boneless gums,
> And dash'd the brains out, had I so sworn
> As you have done to this. (1.7.54–9)

Lady Macbeth's stark image is unmistakably violent. Why should she use this particular language here? *Brains* in this context functions as synecdoche for the head, for the soul, for life. Her speech moves from the broader emotion of maternal love to a close-focus on the body: the baby's smile, her own face, her nipple, the baby's "boneless gums," the brains she will dash out. This movement underscores the physicality of her imagined sworn oath, and it infantilizes Duncan by metaphorically linking him to the baby as the object of the sworn violence. Lady Macbeth's example undoes a purported bloodline via this particular form of brain damage, uncoupling a putative child from both breast and life.

The image of violence towards the brain seems to prime Lady Macbeth for broader considerations of brutality, both with respect to death and its shadow state. When, later in the very same scene, Lady Macbeth speculates on how her husband might murder Duncan, she imagines the king at the end of a long day, asleep and vulnerable. At that point, she says she will approach his two chamberlains with wine and wassail, working on them in such a manner that

> memory, the warder of the brain,
> Shall be a fume, and the receipt of reason
> A limbeck only: when in swinish sleep
> Their drenched natures lie, as in a death,
> What cannot you and I perform upon
> Th'unguarded Duncan? (1.7.65–70)

Memory dissipates from stolid guard of the brain and its associated senses to "a fume … a limbeck only," shifting states with easy liquors both literal and metaphorical. The progression from solid to gas to liquid ends in a drowning, taking with it memory, reason, and life, collapsing sleep and death – and enabling Duncan's murder. Further, damage to the brain extends across the line separating physical from metaphorical in this passage, and an incidental act of violence reveals the deeper embedded intertwining of the brain and its psychological functions.

I will collapse here three further mentions of the word *brain* in *Macbeth*, as all have to do with the brain being physically compromised: Macbeth famously wonders if the dagger he sees before him proceeds from the "false creation" of a "heat-oppressèd brain" (2.1.48–9); his wife then chides him for thinking "[s]o brain-sickly of things" (2.2.44); and Macbeth, upon seeing Banquo's ghost, laments, "The time has been / That, when the brains were out, the man would die, / And there an end" (3.4.77–9). In each of these cases the brain is physically affected in an adverse manner and linked closely to death and delusion; it may be damaged variously, oppressed by heat, sick of thinking, or knocked "out" and therefore obsolete. These examples incorporate and elaborate upon the lines of reference set out by *Hamlet* in that they describe in different ways what happens to the soul, mind, or person when the brain is in some way unsettled.

Finally, and perhaps most tragically, we turn to Macbeth's lamentation upon his wife's mental state prior to her death. He begs her doctor to cure her of her troubled sleep: "Canst thou not minister to a mind diseased," he asks,

> Pluck from the memory a rooted sorrow
> Raze out the written troubles of the brain
> And with some sweet oblivious antidote

Cleanse the fraught bosom of that perilous stuff
Which weighs upon the heart? (5.4.41–7)

Telling here is the collapse of physical "antidote" and abstract "sorrow," "troubles," and "stuff." The underlying implication is that the physician could act materially upon the brain to affect these emotional states – and this notion is a strikingly modern one. Not only that, what Macbeth desires the physician to do is broadly damaging in nature, requiring plucking and razing. Even the apparently benign act of cleansing may be harsh as well. Particularly of note here is the line about razing out "the written troubles of the brain," which will recall for us Hamlet's vow to *wipe away* all the knowledge embedded in his brain, the better to make way for his father's (ghost's) commandment. In both cases a kind of curative blankness is sought – for Macbeth this will ostensibly restore his wife's sanity, while for Hamlet the necessary damage will pave the way for the better remembrance of his father.

Taken together, these references to brains in *Macbeth* signal a disquietude with the organ, a sense of its power but also its fallibility. As we have seen, such a relationship to the brain is shared by *Hamlet* and early modern anatomical texts. Marina Favila productively traces the shift from magical to mortal thinking in *Macbeth*, noting that idealizations in the play lead very quickly to thoughts and actions of mortality, and that these mortal thoughts in turn "rend the thinker."[66] *Macbeth* and *Hamlet* share this crucial connection, of dangerously shifting the imaginary to the physical. Such a shift is not, to be certain, always necessarily problematic, but in these two plays, and in Shakespearean tragedy more broadly, it both follows from and signals a menacing link between *thought* and *action*, between *mind* and *brain*, and between the proper action of thought and its malfunctioning counterpart.

CHAPTER THREE

Labour Pains: William Harvey and the Travails of Conception

Kent. Is this not your son, my lord?
Gloucester. His breeding, sir, hath been at my charge. I have so often blush'd to acknowledge him, that now I am braz'd to't.
Kent. I cannot conceive you.
Gloucester. Sir, this young fellow's mother could; whereupon she grew round-womb'd, and had indeed, sir, a son for her cradle ere she had a husband for her bed.

– William Shakespeare, *King Lear*, 1.1.3–6[1]

Conceiving Conception

The very opening of *King Lear* highlights the double meaning of *conceive* (and its correlate *conception*), playing on an ambiguity fundamental to both the scant humour of the scene and the plot of this chapter. Seeking to confirm the bastard Edmund's paternity, Kent is taken aback by Gloucester's reply. "I cannot conceive you," he replies in apparent confusion, intending the psychological instance of the word (i.e., I cannot *understand* you). Gloucester, amused, spins his answer using its biological meaning instead (i.e., Edmund's mother could *get pregnant* by Gloucester). However, beyond the obvious pun it allows, Kent's use of *conceive* to denote a mental act is compelling because it actually can work here on a level closer to the biological as well: for the early moderns, to conceive something in one's mind implied an essentially creative act, one that conjoined sensory information, memory traces, and imaginative force to produce a novel idea not previously accessible to the thinker. Biological conception is, of course, likewise a creative act, resulting in the production of a new being.

The double valence of *conception* remains very much alive in our language today, but in this chapter I examine a moment in the seventeenth century when the word managed to forge a peculiar and especially productive link between intellectual and biological processes, and between the very organs in which those processes took place.

The seventeenth century witnessed a veritable overflux of metaphorical connections between brain and womb, which were bound through theories of conception and imagination, both of which might occur inside or act upon the vessel in question.[2] The metaphorical ties between these two organs went in both directions: brains-as-wombs and wombs-as-brains. In this chapter, we will see the former connection and its ramifications most readily in (predominantly) male writing about the process of writing – prefaces, dedicatory epistles, and the like. That is, the writer's brain conceives an idea, gestates the argument, grows big with child, and eventually suffers the labour pains of writing (with which we are all doubtless familiar) in order to bring forth the finished book. Sometimes helped by midwife-translator or midwife-editor, the babe-book can then be handed gently to the reader or else left to make its own way in life. The conceit is more often than not rhetorical, and intended to forestall criticisms about the actual written work – but I will argue that the metaphorical link reveals an unusual generative capacity for the indubitably physical brain, and that in actuality the metaphor may be far more literal than it appears at first sight.

In this and the following chapter, I will also detail how the latter connection helped shape a perception of the feminine body as a provocatively and dangerously creative space for the generation not only of other literal bodies but also of metaphorical book-bodies. Furthermore, a consideration of imagination's role in conception sheds new light on the problem of maternal imagination, a dimension key to the converse metaphor of the womb as a kind of brain. It houses the conception, protects it as it grows, and seemingly has a direct line to the mother's wants, desires, fears, and fancies. Stories abound, from the innocuous (a birthmark in the shape of a strawberry, because the mother had craved strawberries) to the worrisome (a black child born to two white parents, explained away by the claim that the mother had been looking at a picture of a Moor at the moment of conception) to the disastrous (the monstrous birth). As men's brains become wombs to create lively written work, women's wombs become brains that create a thinking organism, and in both cases imagination takes an active role in shaping the finished product.

This chapter takes as its starting point William Harvey's strange and often under-noticed short essay on conception, which is tacked on to his more widely circulated *De generatione*. Today, the renowned seventeenth-century anatomist is largely remembered for having established the theory of the circulation of blood in the human body, but his interests in biology were always more comprehensive. In seeking to understand the very principle of generation, namely how the child is formed in the womb, Harvey sets the terms for thinking about the womb as a kind of brain, and begins to explore how the two organs corresponded on an anatomical level. As we shall see, their connection would be testified to by the eerily similar visual representations of both organs, as well as by the curious decision of anatomists to name many parts of the brain after reproductive structures in the human body. After tracing the lines of filiation, both anatomical and literary, that feed into and grow out of Harvey's version of the womb-as-brain metaphor, this chapter turns to a range of paratextual instances drawn from early modern texts to examine more closely how male writers used the metaphor of the womb to describe their own creative efforts. Their womb envy may seem at first simply a form of appropriation, the metaphor serving to co-opt a fundamentally female space (what could be more essentially a female preserve than the womb?). Nonetheless, their practices also usefully highlight the complex role that imagination, as a faculty of the soul, played in the process of conception, be it textual or biological.

A "plastick generative power"

To the end of his 1651 *Exercitationes de generatione animalium* (translated in 1653 as *Anatomical exercitations, concerning the generation of living creatures*), Harvey appends the brief essay "De conceptione" ("On Conception"), in which he puzzles over the problem of the intangible (and potentially immaterial) but nonetheless structurally necessary presence of some "agent" of conception. As he points out, "I plainly see that nothing at all doth remaine in the *Uterus* after *coition*, whereunto I might ascribe the *principle* of *generation*" (p. 546, sig. Nn_1^v).[3] The uterus has, to be certain, received "spermatical contact" or "contagion" (in the sense of contiguity rather than spreading of disease) from the masculine member, but how exactly it exercises the "plastick generative power" necessary to drive the ensuing formation of the fetus remains a mystery (p. 539, 540, sig. Mm_6^r, Mm_6^v). The desire for an experimentally verifiable entity responsible for converting seed into child and impressing the image of the father onto the matter supplied by the mother had driven other philosophers to a variety of conclusions, all of which Harvey weighs as equivalent in their unlikelihood: some "have furbushed [*sic*] over the old opinion concerning the *Atomes* ... Some againe raise up certaine *incorporeal spirits*, like so many *Agents*, *Angels*, or *Demons*. Others understand a *Contagion*, like to a kinde of *ferment*, or *sower levening*" (p. 547, sig. Nn_2ʳ). That these theories reach back ("old opinion," "againe raise up") to prior – and discarded – philosophical strategies underlines their ineffectuality for the neo-Aristotelian Harvey. Thus the paradox of conception still begs a resolution: "I say, there is no *Sensible thing* to be found in the *Uterus*, after *coition* and yet there is a necessity, that something should be there, which may render the *female fruitful*; and that (in probability) can be no *corporeal essence*" (p. 547, sig. Nn_2ʳ). What might this entity be, then?

For Harvey, the real solution to what this "something" is emerges from seeking to answer a different, related question: how should one apprehend the force that drives female fertility? In other words, rather than search directly for a material cause, for some "sensible thing" to which generation must be attributed, Harvey attends instead to the process of fructification. His response proceeds by way of analogy and metaphor, a technique that has served us well in the past, and here he is gloriously transparent about his modus operandi. The hinge-point for him is the double-sided idea of *conception* itself. As the brain *conceives* an idea, so the uterus *conceives* a fetus. The commingling of physical and psychological processes in the very word drives Harvey's analogy from the outset: could it be, he wonders, that just as we "think with our *braines*, so a *female* doth conceive with her *Uterus*?" (p. 540, sig. Mm_6ᵛ). Since "both their functions are equally called *conceptions*, and both are *Immaterial*,"[4] thought-conception and impregnation-conception share a common designation and a comparable absence of a material basis. This in turn suggests a functional resemblance between them, allowing our understanding of how the brain works to be transferred onto the womb (p. 543, sig. Mm_8ʳ). Rather than outright causation, the relationship becomes one of teleology and structural analogy. Benjamin Goldberg has suggested that Harvey's dependence on functional likeness was probably the reason why the theory did not take full hold in his time.[5] But, as we shall see, even if the explicit analogy posited by Harvey failed to self-propagate in the domain of anatomical writing, a milieu of contemporary thought nonetheless existed that enabled and justified such a connection: for the relationship between brain and womb was repeatedly taken up in the period as a means of illuminating the process of creation more broadly and of artistic creation, especially writing, in particular.[6]

Moreover, the functional correspondence between brain and womb underpins the emergence of a visual connection between these two as biological organs, which in turn reinforces the sympathy between them. Consequently, Harvey points to the physiological similarities between brain and womb post-"coition," whether of idea or of fetus, even as he again emphasizes the invisibility – and thus (for him) immateriality – of the engendering agent at play:

> But because there are no manifest signs of Conception visible, before the *Uterus* doth begin to open, and the albugineous liquor, or slender threads, (like the Spiders web)[7] and the first *rudiments* of the future Egge or Conception appear: and seeing the substance of the *Uterus*, now ready for *Conception*, doth so neerly resemble the Constitution of the Braine: why may we not imagine, that both their functions are also alike; and that something like, if not the selfe same thing that the *phantasme*, or *appetite* is to the *brain*, is excited in the *Uterus*[.] (p. 542–3, sig. Mm_7^v–Mm_8^r)

Several things are striking about this parallel. First, Harvey emphasizes the utter absence of tangible phenomena connecting the act of insemination to the sudden appearance of the embryo. *Something* must drive the process, because there is clearly a causal relationship between the joining of seed and the creation of new life – but this *something* remains outside of our sensible grasp. This quandary iterates the very soul-body problem that Cartesian dualists would have to grapple with: how could it be that the immaterial soul actually interacts with the material body? One could winnow the interface down to the atomic level, but, given the fundamental duality of the entities involved, the mechanics of the process still needed to be accounted for. Harvey spans the gap by analogy: we should look for a solution to that problem in the functions of the brain.

Second, and especially germane to my argument, Harvey connects the "substance" of the uterus and the brain at the critical moment of preparation for conception. In so doing, despite his insistence on immateriality, he grounds the analogy in the physical realm, a crucial move that protects him from criticism that he could be searching for something so intangible as not to exist at all. To this end, Harvey points his reader in the direction of a particular feature of the brain, namely its ventricles, offering these as the closest visual analogue to the womb. As he puts it, "[T]he *uterus* appears thicker and more fleshy: and afterwards … it groweth more tender, answering in lubricity and softness to the internal ventricles of the *Braine*" (542). These holes inside the brain had been seen as useful caverns for manufacturing and refining the animal spirits, or, alternately, for housing the various functions of the soul. This latter idea, known as ventricular localization theory, posited that the soul's different aspects – usually commonsense, cognition, imagination, and memory – each resided in one of the ventricles, communicating mutually through the various connecting channels. We will explore ventricular localization further below, but we need to keep in mind the generative potential of the brain's ventricles in reading this fragment of Harvey's philosophy, since it constitutes a key link between the production of ideas in the abstract and the brain's biological-anatomical processes in the particular. That is, Harvey tacitly links the possibility of generation of *thought* and *idea* in the brain to its especial local point, the ventricles, on the basis of their putatively physical similarity to the womb upon conception: their membranes are subject to similar thickenings, their body simply *fleshier*, the entirety softer and better lubricated.

It is a mistake to assume, as Goldberg and others do, that in seeing them as similar Harvey has made some error in his descriptions of womb and brain. Rather, biological likeness – and, subsequently, the visual representation of such a likeness – is consequent upon their *metaphorical* similarity in Harvey's configuration.[8] In short, we should discard the assumption of modern anatomical correctness, because it does a disservice to the organic analogy that Harvey develops here. Whether it was actually true, then or now, that the recently impregnated womb and the brain's ventricles appear anatomically similar is less the issue than the very fact that Harvey insists upon their likeness, seeing a physical parallel that both emerges from and serves to justify the metaphorical equivalence between their generative functions.

The beauty of the metaphorical connection is that it needn't be *physically true*, however physical the organs at issue might be. But at the same time it is the *matter* of Harvey's analogy – the assertion that the interior of the womb and the ventricles of the brain share an anatomical thickness, fleshiness, lubrication, upon conception, whether of embryo or of idea – that renders real the very analogy that permitted the physical likeness to be seen in the first place. The logical dependence of physical similarity upon a prior metaphorical connection is especially crucial in an era that has no recourse to microscopy (or at least, microscopic studies are incipient rather than given), which means that, despite the increased importance accorded to experimentation, much of anatomical theory continues to rest upon *reasoned* principles rather than observed facts – to the extent that the observations intended to ratify the theory independently are often ultimately effects of the theory itself. The evident circularity obviously occurred to Harvey and his associates as well – hence his complaint about the failure of empiricism and experimentation. Anticipating that "some scoffing persons will laugh at these conjectures" because (as he has already pointed out) "there are no manifest signs of Conception visible," Harvey explains that when a philosopher "cannot clearly discover how things themselves are brought about" he must "conceive some way consonant to the *course* of *nature*" until such time as observation may ratify or correct such a theory (546, 542, 546). All our philosophical opinions were "at first mere *figments*, and *imaginations*; untill they were wrought a solid *credit* in us, by sensible experiment," but, as Harvey argues again and again, experiment alone cannot account for the intangible but structurally necessary power enabling generation (546). Connecting brain and womb via the notion of conception therefore allows both to be better understood without direct observation.

Finally, the matter of the connection leads to (and indeed grounds) the possibility that the same force operates in both entities: imagination. The "appetite" or "phantasme" – terms used as synonyms for imagination more broadly – was often seen in the period as a generative power, one that drew from the other three faculties of the soul (commonsense, cognition, and memory) to form new thoughts. The brain's appetite impels the creation of the intellectual conception – it is a force that drives creation, and despite our not being able to see or touch it, we know it exists. Harvey elaborates:

> For as we, from the Conception of the Form, or Idea, in the Braine, do fashion a form like it to our works, so doth the Idea or Species of the Genitor, residing in the *Uterus*, by the help of the formative facultie, beget a *Foetus* like the Genitor himself; namely by implanting that *Immaterial species* which it hath, upon its Workmanship. In like manner … *Art*, which is the … Species of the future work, doth produce a Like in its operation, and generate it in the *matter*. (543)

Thus as we conceive an idea and bring it into being in our works, so does the male "genitor" impart a correspondent form, which the uterus uses to bring the fetus into being. A key component of the analogy concerns the active nature of the idea or the form: the "species" of the father assists the "formative facultie," just as the idea in the brain fashions itself in our work.

Harvey's reasoning works both anatomically and linguistically: not only do brain and womb look alike, we also use the same language to talk about them, *and* what they produce is directly related to (and looks like) us.

Crucially, Harvey explains the specific operation of biological reproduction through its likeness to artistic generation, our creative intellectual output.[9] He acknowledges that his propositions might seem outlandish, particularly were the critic (as the narrow-minded and backwards-thinking philosophers above) unreceptive to the idea in the first place: "Let the Learned and ingenious stock of men consider of it, let the supercilious reject it: and for the scoffing ticklish generation, let them laugh their swinge" (p. 546–7, sig. Nn_1^v–Nn_2^r). Doubtless, Harvey goes to great and often bizarre lengths to justify the connection between brain and womb, pointing out that just as the brain of an artist can create things the artist himself has never encountered sensibly before – just as "a litle *bird* will most *artificially* contrive a *Nest* (whereof shee never saw any *platform* before) and that not from her *memory*, or any *habit* implanted in her, but onely by meere *phansie*" – so, too, a woman may become "the *efficient cause of Generation*, being *impregnanted by the conception of a generall, immateriall Idea*" (p. 545–6, sig. Nn_1^{r-v}; Latin: "foeminam ex conceptu ideae generalis fine materia impraegnatam, generationis opisicem evadere"). The logical progression from artist/brain to "immaterial Idea" / womb would seem unclear but for the inclusion of the avian example drawn from nature's art,[10] the important words "memory," "habit," and "phansie" priming the reader for an intellectual transformation occurring in a material substance, with the Harveian "*Idea*" itself constituting the generative power.

As we will see in more detail later, this detour through nature's artful processes reveals, too, Harvey's implicit dependence upon the handy metaphor through which masculine authors co-opted a fundamentally feminine space by claiming to *labour* over the children of their brains. The puissance of the "idea," both intellectual and biological, makes the brain an appealing location for thinking about a kind of metaphorical motherhood. And so, writers in the early modern period would often describe the brain as being pregnant with the "child" of fantasy, likening the generation of ideas to the growth of the fetus, and the act of writing to labour. The newly birthed text they then present to the hands of the reader, often in the form of a preface or dedicatory epistle, which functions in turn as a sort of birth certificate, documenting the process. If, as Harvey argues, man is the "author" of his desires, arts, and principles, then the brain-womb is equally the author of man's sensitive faculty and virtue.

Ventricular Localization

Harvey's "De conceptione" thus reveals a dual emphasis. The initial analogy between brains and wombs based on their shared task – conception – leads him, on the one hand, to assert the physical likeness between the two organs. On the other, his text explains the generative process common to both through another metaphor: by invoking nature's imaginative artfulness, of which man too partakes. The two directions opened up by his treatise point in turn to two relatively autonomous intellectual traditions that feed into and ground Harvey's analysis, the one anatomical and the other mythological. Let me begin by treating the former. For the physical correspondences that relocate in the very matter of the brain the metaphorical relationship between these two organs of conception were not posited by Harvey *ab novo*. Rather, they derive from early modern anatomy and the traditions from which it grew. Our focus here is the brain's ventricles, caverns in the matter of the brain that contemporary neuroscience understands as lined with cells to manufacture the cerebrospinal fluid cushioning our brain and spinal column.[11] Imagine yourself, however, as an early modern anatomist encountering an actual brain for the first time. You know from your training that this organ is the physical instantiation of the soul, the best and highest part of man's body, our closest link to God. Yet, when you finally turn your scalpel to the cadaver to reveal this precious organ, you find that it contains great hollow vaults – there are holes in the brain. How would you explain this curiosity? Does the soul reside in the brain's matter or in its ventricles?

Galen was perhaps the first to speculate upon the ventricles' role. Using animal vivisection, he was able to demonstrate that obliterating the rearmost ventricle resulted in the animal's death.[12] However, opening the ventricles did not necessarily cause death or even serious injury, a fact that led Galen to consider it unlikely that the soul actually took up its residence there. Furthermore, the liquid inside the ventricles, which he termed *pneuma psychikon* (the animal spirits), was not itself the soul but merely a sort of instrument or handmaiden to the soul.[13] Consequently, he suggested that the soul must be situated in the very matter of the brain. Following this initial foray, two controversies arose in subsequent Classical anatomical studies: the first concerned the nature of the *pneuma psychikon* and its exact relationship to the soul, and the second the actual number of ventricles in a human brain.

Even if the Galenic account predominated, there were always others who dissented from that view. Nemesius, for example, writing two centuries after Galen, was an early proponent of what came to be called ventricular localization theory – namely, that the different aspects comprising the soul were each lodged in a separate ventricle.[14] "The organs of imagination," writes Nemesius, "are the frontal cavities of the brain, the psychic pneuma within them, the nerves from them soaked with the psychic pneuma and the apparatus of sense organs."[15] As the phrase "organs of imagination" suggests, the problem of the *pneuma psychikon* and the number of ventricles had to do with what would come to be known as the faculties of the soul, among which was imagination. If, like Nemesius, you subscribed to ventricular localization theory, you had to consider what exactly was localized in each ventricle. The two major camps in play at the time (and extending into the early modern period) were, as we might expect, Platonic and Aristotelian in their respective affiliations: the former divided the soul into three key faculties (intellect/*nous*, affection/*thumos*, and passion/*epithumetikon*), whereas the latter found five faculties (reason/*dianoetikon*, appetite/*opektikon*, sensory perception/*aisthetikon*, a vegetative faculty/*threptikon*, and a locomotive faculty/*kinetikon*).[16] Thus the twelfth-century writer of the *Anatomia Magistri Nicolai Physici* would argue that there were three ventricles housing three faculties of the soul – the *cellula phantastica* (imagination), the *cellula logistica* (rational thought), and the *cellula memorialis* (memory) – while his Persian contemporary Avicenna held that there were five ventricles housing five faculties – the *communis sensis* (commonsense), *fantasia* (fancy), *imaginatio* (imagination), *cogitatio* (rational thought), and *memoria* (memory).[17] Albertus Magnus took a middle road, subdividing the ventricles themselves, so that commonsense was housed in the first ventricle; imagination, instinct, fancy, and rational thought shared the middle ventricle; and active memory and memory storage resided in the third ventricle.[18] Figure 3.1, an illustration inserted into a manuscript of Avicenna's *De generatione embryonis*, shows one variation in which the brain is divided into three large compartments (the vertical lines which run from just behind the eye and where the ear would be), but the smaller circles represent the subdivisions; the first circle represents commonsense, as the several nerves running from the throat (touch), tongue, nose, ear, and eye join there. Early Church Fathers would offer another, more streamlined tripartite division of the soul, also derived in part from Galenic theory, in which imagination, reason, and memory each inhabited one of the "cells" of the human brain.[19]

While in actuality the human brain has four ventricles, confusion about their number arose during the medieval period because the two

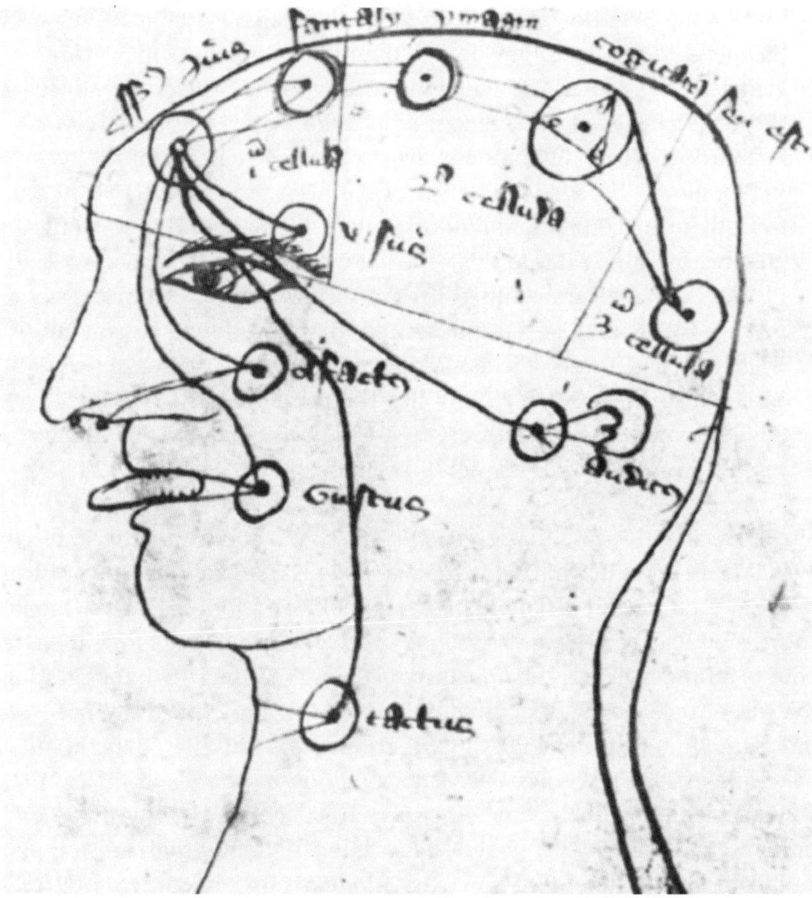

3.1 Avicenna's ventricular localization. Ibn Sina (Avicenna), *De generatione embryonis*, ms. Dated 1347. Walther Sudhoff, "Die Lehre von den Hirnventrikeln in textlicher und graphischer Tradition des Altertums und Mittelalters." *Archiv für Geschichte der Medizin* (Stuttgart, Germany: Franz Steiner Verlag, 1913), figure 5, 189–91.

anterior ventricles were often paired and thus treated as one. Moreover, for much of the time anatomical knowledge was gathered less from human bodies than from animal dissection.[20] Consequently, even through the early modern period the precise number of ventricles and the nature of what they contained was up for debate. Vesalius, arguably the period's most influential anatomist, was himself taught ventricular localization theory. As he recalls in the *Fabrica*,

> I well remember how when I was following the philosophical course [at the University of Louvain] … the brain was said to have three ventricles … [which] had names according to their function. Indeed those men believed that the first or anterior, which was said to look outwards towards the forehead, was called the ventricle of the *sensus communis* because the nerves of the five senses are carried to it from their instruments, and odors, colors, tastes, sounds and tactile qualities are brought into this ventricle by the aid of those nerves … [These were then transmitted to] the second ventricle, joined by a passage to the first so that the second might be able to reason and cogitate about those objects; hence cogitation and reasoning were assigned to the [middle] ventricle. The third ventricle was consecrated to memory … as it were moist or dry, either more swiftly or more slowly, engraved [memories] as into [softer?] or harder stone.[21]

However, the three-ventricle schema does not appear in the *Fabrica*, because, as Vesalius himself emphasizes, his illustrations come from first-hand observation rather than rote memorization of the traditional sources he had been taught in school. Instead, he accounts for the presence of animal spirits or *pneuma psychikon* in the ventricles by suggesting that they might be manufactured there via the choroid plexus (a network of vessels lining the ventricles). By ascribing this purpose to them, he in effect fills the empty holes that ostensibly housed the soul's three inner faculties, denying them any other function.[22] That both animals and humans had ventricles further suggested to Vesalius that the rational soul could not reside there, since this mode of ensoulment was precisely what animals lacked.[23] Ultimately, even this great early anatomist had no systematic positivist explanation for where exactly in the brain the soul resided, but he nonetheless asserted with confidence that one would find neither the soul nor her faculties in its ventricles.

Yet, despite the authority exerted by the *Fabrica* over anatomical studies for the following two centuries, not all of Vesalius's ideas actually took hold in European popular thinking. As with the *rete mirabile*, the argument against ventricular localization met with resistance from his translators/purloiners. For example, while Thomas Geminus and his translator Nicholas Udall, who were among the first to transport the *Fabrica* to England in a vernacular translation, generally remained faithful to that original inspiration, they could not entirely resist the lures of ventricular localization theory. So, Geminus assigns commonsense to the first ventricle, which serves as a meeting-ground for information from the five senses and fantasy and imagination (which he sees as essentially a recombination of sensory data). The second ventricle "ponderith and weigheth, rehersith [*sic*] declarith and demeth" what is fed to it from the first ventricle; thus its office is a cognitive one. Finally, the third ventricle houses the memory, where "are registered, reserved & kept, al such thinges as are done or expressed by the wittes and senses before rehersid, & here be as treasure preserved, unto the … instrumentes of the soules operations."[24] No doubt, Geminus's description does not specify exactly where the soul is located, but in a signal departure from the Vesalian model, the burden of the soul's action nonetheless falls upon the brain's ventricles, for these are what accomplish the pondering, weighing, and so forth of ideas, the preservation of treasure, and the presentation of the "wittes and senses" to the "instrumentes of the soules operations." In other words, the ventricles, rather than the grey matter, are seen as containing the soul's power, its ability to act, and this was broadly the model that, with various alterations, would prove difficult to shake from the British imagination.

Later in the seventeenth century, as the structure and function of the brain came to be more closely aligned, what the ventricles could signify psychologically began increasingly to depend upon what they could accomplish physically. Vesalius's argument that the ventricles produced and stored the animal spirits became more commonly cited as their likeliest role – but, in turn, the animal spirits themselves became more readily identified as the agents of the soul's faculties. An hydraulic model began to emerge: the ventricles manufactured and housed the animal spirits, which the soul pumped out through the nerves in order to gain sensory information about the world around it; when the animal spirits returned to the brain, they conveyed this information, which the soul then used to make judgments, create fantasies, and craft memories.

Thus, despite Vesalius's reservations, the ventricles became the focus of anatomical thinking about the nature of the soul – even if commentators remained unsure as to whether the soul could inhabit a ventricle, like an empty womb, or whether it needed to innervate the physical matter of the brain as well. For, seen as part of an hydraulic system, the ventricles were still involved in the creation and action of thought, and thereby essential to the soul's functions. As Harvey makes clear in "De conceptione," the temptation to link the brain to the womb in terms of the generative power of each organ was strong. Etymologically the link had a solid basis: *ventricle* derives from the Latin *ventriculus*, which in turn is the diminutive form of *venter* – and *venter* can mean *womb*.[25] The use of *venter* as a direct synonym for *womb* was rare and is now obsolete, but *venter* also frequently meant (in the fourteenth to seventeenth centuries) a wife as a child-bearing figure, or one's mother insofar as she possessed a womb. In Aristotelian nomenclature, *venter* described the three womb-like pockets in a human, where the three spirits (vegetable, vital, and animal) were formed and refined.[26] Moreover, the choroid plexus, which Vesalius marked out as the likeliest structure in the brain to manufacture the animal spirits, shares an etymological root with the *chorion*, the afterbirth or membrane which most directly covers the fetus.[27]

Not surprisingly, these etymological connections would be transposed to explain biological conception, supporting the notion that its physical makeup could affect what the womb conceived. Recall from chapter 2 that the part of the brain responsible for memory (usually the third or fourth ventricle) could vary in its temperature and softness; a soft, moist ventricle would allow the easier impression of ideas into the memory, whereas a hard, dry ventricle might impede the memory process. A distempered womb would likewise alter the ability of the woman to conceive and to gestate properly. "Women whose Wombs are too thick and cold, cannot Conceive," writes an author going by the pseudonym Aristotle, "because Coldness extinguisheth the Heat of the Humane Seed."[28] Along these lines an overly moist womb could also destroy seed from the man, while excesses of heat or dryness would be equally problematic.

Indeed, the womb itself is often described in language strikingly like that used to describe the brain, and in particular its ventricles. Jane Sharpe, one of the only female midwife manual authors in the period, observes that

> [i]t hath been much and long disputed how many Cells are in the womb: *Mundinus* and *Galen* say there are seven severall Cells, and that a woman may, by reason of so many places distinct one from the other, have seven Children at a birth, and many midwives are of this opinion, but none that ever saw the womb can think so; for there is but one hollow place, unless Men will say that those holes where the seed vessels come into the womb are places for Children to be conceived in.[29]

Sharpe makes a properly Vesalian divide, between those whose knowledge (or, indeed, ignorance) rests on earlier canonical anatomical texts, and those who draw their understanding from empirical, tangible evidence. On the basis of the latter, she rejects early authorities Galen and Mundinus, along with those (presumably unread) midwives who share their "opinion." Those whose experience has led them actually to see the womb first-hand will know the reality – unless of course they bow to the authority of what "Men will say" regarding (what are presumably) the fallopian tubes.[30] At the same time, her very concern with the number of hollow cells in a given organ reveals her indebtedness to the anterior debate regarding the number of ventricles in the brain.

Penis cerebri

But there was another set of anatomical features within the brain that made a connection between it and the womb even more attractive. Among these was the pineal gland, which was central to Descartes's psycho-physiological system as the organ used by the soul to direct the flow of animal spirits through the nerves. The renowned anatomist Helkiah Crooke describes the gland thus:

> The Figure of it is like a Cone, that is a round turbinated figure, much like the fruit of a Pine-apple, broade and round in the Basis, and growne smaller but keeping round to the top; and hereupon it is that it is called … *Glandula Pinealis*: some have resembled it to the end of the Virile member, and therfore call it *penis Cerebri*, the yarde of the brain.[31]

This description, by no means the only one of its kind, might have troubled the French philosopher. Yet, as we have seen, so much of anatomy in the period is predicated upon physical and metaphorical correspondences, and what object was more readily available, when the time came for naming a given structure within the human body, than the human body itself? In short, because the pineal gland ostensibly *looks* like a penis, it can be called one.

This brain phallus was not, however, the only extra-cerebral figure contemporary anatomists were to see in the brain, for finding it led, almost irresistibly, to further such discoveries. Continuing his description, Crooke goes on to tell us that we see, posterior to the gland itself, an arch separating the two ventricles surrounding the conarium. Next to the arch are "foure small bodies, swell up round and are somewhat hard"; "because they carry the representation of two buttocks joyned together, *Galen* … [and others call them] small buttocks." If you glance over these at the spinal marrow that runs between them, "you will readily compare them to a mans thighes set close together. Others compar[e] the*m* to Testicles." If you then remove the afterbrain, "the passage out of the third ventricle into the fourth conspicuously appeareth … which passage resembleth the very fundament between the buttocks."[32] Crooke's account renders the brain a space that has been gendered masculine, with testes, a penis, and "mans thighes," along with buttocks and anus. The male genitalia have been reproduced in miniature in the brain – at least, from a structural standpoint. But, curiously enough, even here some element of the mother inheres in the brain; in describing the olfactory bulbs, Crooke calls them "two swelling Pappes," or breasts.[33]

Other accounts of the brain's internal architecture muddle the boundary between it and the body even further. Anatomist John Banister – upon whose medical authority Crooke often relies and from whom he probably lifted portions of his own description – would also sketch out the brain's geography in a way that included members of both sexes:

> Behind this vaulted part in the extreme part of the brayne towards Cerebellum, and in the uppwer part of the thyrd ventricle, Nature hath feyned certaine eminent partes, which in their upper partes, represent the likeness or Image of Testicles, and so called therefore of Anathomistes Testes: neare unto the which, two other particles yet somewhat greater are to be discerned, called accordyng to their figure clunes, the haunches or buttocks. Betwene which lyeth that hole, which is already noted to from the third, to the fourth ventricle, and seemeth like unto the fundament. Furthermore in the forepart of these Testicles (as we call them) stretchyng to the thyrd ventricle, an other part of the brayne appeareth, which not unaptly, but very elegantly expresseth the shape or privye part of a woman.[34]

Here the blame for the likeness lies with Nature, who "feigns" a visual connection between structures in the third ventricle and testes and buttocks. Banister is careful to let his reader know that these parts of the brain only "represent the likeness or Image" of the "eminent part" in question, and that anatomists ("we") are the ones who *call* the structure by such a name. The artificiality of the construct seems crucial – Banister wouldn't have us believe that the brain literally contained a penis, let alone the "elegant[]" formation "that expresseth the shape or privye part of a woman." But, despite Banister's reminder that the resemblance is a constructed one, he is not immune to naturalizing it: following their presumably innate desires, the testes strive to join the woman's "privye part," for they "stretch" towards the other part of the brain that "appeareth" under the anatomist's hand.[35]

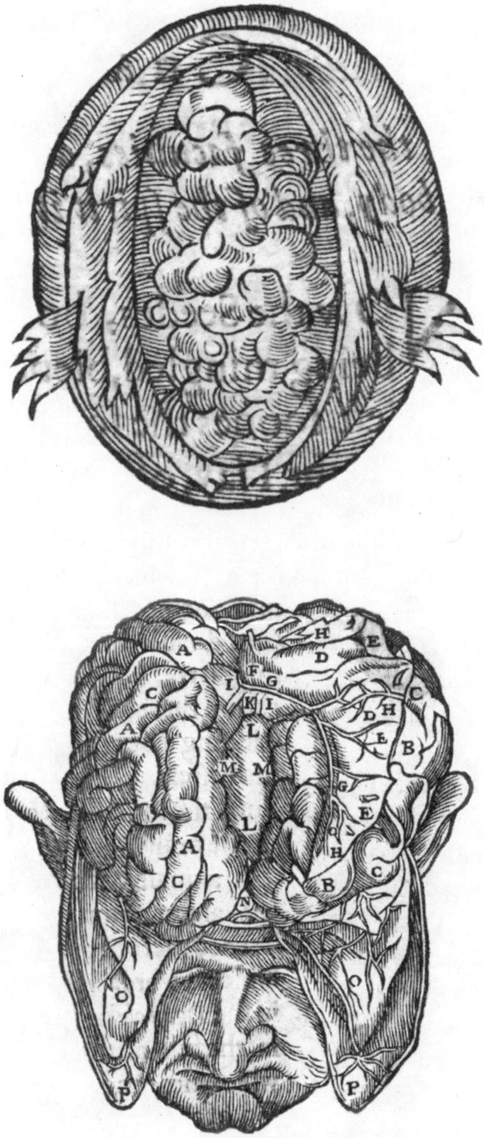

3.2 Top: The womb opened to show the inner membranes; from Jakob Rüff, *The Expert Midwife*, trans. E. Griffin (London, 1637), p. 15. Yale University, Harvey Cushing / John Hay Whitney Medical Library, New Haven, CT.

Bottom: The cranium opened to show the cerebral cortex; from Helkiah Crooke, *Mikrokosmographia* (London, 1615), sig. Rr$_1^{v}$. Yale University, Harvey Cushing / John Hay Whitney Medical Library, New Haven, CT.

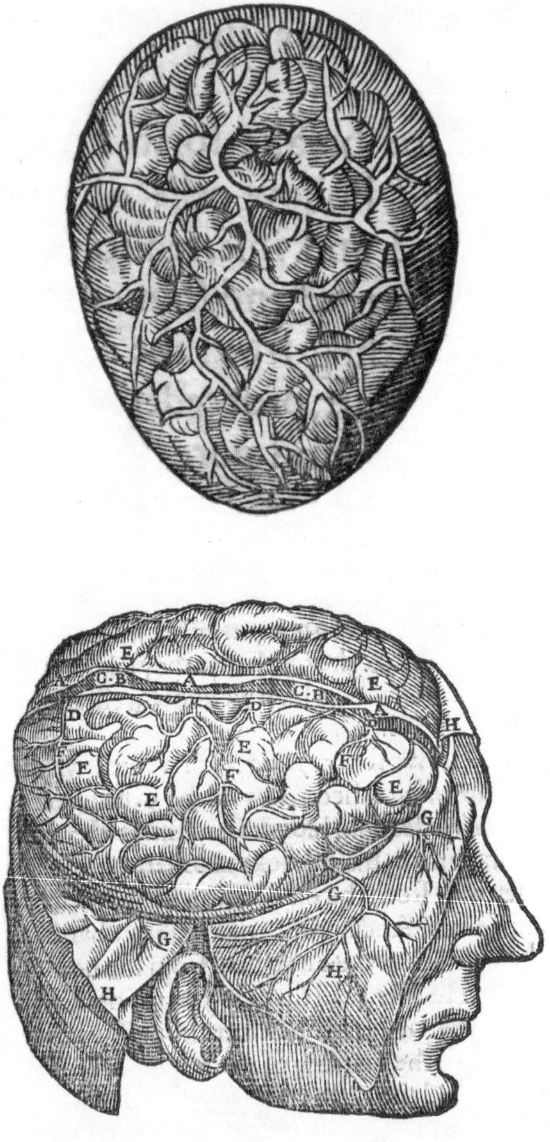

3.3 Top: The womb at the time of the growth of the brain; from Jakob Rüff, *The Expert Midwife*, trans. E. Griffin (London, 1637), p. 34. Yale University, Harvey Cushing / John Hay Whitney Medical Library, New Haven, CT.

Bottom: The brain dissected to show the blood vessels supplying it; from Helkiah Crooke, *Mikrokosmographia* (London, 1615), sig. Rr_1^r. The Huntington Library, San Marino, CA. RB 53894.

Indeed it is striking that there is no alternative nomenclature for these sections of the brain; that is, they are what they are thought to resemble, functionally and visually. No wonder, then, that visual representations of the brain and the womb often seem to overlap, to the point that it sometimes becomes difficult to distinguish cortical folds from uterine furrows. Figure 3.2 presents a womb and a brain in various states of dissection, the view in both cases coming from the top of the structure. The membranes covering both the womb and the brain have been peeled back in similar fashions, and the seed from both mother and father (which Rüff explains has been "congealed and curded together, after the manner of a tender Egge" as the fetus develops) and the cortical matter are rendered in comparable sets of infoldings.[36] The shape of the depiction of the womb echoes that of the brain, down to the details such as the blood vessels interlacing over the interior, as in Figure 3.3. Rüff's womb diagram here shows the fetus at the point at which the brain is beginning to develop: "the highest and most principall part of this beautifull and admirable frame and Architecture of Nature, is formed and produced, that is to say, the Braine."[37] Again, the cortical folds and the tissues of the fetus, along with the veins and arteries, are strikingly akin to those folds and windings of the brain, revealing a realized version of the more metaphorical associations amongst body parts. The visual correspondence thus reflects the desire of anatomists to read the body in terms of the closest thing itself: namely, other parts of the very same body.

Even names of brain structures that derive more from functional similarity rather than from the visual recollection of other parts of the body reveal the importance of gendering the brain. The two membranes covering the brain, for example, are named "hard" (*dura mater*) and "soft" (*pia mater*) "mothers" because of their role in protecting the brain. Even as Crooke likens their function to that of the pleura that holds in the organs in the torso, he points out that the *pia mater* is called "the deere or neere Mother, because it immediately incompasseth and imbraceth the substance of the braine."[38] Contained and embraced by the membranes that protect it, the brain is here "mothered" here quite literally.

Not of Woman Born

It may appear from the foregoing that the broader cultural phenomenon of seeing brains as wombs – that is, of explaining intellectual creation on the model of biological reproduction – rested entirely on the prior labours of anatomists and physicians, whose metaphors and analogies

3.4 "Athena born in a shower of gold." Michael Maier, *Atalanta fugiens, hoc est, emblemata nova de secreties naturae chymica…* (Oppenheim, 1617), sig. N_3^r. Boston Medical Library, Francis A. Countway Library of Medicine.

were absorbed by other writers of the period. However, this would be to ignore influences in the other direction – namely, from the aesthetic to the anatomical. Arguably, behind the anatomists' desire to connect brains and wombs, and forming the crucial bridge that would lead to these connections being taken up more generally, lay a mythological *locus classicus*. That thought, like a child, springs from the mind, is of course, the central action in the widely cited story of Athena's birth, depicted emblematically in Figure 3.4. In Maier's rendition, Hephaistos uses an axe hot from the forge to split Zeus's head, allowing a naked Athena to emerge, while an eagle supervises (prompting the viewer to imagine Ganymede and thus exploiting the potential sexuality of the scene). In the background Aphrodite and Ares are depicted in a severely compromising position, with Eros to egg them on – and that this tryst takes place within view of Hephaistos emphasizes both the wrongness of Athena's generation and the eroticism of such an odd pregnancy.

According to Hesiod's version, Zeus, having slept with his first wife Metis, fears a prophecy that one of his children will become more powerful than he himself. To circumvent the problem, he tricks the newly pregnant Metis into turning into a fly, and then he

> seized her with his hands and put her in his bellow, for fear that she might bring forth something stronger than his thunderbolt: therefore did Zeus … swallow her down suddenly. But she straightaway conceived Pallas Athene: and the father of men and gods gave her birth by way of his head.[39]

While inside Zeus, Metis creates armor and weaponry for her daughter; the hammering causes Zeus to be so troubled by a headache that he calls upon Hephaistos to split open his skull, the very moment captured foregrounded in Michael Maier's engraving of the story.

One of the Homeric hymns recounts this violent "birth":

> Metieta Zeus gave himself birth to her [Athena] out of his terrible head, arrayed in warlike arms, golden, gleaming. Astonishment seized the gods as they watched. She sprang forth at once from the immortal head and stood before Zeus who holds the aegis, shaking a sharp spear. Great Olympos began to quake dreadfully at the might of Glaukopis [Athena], and earth all about screamed horribly, and the sea moved and frothed with dark waves, while foam suddenly burst forth. The brilliant son of Hyperion stopped his swift-footed horses for a long time, until the girl, Pallas Athena, stripped the godlike armor from her immortal shoulders, and Metieta Zeus rejoiced.[40]

Without a doubt one not of woman born – the very earth seems to rebel at her unnatural birth and immediate readiness for battle – clear-eyed (*Glaukopis*) Athena represents the union of cunning (from Metis) and power (from Zeus): she is wisdom embodied. She enters the world from a double womb, first from Metis's literal womb and then from Zeus's more metaphorical womb, his head and brain. In this story, Zeus becomes the author who must consume wit or cunning in the form of Metis in order to produce his "text," Athena, hence the new name he acquires here, Metieta

Zeus. The episode proves ripe for metaphorical transfer, particularly given Athena's designation as the goddess of wisdom, as writers could easily transpose the story of her birth onto a larger discourse of the generation of thought itself.

That this account fascinated the intellect of the early modern period can be seen in the wide-ranging adoption of the formal possibilities of imagining the brain as a creative intellectual womb, in a variety of genres, from poetry to natural philosophy and beyond. For example, in a preface to a treatise elaborating upon Cartesian mechanism, Antoine Le Grand philosophizes, "Hence it is the *Poets* tell us, That *Pallas* or *Minerva*, the *Goddess* of *Wisdom*, was the birth of *Jupiter's Brain*; intimating thereby that all *Knowledge* comes from GOD, and takes its rise from his *Understanding*."[41] Here the brain-birth story has been transposed onto a theological account of the production of knowledge, and we are asked to figure our own knowledge in terms of an outpouring from God's "understanding." The comparison stems from what the poets have to say, but for Le Grand the poetical association is useful for comprehending the philosophical implications of knowledge birthed from the brain.

Neither was Metis, the original mother of Athena, ignored for her role in the birth of knowledge. In an heraldic treatise based largely on an essay by Francis Bacon, Thomas Philipot instructs royalty on the benefits of taking outside counsel: "*Jupiter* espoused *Metis* which signifies Counsel, and after by swallowing her, conceived *Pallas* in his Brain; so Princes must wed themselves to sober Counsellors, and by swallowing their sage advice, their Heads shall be pregnant with Wisdome."[42] Philipot emphasizes the interaction between wise counsel and the (re)production of such wisdom in the mind that "swallows" it. Metis apparently dissolves in Jupiter's brain, becoming something like Harvey's "plastick generative power," thereby allowing Jupiter himself to conceive both physically and intellectually. The line between the two actions of conception is narrow, and the metaphor has been so rubbed through that one may peer from the side of the biological and maternal to the side of the intellectual and paternal.

While Philipot focuses on the image of pregnancy, other authors borrow terms from the birth itself to make vivid their analogies. Thus in Richard Brome's play *The Love-Sick Court*, Matho, accomplice of the schemingly ambitious Stratocles, confides in his partner:

I can perceive
That now the labour of your *Jove*-like brain

Is bringing forth the *Pallas*, shal in spire
Me, to perform the work of my advancement.

Stratocles replies, not without sarcasm, that his own Pallas is "not yet ripe for delivery."[43] The ideas require further gestation, and just as a real child should not exit the womb before his time, so too the intellectual child must await the pangs of labour. A mechanism similar to the one described by Philipot is at work here: the ingestion of an idea, here the very idea of labour itself, bestows upon the ingestor the powers necessary to accomplish a similar task. The act of birth – labour, bringing forth – taking place in Stratocles's mind inspires a similar response in Matho's. Intellectual pregnancy, at least, can be contagious.

The poetics of the metaphor likewise worked their way into even Thomas Willis's later neuroanatomical philosophy. "The Poets feigned *Pallas* to be formed within the Brain of *Jupiter*, and from thence to be born," writes Willis (thereby excising Metis from the story entirely): "In truth, within the Womb of the Brain all the Conceptions, Ideas, Forces, and Powers whatsoever both of the Rational and Sensitive Soul are framed; and having there gotten a species and form, are produced into act."[44] Willis here proposes an artistic justification of his story of the brain's architecture – and, crucially, a justification that takes as its basis a comparison originated by poets, and given by Willis the imprimatur of truth. The elision of the boundary between womb and brain underscores his distinctive use of the metaphor: the "framing" that takes place "within the Womb of the Brain" unites all the powers of rational and sensitive souls with "a species and form," just as Harvey's "plastick generative power" translates the immaterial to the material, producing both biological and intellectual offspring.

As the variety of these examples suggests – and they could be multiplied almost indefinitely – a pre-existing mythical narrative, the story of Athena's quasi-parthenogenetic birth, was re-functioned in the early modern period to legitimate male intellectual and literary production, and this appropriation rebounded upon anatomy itself, inspiring its own theoretical and empirical accounts of human generation. Their functional analogies would even inform what they took to be the independent evidence of their eyes. In turn, coupled with the poetic vision of male fertility embodied in the myth of Athena's birth, anatomical discussions of brains and wombs would echo more widely, as we shall see below, spurring a set of metaphors and analogies that allowed male authors of the period to arrogate to their own creative needs the productive, generative powers of the female body.

My Brain, the Womb

If, following Harvey, the womb could be conceived of as a brain, so too, then, could the brain be a womb – and, specifically, a writerly womb, in which sensation and perception create a conception, nourished by imagination and birthed by the author through his pen. Prepared by both anatomical and mythological precedents, a popular trope, especially in prefaces and letters to the reader, described the toils of the author's gestation, and the labour pangs required to produce the child of his writing. An especially fine example may be found in the striking image in the sonnet that opens Sir Philip Sidney's *Astrophil and Stella*, representing the speaker's inability to produce something new, through which to win the eponymous woman of his desire, as a thwarted pregnancy. This dire condition, which leaves him "Byting" his "trewand penne," comes from seeking too hard to model his poetic production on the works of others who have preceded him:

> Studying inventions fine, her wittes to entertaine:
> Oft turning others leaves, to see if thence would flowe
> Some fresh and fruitfull showre upon my Sunne-burnt braine.
> (Sig. $A_2{}^r$, ln. 6–8)

But far from fructifying his brain-womb, all his "study" only incapacitates it, words come "halting out," invention flees, and "others feete, still seem'de but strangers in my way" (ln. 9–11). His brain thus stymied, the would-be poet becomes trapped in a potentially never-ending labour: "great with Childe to speake, and helplesse in my throwes" (ln. 12). Indeed, the only available egress for Sidney's nom de plume, Astrophil, demands breaking with the analogy between brain and womb, reconceiving poetic creation not as an intellectual but as an emotional enterprise that bypasses rational calculus. "Foole saide my Muse to mee, looke in thy heart and write" (ln. 14). What is thereby birthed, the speaker would have us believe, is the sonnet sequence itself.

But, as with the use of the Athena legend, the metaphor of brain as womb goes well beyond poetry, the privileged domain of metaphor as such. It traverses a range of genres, and can be found in atlases, scriptural commentaries, magazines of melodies, and works of philosophy, history, and, of course, anatomy. In one of the metaphor's more curious instantiations, a prefatory poem to his tract describing the colony of Maryland and extolling its natural virtues, George Alsop imagines himself having been ravished, and subsequently abandoned, by Apollo:

When first *Apollo* got my brain with Childe,
He made large promise never to beguile,
But like an honest Father, he would keep
Whatever Issue from my Brain did creep:
With that I gave consent, and up he threw
Me on a Bench, and strangely did he do;
Then every week he daily came to see
How his new Physick still did work with me.
And when he did perceive he'd don the feat,
Like an unworthy man he made retreat,
Left me in desolation, and where none
Compassionated when they heard me groan.
What could he judge the Parish then would think,
To see me fair, his Brat as black as Ink?[45]

The naive author retains his own gender while turning his brain into a place of generation, the ensuing "Brat as black as Ink" a cheeky reminder of the materiality of the book itself. The "Issue" from the author's brain is the product of Apollo's workings – the "new Physick" and strange doings, which may take a rather vivid imagination to picture properly – upon the author's heretofore innocent brain. Here the Roman god, as an emblem of divine (though, in this case, only minimally successful poetic) inspiration, stands in as an external agent who fertilizes the author's active imagination in the womb of his brain in order to produce the aforementioned offspring.

More generally, the creation of brainish children is rendered as a process of parthenogenesis, with no obvious impregnator and only the author involved in the act of creation. In an epistle to a female patron prefixed to a treatise of medical observations, a writer explains:

> Some Authors call their *Books* their *Children*, all sure must own them for their *Conceptions*. This of mine is but a *Daughter*, begotten then, when your [*sic*] lay sick at *Dorchester*; and born, by the help and Midwifery of your obliging and *Testimonial Letter*, from *Spargrave*. But, having been sick almost as long as *yours* hath been well, she is but slenderly grown, and none other but a plain *English Girle* ... this *Child* stuck at the Birth above twelve months.[46]

Ingratiating himself by likening his creation to his patroness's, the author here takes control of the process of generation: "all sure must own them for their *Conceptions*," he writes, indicating that the parental/paternal relationship between author and book is one of self-creation and then separation. He plays on the term *conception*, neatly collapsing its two operative senses (i.e., *idea* and *child*). The patroness's illness caused the author to want to write for her – and there the idea/child was conceived by the author himself – and the "Testimonial Letter" serves as midwife to the difficult birth. Obviously the birth was a problematic one, given that it (rather than the gestation, which would have been bad enough!) went on for twelve months. Yet the point here is also to distinguish writerly gestation from writerly birth: the one involves the development of the concept in the author's mind, whereas the other is the physical labour of the writing itself.

The above example further showcases a particular way in which such letters operate as a preemptive undercutting, a kind of shield of false modesty protecting the author from foreseeable critics. Part of the charm of the metaphor stems from its giving authors space to present their works as potentially imperfect – inky, female, ill – the better to garner reassuring praise from the reader. If I say that my work is "but a Daughter," "slenderly grown" because she has been sick, and "plain," at that, you (the reader) are presumably less likely to find fault with my arguments. A similar idea comes across even more explicitly in the preface of an eighteenth-century satirical essay on nothing, by Hugo Arnot:

> I Acknowledge without affectation, that I think this child, whom I am about to usher into the world, well made, handsome and sprighdy [*sic*]. I am sensible however of the blindness and fond partiality of a father, and that the public alone are proper judges of his merit. Lest, therefore, it should be discovered that my favourite child is lame and deformed, my parent affection has furnished him with a couple of stilts, to enable him to hobble through the world with tolerable grace.[47]

While the "child" in question here is a boy, and thus not immediately denigrated as a daughter might be, the author admits nonetheless that his own paternal partiality may render him incapable of seeing whether the child, his favourite, is "lame and deformed." Only the reader can judge the work, yet the stage is so strewn with curried favours that the reader cannot help but see it as a "well made, handsome and sprighdy" issue of the author.

The image of labour that Sidney skillfully deploys would in other cases be extended even more literally to take in the actual circumstances of childbirth in the early modern period. For example, a friend writing an elegy to the author Richard Vines emphasizes his own intermediary role in presenting the book to the author's patroness: the friend describes the book as "the fruit of [Vines's] studies and pen; the more lively characters of his heart and spirit … [and] neither did it need any thing but midwifery to bring it forth into the World, and present it unto you as *his*, or (if you please) as *him*."[48] Here it lies in this specific reader to identify the *fruit* as belonging to the author – or, more strangely, as actually standing in for the absent author, so strong is the family resemblance between the man and the "characters of his heart and spirit."

This is but one example of calling the office of the midwife into play, in order to extend and elaborate the metaphor as well as ground it in the actual circumstances of childbirth. Because it is usually a male writer referring to himself as a midwife, the choice of office (rather than, perhaps, physician) is both curious and shrewd: curious because it extends further the gender-reversal already occurring in the metaphor, and shrewd in the context of contemporary debates concerning the "male-midwife."[49] Midwifery manuals were themselves usually written by men as prescriptions for female midwives during birth, but a growing contingent of male physicians was arguing for a masculine presence during the birth – which was generally seen as predominantly women's work – as a means of authorizing and legitimating any actions taken during what was ultimately an extremely dangerous process.

Beyond being simply a friend and endorser of an author's work, the writerly midwife could equally stand in for a translator or editor. George Ent, who performed both these offices in the case of William Harvey's *Anatomical Exercitations, Concerning the Generation of Living Creatures*, thus positions himself as a "meer" midwife, working under the shadow of the great physician Harvey:

> But, of *My selfe*, I shall add onely thus much; that in this [great] Business I have performed no more than the meer office of a Midwife: producing into the light this noble Issue of His [Harvey's] Brain, in all its parts and lineaments perfect and consummate, as it is now presented to your Veiw [*sic*]; but staying long in the Birth, & fearing, perhaps, some injurious Envy or Detraction.[50]

Ent's worries stem from the fear that the long birth of the English translation of *De generatione* could be disastrous, the longer the Latin edition circulated amongst those who might criticize it. An English edition would be subject to the same "injurious Envy or Detraction." Humbly, then, Ent brings "into the light" the "noble Issue" of Harvey's brain, asking us to examine its "parts and lineaments," as one would do with a real child.

Indeed, the desire to read in the child the stamp and character of the parent was yet another a crucial component of this metaphor, as the likeness signifies authority, legitimacy, and, eventually, the perpetuity of the author-parent. In a letter to the author of a satirical typology of "intrigues," the anonymous letter-writer at first thinks to chastise the author (identified on the title page as "one of the fair Sex," and called Idalia by the letter-writer) "in concealing a legitimate Birth of your Brain from me," but then considers that it may have been Idalia's modesty that "deny'd me the Privilege of assisting your Labour," and so the letter-writer decides instead to "congratulat[e] you upon so hopeful an Issue":

> Had I suspected you so near your Time, I should have attended the first Cry, and look'd upon the Offspring with all the forward and officicious Sentiments of a Friend … Methinks I trac'd you in every Feature; each beauteous Lineament confessed the happy mother … The kindly Stamp betray'd the innocent Original.[51]

In another variation, a publisher might step in to explain the origin of a work that is the product of a putatively reluctant author, as does Edward Blunt in describing his role in helping bring John Earle's *Micro-cosmographie* into the world. He writes that he has "adventur'd to playe the Mid-wifes part, helping to bring forth these Infants into the World, which the Father would have smoothered: who … left them lapt up in loose Sheets, as soon as his Fancy was delivered of them."[52] This iteration of the metaphor calls attention to the very means of imaginative production, for it elides womb-brain and the creative force of fancy. The troubled and dissatisfied author would have crumpled up his pages (the "loose Sheets"), smothered his own creation, had not the publisher intervened to rescue the child for the reader and for posterity.

Ultimately the purpose of the brain-as-womb metaphor would appear to be twofold. First, the likening of the creative process to generation allowed the male author to imagine and even co-opt the experience of birth: if a woman may produce, so, too, can a man. Second, and perhaps more recognizable to us, is the urge to read in the "birthed" book evidence of a lineage, the creation of an heir to an author that will (it is hoped and fervently desired) outlive its parent, entering into new minds and breeding new ideas, while also serving as a monument to the author himself. To cite again one of the epistle writers I have referred to above, a reader should receive the work not only as belonging to but as actually *being* the author: "as *his*, or (if you please) as *him*."

CHAPTER FOUR

The Mechanics of Reproduction in the Art of Cavendish

Yet *Fancy* cannot be without some *Braines.*
– Margaret Cavendish, *Poems and Fancies* [1]

The Disease of Wit

To be "Authoress of a whole World," one must first be willing to make sacrifices: the "Paper Bodies" of one's drafts need to be burned, "like Parents which are willing to Die, whenas they are sure of their Childrens Lives."[2] With this image, Margaret Cavendish presents a version of creative lineage that runs counter to a male-oriented fantasy of writing being birthed from the brain-womb. Rather, she sidesteps the question of how authorial thought acquires a "paper body," with the written drafts themselves instead providing the intellectual parentage heretofore supplied by the author. A kind of collapse has taken place, the line between author and her works having been all but cancelled out – while her writing itself begets more writing. Indeed, through this conceit, the author is herself freed to proliferate, her written productions becoming ultimately the containers her surrogate self inhabits, and from which her selves may then depart to take up residence elsewhere. Thus, Cavendish hatches a curious plan to achieve "some Work, wherein I may leave my *Idea*, or Live in an *Idea*, or my *Idea* may Live in Many Brains, for then I shall Live as Nature Lives amongst her Creatures" (SL 178). Her self-multiplication depends upon eliding distinctions among herself, her ideas, and her works, and, in a reverse-birth, this hybrid of three is reborn in the brains of her readers. For Cavendish, this process would be a path to eternity; to her critics, it reeked of infection.

That the fecundity of Cavendish's writing had potentially dangerous consequences was lost neither on her contemporary critics nor on Cavendish herself. Pepys, for example, details over several diary entries the ineluctable curiosity her work kindled in him. He grumbles that he went to see "a silly play of my Lady Newcastle's, called 'The Humourous Lovers'; the most silly thing that ever came upon a stage. I was sick to see it, but yet would not but have seen it, that I might the better understand her."[3] The play is by his account "silly" twice over, and yet Pepys's expressed reluctance is an ambivalent illness: he is sick both at seeing and with the desire to see the play. The felt necessity of "better understand[ing]" Cavendish (whether the woman or the phenomenon) compels him to return, directly and repeatedly, to experience the effects of what she has written. While disparaging her writing as ridiculous, he is drawn through her play to the woman herself, a figure both obscure and obscured, and one who seems always just beyond Pepys's reach, as we read in ensuing diary entries. He is frequently going somewhere in the hopes of meeting Cavendish, the better to know her – only to find that he has just missed her, or that she is too surrounded by baffled onlookers for him to do more than catch a glimpse of her. Eventually they do meet, but not before Pepys observes, witheringly, that "[t]he whole story of this lady is a romance, and all she do is romantick ... [H]erself in an antique dress, as they say ... There is much expectation of her coming to Court, so that people may come to see her, as if it were the Queen of Sheba."[4]

Others would echo the diarist, testifying to the unsettling impact of Cavendish's writing as well of her persona, conflating the two as if responding to her own encouragement. Upon reading Cavendish's first book of poetry, the 1653 *Poems and Fancies*, Dorothy Osborne transfers judgment onto the authoress, averring that "there [are] many soberer people in Bedlam."[5] Cavendish made such an impression upon Mary Evelyn (wife of diarist John Evelyn) during a visit between the two women that Mary hoped, "as she [Cavendish] is an original, she may never have a copy."[6] In Katie Whitaker's biography of Cavendish we see that this reaction was certainly not intended as laudatory, since Evelyn would later "rush[] ... to suggest insanity" as an explanation for Cavendish's apparent oddness. Whitaker elaborates that "Mary Evelyn was not only offended: she also felt decidedly threatened."[7] Reflecting on her encounter with "Mad Madge," Evelyn would write: "I acknowledge, though I remember her some years since and have not been a stranger to her fame, I was surprised to find so much extravagancy and vanity in any person not confined within four walls ... At last I grew weary, and concluded that the creature called a chimera which I had heard speak of, was now to be seen, and that it was time to retire for fear of infection."[8]

Not only did Cavendish merge herself with her writing in a way that licensed such responses, but she anticipated, too, a readerly recoiling, as is evident in how she repeatedly and self-consciously defends herself, particularly in her prefatory writing. In the preface to her *Observations upon Experimental Philosophy* (hereafter OEP), for example, she envisions her critics interpreting her constant generation as worrisome and epidemic, much as Evelyn had done:

> It is probable, some will say, that my much writing is a disease; but what disease they will judge it to be, I cannot tell; I do verily believe they will take it to be a disease of the brain; but surely they cannot call it an apoplectical or lethargical disease: Perhaps they will say, it is an extravagant, or at least a fantastical disease; but I hope they will rather call it a disease of wit. (*OEP* 7)

No one can accuse her, Cavendish insists, of sloth – her disease could never be called "apoplectical or lethargical," for she is constantly putting forth work. But it is less the quality of her writing, its ideational content, that is at stake here than its quantity: the "muchness" that she believes will cause her readers to wonder if she is ill. The voluminosity of her prose might be read, she claims, not as appropriate production; it might instead be seen as a sign of something pathological, dangerous, and potentially *communicable*, even if it is hidden in the inner recesses of the mind (be it anatomical or metaphorical).

In fact, Cavendish takes almost for granted that her overabundance of literary production is a disease, but the question that remains is: what is the nature of this disease? How should it be judged? A clue may be found in her hope that her "much writing" would be taken as resulting from "a disease of *wit*," a key seventeenth-century term which we might gloss variously as rational intelligence or imagination, both faculties of the soul and both thought to have their seats in the brain (whether in the ventricles or elsewhere). In the preface to the second book of *The World's Olio*, Cavendish elaborates upon the locus of this generative source in a remark concerning what Kate Lilley calls "the transgressive potential of grammatical singularity."[9] Pledging to renounce the rules of grammar because "those that are nobly bred have no rules but honour, and honesty," Cavendish argues that the writer should not be "hidebound with nice and strict words, and set phrases, as if the wit were created in the inkhorn, and not in the brain."[10] Rather than be so bound, wit ought to remain free, for in its freedom lies its creative power: "As for wit, it is wild and fantastical, and therefore must have no set rules" (*WO* 94). Wit, then, contains in itself the potential for productive transgression, and a disease of wit can create work which likewise can be read as wild and fantastical, from both positive (Cavendish's) and negative (e.g., Evelyn's) perspectives. As Brandie Siegfried observes, Cavendish invokes a "passive *state*," i.e., of *much writing*, "as a teasing contrast to what she considers the real character of her writing persona, the daring *act*."[11] A "wild and fantastical" wit thereby overwhelms the disease of excessive writing which Cavendish's critics ascribe to her.

These descriptions hint at the metaphor governing this chapter: the brain as *writing machine*, an overactive, automatic entity hell-bent on producing writing. While not explicitly articulated in these terms by Cavendish herself, the image of the writing machine nonetheless seems the best way to encompass Cavendish's apparent compulsion to write, to translate her restless thoughts, born of fancy in her brain, onto the printed page. Implicit in her philosophical tracts – and suggestive of her complex relationship to the mechanistic philosophies of her day – this sense of the brain as both fecund and automatic feeds into how Cavendish redirects inherited discourses surrounding childbirth – along with their associated metaphors – to characterize the motive forces driving the brain's activity, and to convey its almost compulsive power of generating words, ideas, and selves. This chapter first explores a range of different metaphors that Cavendish deploys to describe thought at work in the brain, and situates these in relation to her idiosyncratic natural philosophy. Dissenting from the Cartesian dualism of mind and matter, she develops a very different, far more dynamic understanding of their relationship, which is exemplified in how she conceives of the brain as an entity thoroughly physical yet also a direct participant in mental activity. Key to her distinctive translation between intellectual and corporeal production are Cavendish's philosophies of perception and intellectual activity, which she understands through the related metaphors of *patterning* and *figuring*. Allied to these is a further important distinction between different modes of thinking: between the ordered precision of rational thought, on the one hand, and, on the other, fancy, which she aligns specifically with writerly creativity.

Yet, as already intimated, Cavendish remains fully aware that this writing machine can very easily go haywire, that the disease affecting her brain may not just be said to be but may actually be, like herself, "extravagant, or at least … fantastical." It is significant that she expresses concern for her over-production by describing it in terms of disease and monstrosity. As the previous chapter has argued, for male writers such as Harvey, likening the creative process to childbirth enabled a sort of imaginative co-optation of the experience of giving birth. But it must also be added that this fantasy of masculine agency, which lends early modern discourses of childbirth their distinctive contours, was not entirely without its limits. With respect to one phenomenon at least, the generative powers of patriarchy were all too willing to cede agency and control: monstrosity. For, as we shall see, monstrosity in particular was thought for much of the sixteenth and seventeenth centuries to be caused by a defect in the feminine (and specifically maternal) imagination. Thus, considering the metaphors Cavendish uses to describe the thinking brain, this chapter will turn to survey the problem posed by the maternal imagination. That excursus lays the grounds for returning to Cavendish to show how she turns the masculine metaphor of intellectual conception on its head, co-opting for her own productive ends the possibilities for the female imagination still left open by the dominant discourse. One early physician described the process of the mother's imagination working on the fetus as a kind of *imprinting*, and with this characterization in mind we will conclude by examining how Cavendish's reconfiguration of this received metaphor in terms of patterning and figuring shapes her expansive authorial claims – to be "authoress" of "whole world(s)" – in her best-known work, *The Blazing World*.

Brain-Pan, Thought-Pancakes

The brain, so aligned with both wit and fancy, takes centre stage in the creative process for Cavendish, but it does so in a roundabout fashion. We can begin by recognizing that her philosophy of animated matter – which in turn underlies thinking matter – leads to blurring the distinction between mind and matter, and concomitantly between *mind* and *brain*. As David Cunning summarizes it,

> Cavendish is in effect asking us what we think we are engaging when we engage the mind of another person. If the person is before us, his mind is before us as well. If the person moves from one place to another, his mind is carried along with him. A mind is a collection of bodies in a human body, and among its properties are the same ideas and volitions that the substance dualist would regard as immaterial.[12]

Properly speaking, the mind animates the matter of the brain; thus the presence of one presumes the presence of the other. By contrast, thoughts as well as the faculties of the soul are for Cavendish more distinct entities, capable of moving about in the mind-brain (as her metaphors below will make vividly clear).

While not strictly monist, then, Cavendish's philosophy nevertheless opposes facets of Cartesian dualism, which distinguishes sharply between mind and matter – between *res cogitans* and *res extans*, the immaterial (because not extended in space) "thinking stuff" and the material "extended stuff."[13] Instead, Cavendish proposes that matter itself is both self-moving and self-knowing. As she argues in *Philosophical and Physical Opinions*, "Whatsoever hath an *innate motion, hath knowledge*; and what matter soever hath this *innate motion*, is *knowing* … *knowledge* lives in *motion*, as motion lives in *matter*."[14] As a consequence, in her later *Philosophical Letters* (1664) she writes that "it is very probable, that not onely all the matter in the World or Universe hath sense, but also Reason," for "Thoughts, Ideas, Conceptions … as also Natural Life, and Soul, are all Material."[15] Thus the entirety of the material world can think – and, conversely, thought is itself in some fundamental sense material. In Stephen Clucas's words, "she collapses the dichotomy between matter and spirit, mechanism and vitalism, by insisting that matter is living and sentient."[16] This is the very dichotomy that Cartesian mechanists worked hard to uphold; for the mind to be composed of thinking matter, and for thoughts to have physical presence, would have seemed nonsensical to them. The implications of thinking matter and material thought are significant, for they colour how we read her descriptions of the brain. For Cavendish the brain is less a passive vessel for the soul than an active agent producing material thoughts.

Nowhere is the strange pliability of Cavendish's thinking more evident than in the "allegories" section of her 1655 *The World's Olio*, which finds the brain and its inseparable companion, the rational soul or mind, trying on and casting off a dizzying array of analogical costumes. The *Olio* precedes her more extended and sophisticated philosophical works, but its ideas clearly seed the later *Philosophical and Physical Opinions*. As its name suggests, the *Olio* is rather a mishmash, and in it the brain is rendered protean through a series of minor metaphors, none of which is ever developed more than locally or indeed consistently at any length – but whose general tenor resonates with the lines of thought that traverse her subsequent, more philosophically oriented writings.

The mind- and brain-related metaphors of *The World's Olio* are a mixed bunch. She disparages the minds of "Most Men" because "their Brains most commonly are like Barren Grounds, which bear nothing but Mossy Ignorance, no Flowers of Wit" (*WO* 95). In the horticultural parlance of the day, moss was thought to be anathema to a healthy garden, with one writer counselling that a gardener should "kill and destroy" moss, for the improvement of the ground.[17] Only slow-growing, creeping moss can take root in everyday brains void of ideas. The "Flowers of Wit" hint at Cavendish's true prize, the vegetation that should instead take precedence in the garden of the brain. Such is the thrust of another allegory extending the horticultural theme in a more positive direction: "The Mind is a Garden where all manner of Seeds be sown," she writes, before listing different beneficial ("Prosperities are the fine painted Tulips, Innocency the white Lillies") and less-than-beneficial ("Poppy is Stupidity; Sloth, and Ignorance are Weeds which serve for no use") assortments of foliage (*WO* 97). Reading these two garden-metaphors together suggests that the cultivation and ordering of the garden must occur through the care of the intelligence, which will cut away sloth and ignorance while ensuring the garden runs neither too wild nor too dry. One needs take pains to avoid gardens sprawled with unruly moss and sense-deadening poppies: these are signs the mental garden has been badly managed.

Even as the *locus amoenus* of the garden collapses mind into brain by fixing them spatially, Cavendish's desire to introduce a different distinction between *mind/brain* and *thought* becomes clear in a set of varied metaphors running through *The World's Olio*. "The Mind is like a Commonwealth," she writes, "and the Thoughts as the Citizens therein; or the Thoughts are like Household-servants, who are busily imployed about the Minds Affairs, who is the Master" (*WO* 95). Elsewhere, Cavendish considers that "[t]he Thoughts are like Stars in the Firmament, where some are fixed, others like wandring Planets; others again are only like Meteors" (*WO* 97). Or: "The Head of Man is like a Wilderness, where Thoughts, as several Creatures, live therein" (*WO* 99). Or yet again, suggests Cavendish, revisiting the horticultural metaphor, "The Brain is like a Forest, and the thoughts as Passengers that travell therein, making Inrodes and beating out Paths" (*WO* 100). My personal favourite, however, is this: "Thoughts are like Pancakes, and the Brain is the Pan wherein they are tossed and turned by the several Objects, as several Hands" (*WO* 101).

These comparisons, ranging widely and wildly, from the heavenly firmament to domestic culinary arts, nonetheless share a distinguishing feature, united as they are around the theme of a container housing jittery, chaotic, constantly moving things. Each example involves an enclosing field – a commonwealth, a firmament, a wilderness, a forest, a pancake-pan – within which something else – citizens, stars, creatures, passengers, pancakes – is in continuous and potentially topsy-turvy motion. The overarching metaphor, acting as genus to the species, is thus of *container* and *moving things contained*. And, in each case, one could substitute *mind* for *brain*, and vice versa (or even *head*, in one example) without altering the underlying conceit, emphasizing central tenets underpinning Cavendish's philosophical attitude even in these early writings: namely, that the matter of a thing is fundamentally inextricable from whatever animates it, lending it a unity; and, concurrently, that such unities are themselves also congeries, containing other things that dynamically interact with them and each other.

The *Poems and Fancies* and *The World's Olio* find Cavendish exploring a more systematic natural philosophy; *Philosophicall Fancies* revisits, anticipates, and revises some of the basic premises, both poetic and fanciful, she had begun to address. (Though *Poems and Fancies* and *Philosophicall Fancies* were published in 1653 and *The World's Olio* in 1655, there is evidence to suggest Cavendish composed the texts around the same time.[18]) Cavendish here first distinguishes between the categories of matter and motion, both conceived as essential, eternal, singular, and capable of infinite variety, only to insist that they are inextricable, for motion acts upon matter, and neither can exist without the other.[19] Motion is thus an animating force inhering in matter, which Cavendish, following an earlier tradition, identifies with spirits. Both matter and motion can vary in degree, "as *thinner* and *thicker*, *softer* and *harder*, *weightier* and *lighter* … *swifter* and *slower*" (PF 1–2). The thinner the matter, the swifter the motion of its spirits – and thicker or "duller" matter is inherently slower: "Those *Spirits* most commonly move according to the matter they worke on. For in spungy and in Porous *light matter*, their *motion* is quick; in solid, and weighty, their *motion* is slower" (PF 17). Spirits create the forms of matter, can "carve the thicker *infinites* [of matter] all into severall *Figures*; like as *Aqua-fortis* will eate into the hardest Iron, and divide it into small parts" (PF 18). A human being, for example, is made by the spirits' crafting of the different degrees of matter into the different parts of the body: "the *Spirits* of life take the solid and hard *matter* for the *Bones*: The *Glutinous Matter* for the Sinews, Nerves, Muscles, and the like; and the *Oyly matter* for Flesh, Fat, and Marrow. So the *fluid* for Blood … And the *Spirits* themselves do give this *dull matter, motion*" (PF 19).

But what of the mind? The spirits which craft and animate the "duller" matter, the *sensitive* spirits, are for Cavendish distinct from the *rational* spirits which animate the lighter matter, and just as matter may vary to infinite degrees, so too may spiritual motion. Thus, "[t]here is a degree of stronger *Spirits* then the *sensitive Spirits*, as it were the *Essence* of *Spirits*; as the *Spirit* of *Spirits*: This is the *Minde*, or *Soule* of *Animalls*" (PF 30).[20] Cavendish here borrows terminology we will find familiar if we recall the process of sublimation of spirits described in chapter 1: in the Galenic model, vegetable spirits (created by the distillation of food and responsible for basic life functions such as digestion, growth, and reproduction) are sublimated into the thinner and more active vital spirits, which drive the motion of a body; the vital spirits are further refined (in the *rete mirabile* of the brain) into animal spirits, which obey the soul and are responsible for sensation, perception, and rational thought.[21] As we have seen, for early modern natural philosophers the animal spirits can be found only in human beings, and even Descartes and his fellow mechanistic philosophers remained true to this classificatory principle, theorizing a sharp divide not just between matter (the human body *and* the spirits, which are created themselves from matter) and soul (the principle which drives the matter, including the animal spirits) but also between man and all other creatures, for only those entities with rational souls are capable of thought.[22]

But for Cavendish – and here she deviates from the main lines of mechanistic philosophy – the "essence of spirits" or "spirit of spirits" still suggests a process of movement and sublimation, and spirit itself is both itself inextricable from the matter *and* constitutive of the mind – in that these rational spirits are held responsible for "*Thoughts*, as *Memory, Understanding, Imaginations*, or *Fancy*, and *Remembrance*, and *Will*" (PF 30). Thus, in a sense, the very animating force of matter *creates* thought. Cavendish proposes a vivid analogy to assist us in picturing this (for her contemporaries probably counterintuitive) generative process: "Imagine the *rationall Essence*, or *Spirits*, like little *sphericall Bodies* of *Quick-silver* several ways placing themselves in several *Figures*, sometimes moving in measure, and in order, and sometimes out of order: this Quick-silver to be the *Minde*, and their severall postures made by *Motion*, the *Passions*, and *Affectations*" (PF 38–9). In this complex metaphor, the quicksilver comprising the mind's matter is already itself endowed with a capacity for movement, the actualization of which as motion animates the mind; motion in turn produces the "several figures" of "passions and affectations," that is, all the emotions we may experience, our fancy, our memory, our sensations. Since all of matter is in motion, all of matter experiences a form of knowledge: spirits make motion, and "motion makes knowledge" (*PF* 49).[23]

Allied to this figurative process of thought production is a complementary one of "patterning out." "The sensitive and rational corporeal motions in one body *pattern out*," writes Cavendish, "the figure of another body, as of an external object, which may be done easily without any pressure or reaction" (*PL* 70). Hence, the means by which sensory information is taken up by the intellect and imagination are also fundamentally material and motive: Cavendish's metaphor of "patterning" involves the transmission of sensory data from the thing perceived to the senses of the person perceiving. These data "pattern out" an impression on the sensory organ, essentially copying the particular sensory attributes into the sensitive organ, which in turn then conveys this pattern to the mind. Susan James offers a vivid example of what Cavendish seems to posit here:

> [A] dog has a figure. When a woman sees the dog, its figure is copied, imitated, printed or "patterned out" by the sensitive matter of her eye which senses and perceives it; when she hears the dog bark, the figure of the sound is patterned out in her ear, and so on. At the same time, each of these figures is patterned out by the rational matter in the woman's body to form an integrated figure of a dog with several sensory properties.[24]

As ever, there is the repetition of metaphor for each stage of the transmission: "The sensitive and rational corporeal motions in one body pattern out the figure of another body, as of an external object," writes Cavendish, "which may be done easily without any pressure or reaction" (*PL* 70).[25] She emphasizes the extent to which such ideation remains through and through material: as with the initial sensory information, the mental figures associated with the dog are fundamentally material, too, having been patterned out in the rational matter of the brain.

Returning to the allegories of *The World's Olio*, we encounter a range of similar metaphors, emphasizing not only how intertwined Cavendish's conceptions of matter and motion are, but likewise the interdependence between her poems and her philosophy. "Wit," for example, "is like a Pencill that draws several Figures, which are the Fancies; and the Brain is the Hand to guide that Pencill" (*WO* 101). Whereas a mechanistic reading would flip and contort the analogy to make wit (or thought) the guiding hand and the brain (matter) the thing acted upon, Cavendish's metaphors suggest instead that wit and brain act together, the pencil being no less material than the hand – implying in turn that matter itself can be an organizing principle, because it is infused with spiritual motion. Shortly afterwards, the language of figuration shifts from writing or drawing to printing, and Cavendish writes that "the Mind is like a God, that governs all; the Imaginations, like Nature, that created all; the Brain, as the onely Matter on which all Figurative Thoughts are Printed" (*WO* 103). While the technology through which the governing "matter" of the brain takes on the figures of thought has changed here, as has the agent (it is not the brain that acts here; rather, the brain is acted upon), the two analogies nonetheless share a crucial component: both picture the interaction of mind – and by extension its faculties of wit and imagination – and brain as a *physical* enterprise. Cavendish thus continues to elide the distinction between what the mind *is* and what it *does*; even though these are linguistically differentiated, the brain remains an active agent with respect to mental faculties, rather than a passive, unknowing material substrate that is visited by an external force separate from it.

Admittedly, Cavendish's natural philosophy is both labile and often murky, but the two main impulses identified above remain remarkably stable through all her divagations: first, thoughts are treated as material and as such capable of interacting with body-matter; second, the brain is far more than just a passive receptacle for the soul – rather, it participates in mental actions, thus binding itself to material thought as an active, self-knowing force. Both strands are in line with the broader continuities of Cavendish's vitalism, despite her shifting stance on atomism and her often confusing revisions to her own philosophy. The necessity of the brain as physical locus for the mind to be engaged and functioning at all, and for the figuring forth and patterning out of thoughts and emotions, thus remains a crucial determinant in her varied writings. After all, as she herself points out, even fancy cannot be without *some* brains.

It is against the background of this broader, somewhat messy vitalist picture that Cavendish differentiates the brains of men from those of women. As several modern critics have noted, her early work in particular tends to describe women as the weaker sex on a variety of different levels, leading to some vexation regarding her purported status as an early feminist.[26] The question of whether or not her writings anticipate modern feminism may be set aside here, if for no other reason than that it is hard to extract a consistent, systematic position from Cavendish's philosophical writings. As many recent critics have noted, her writing resists easy categorization, her theories often doubling back on or cancelling themselves entirely.

Indeed, it is often the case that what seem at first variations on traditional, gendered oppositions become destabilized through the metaphors used to elaborate on or elucidate difference. For instance, in the "Preface to the Reader" of *The World's Olio*, Cavendish excuses herself from writing "so wisely or wittily as Men," because

> Nature hath made … Mans Brain more clear to understand and contrive than Womans; and as great a difference there is between them, as there is between the longest and strongest Willow, compared to the strongest and largest Oak; though they are both Trees, yet the Willow is but a yielding Vegetable, not fit nor proper to build Houses and Ships, as the Oak, whose strength can grapple with the greatest Winds, and plough the Furrows in the Deep; it is true, the Willows may make fine Arbours and Bowers, winding and twisting its wreathy stalks about, to make a Shadow to eclips the Light; or as a light Shield to keep off the sharp Arrows of the Sun … (*WO*, sig. A_4^r)

The set-up of the metaphor seems clear enough: men's brains and women's brains differ by degree as the oak and the willow, with men's brains and oaks being superior in strength and functional utility (i.e., capacity for understanding, and multipurposeness in use, respectively), whereas women's brains are deprecated as "yielding Vegetable[s]." At first blush, then, Cavendish defers to the masculine brain as indicative of the more capable sex. However, with wit and some slyness, the passage then undercuts the very metaphor she has constructed: willows offer excellent hiding-places, they can wind and twist all around, and their shadows may even "eclips the Light." The oak may be the stronger and more fungible of the trees, but the willow is by far the more subtle. Like the willow, the brain of the woman is labyrinthine, proving difficult to trace; it nonetheless creates concealed spaces and holds that can ward off the power of the sun itself. Thus the analogy ultimately refuses to discount the value of a woman's brain; by pointing out the distinct virtues of each gender's brain, it even arguably undoes the very hierarchy that has been set up in the first place.

A similar reversal may be seen in how Cavendish subverts male intellectual conception even as she seems to praise it. A dedicatory epistle to her brother-in-law Charles Cavendish at the start of *Poems and Fancies* describes her own compulsion to write thus:

> True it is, *Spinning* with the *Fingers* is more proper to our *Sexe*, then studying or writing *Poetry*, which is *Spinning* with the *braine*: but I having no skill in the *Art* of the *first* … made me delight in the *latter*; since all *braines* work naturally, and incessantly, in some kinde or other; which made me endeavour to Spin a *Garment* of *Memory*, to lapp up my *Name*, that it might grow after *Ages*. (PF, sig. A_2^{r-v})

While acknowledging that women's brains are generally more suited to labouring with textiles than with texts, Cavendish's metaphor at the same time domesticates poetry-making in order to appropriate it for herself. Emphasizing the materiality of intellectual production by associating it with handiwork, she spins a garment of memory that will ensure not just her continuance but the organic growth of her reputation. A poem at the end of *The Animall Parliament*, which is appended to *Poems and Fancies*, offers a variation on this description of intellectual labour: "Since *Men* doe spin their *Writings* from the *Braine*, / Striving to make a lasting *Web* of *Fame*, / Of *Cobwebs* thin, *high Altars* do they raise, / There offer *Flyes*, as sacrifice of *praise*" (213). Rather than defending her own practice here, she instead undermines the valiant efforts of the masculine brain: for all its efforts to create "a lasting web of fame," it only succeeds in fashioning a cobweb for fly-catching.

In her poem "The Circle of the Brain cannot be Squar'd," the distinctive, intermediate status of mathematical objects, suspended between the ideal and material, opens up yet another way to raise the problem posed by the brain, and underscores the impossibility of distinguishing between what the brain is and what it does:

> A *Circle Round* divided in foure Parts,
> Hath been a Study amongst Men of Arts;
> Ere since *Archimedes*, or *Euclid's* time,
> Hath every Brain been stretch'd upon a *Line*.
> And every *Thought* hath been a *Figure* set,
> Doubts *Cyphers* are, *Hopes* as *Triangulars* meet.
> There is *Division*, and *Subtraction* made,
> And *Lines* drawne out, and *Points* exactly layd.
> But yet *None* can demonstrate it plaine,
> Of *Circles round*, a just *Foure square* remaine.
> Thus while the *Braine* is *round*, no *Squares* will be,
> While *Thoughts* are in *Divisions*, no *Figures* will *agree*. (PF 47)

Given the close connections between the philosopher Thomas Hobbes and the Cavendish family, the choice of the mathematical instance – how to construct a square of the same area as a circle, using only a compass and a straight-edge – cannot be accidental, as the search for a solution to this very problem would consume Hobbes for many years. Moreover, Hobbes's thought is especially congenial to Cavendish not only because it insists upon the primacy of motion to the sciences, but because he sees even putatively ideal mathematical objects such as points and lines as being entirely material and generated through material processes. For Hobbes, definitions must "consist of such names as express the cause or matter of generation," leading, for example, to his defining the circle as "the circumduction of a body whereof one end remains unmoved."[27] Merely seeing a circle cannot confirm us in our knowledge that it is a circle unless we actually generate the figure.[28]

In Cavendish's poem, the problem of knowing the brain is transposed onto the attempts by "men of arts" to construct the square that equals its circle, torturing the brain to uncover its secrets by stretching it upon a line and applying to it the arithmetical and geometrical techniques. The contents of the brain – its thoughts, doubts, and hopes – become mathematical objects, turned into figures, ciphers, and triangles. And yet roundness remains resolutely irreducible to squareness, the brain firmly resisting all attempts to "demonstrate it plaine." While echoing the frustrations anatomists and natural philosophers must have felt when they tried to delineate the precise relationship between soul and body, Cavendish critiques a masculine tradition that obsessively seeks to set forth rationally the myriad mysteries of thought. No one can demonstrate the operations of thought, she argues, because a thought cannot be separated from its thinker, and dividing in order to conquer results only in the failure to grasp the brain's dynamic, material interactions: "While Thoughts are in division, no Figures will agree." If the physical grey matter is being stretched, drawn out, laid on a line, or divided, so, too, are the faculties of the soul – and so divided, neither thoughts nor their figurations match up, neither among themselves nor with each other.

While in these examples, Cavendish does not take up the metaphor of gestation, that crucial analogy returns elsewhere, as may be seen in a "Dialogue-Discours[e]" amongst several characters from *Wits Cabal* (1662):

> *Vain-Glorious.* Wit neither fails, weakens, decays nor dies,
> *Tranquillities Peace.* Though bred and born, as other creatures are,
> *Ambition.* Only in the Brain, the Womb wherein it lies:

Heroick. But when 'tis born, Fame nurses it with care,
Frisk. And to Eternity doth it prefer.
Pleasure. Wit makes the brain sick when it breeding is,
Tranquillities Peace. And painful throws before, and at its birth;
Ambition. But when 'tis born, if good, a Comfort 'tis,
Heroick. The Parent Poetry creates with mirth[.][29]

Unlike other uses of the brain-as-womb metaphor, her account focuses on wit's emergence and breeding – that is, it stresses the birth not of ideas but of the very capacity for generating ideas. Moreover, in accentuating the pain and sickness accompanying the labour of childbirth, her use of the metaphor diverges from that of male writers, who more often than not gloss over this aspect or subordinate it to the result. By contrast, Cavendish stresses the sickness and the "painful throws" that are part and parcel of the entire gestational process, and even the birth itself yields "comfort" only if it is "good."

In one of her *Philosophical Letters*, she further tackles the principle of intellectual conception by taking on Harvey's "De conceptione" directly, playing as he does on the word *conception*. Cavendish's imagined interlocutor has posed the question, "*Whether the Production or Generation of animals is as the Conceptions of the Brain, which the Learned say are Immaterial?*" (*PL* 421–2). The construction of the question is at once a citation of the problem Harvey poses – that there must be some intangible force that causes biological conception to occur – as well as a more direct posing of the consequences of Harvey's solution – that we can reason towards the nature of this intangible force by likening biological to intellectual conception, the womb to the brain. So posed, the question leads Cavendish to compare the two processes (biological and intellectual conception) as well as to weigh the fallout of supposing both processes to be in some sense immaterial, driven by impulses that have no tangible material form.

Given Cavendish's materialist philosophy, she cannot but reject this supposition: "The Conceptions of the Brain," she writes, "are not Immaterial, but Corporeal" (*PL* 422). Defending this counter-assertion, she forwards another component of her philosophy: namely, that the non-existence of an entity cannot be inferred from the fact of its being unintelligible to our senses. "[T]hough the corporeal motions of the brain, or the matter of its conceptions, is invisible to humane Creatures," she writes, echoing Harvey, "and that when the brain is dissected, there is no such matter found, yet that doth not prove, that there is no Matter, because it is not so gross a substance as to be perceptible by our exterior senses" (*PL* 422). In fact, Harvey's point had been similar (even if Cavendish misconstrues his position somewhat in order to assert her own): whatever causes the biological conception, whether visible to the anatomist or not, must exist, given its very material consequences. But to this Cavendish adds that this existence must take a material form, even if it be insensible.

Questioning whether biological conception can in fact proceed in the manner of intellectual production, Cavendish further wonders whether a child that results would be the property of the mother (who imagined the child into being) or the father (who presumably had little to do with the matter itself). Noting the limits of the analogy, she points out that if "amorous Lovers" could reproduce through the intellect, "there would be more Children than Parents to own them" (*PL* 422). However, this limitation does not mean that she sees biological and intellectual conception as fundamentally dissimilar. Despite particular differences between them, she insists that presuming they are similar requires that we accept the materiality of both kinds of conception. The example she offers by way of proof has profound consequences for a reading of her fiction:

> Neither will your *Authors* example hold, that as a builder erects a house according to his conception in the brain, the same happens in all other natural productions or generations; for, in my opinion, the house is materially made in the brain, which is the conception of the builder, although not of such gross materials, as Stone, Brick, Wood, and the like, yet of such matter as is the Rational Matter, that is, the house when it is conceived in the brain, is made by the rational corporeal figurative motions of their own substance or degree of Matter. (*PL* 422)

The metaphor does not deny differences between the two modes of conception; rather, it subsumes the differences under a broader understanding of materiality and the shared logic of the conceptive process. Deeply embedded in Cavendish's metaphor here is a belief that there may be different degrees of matter: "Stone, Brick, Wood, and the like" are distinct from, but no more material than, the matter in which the "rational corporeal figurative motions" operate. Intellectual conception and biological conception are both material processes, albeit differentiated by the degree or type of matter utilized in each case.

Against what she perceives as Harvey's suggestion of immateriality in both kinds of conception, she insists upon the irreducible, if distinguishable, material forms necessary. In fact, her philosophy offers a sort of solution to Harvey's problem concerning the "plastick generative power," for instead of searching for it in an invisible realm of mysterious forces, we find that it inhabits the very matter that it animates. As we shall see, in a real sense, Cavendish's materialist philosophy allows the brain to become the fecund intellectual space we understand it to be today.

Perturbations of the Mind

The sixteenth-century French physician Ambroise Paré relates a story that seemed to have taken especial hold in the early modern imagination. Not only authors of medical treatises but writers such as Montaigne were fond of recalling it.[30] Hippocrates, Paré says, once helped a noblewoman who was under serious suspicion of adultery: "who being white her selfe, and her husband also white, brought forth a childe as black as an Æthiopian." Defending her virtue, Hippocrates proposed that "in copulation" the noblewoman "strongly and continually had in her minde the picture of the Æthiope." To this anecdote attributed to the father of medicine, Paré adds another, of a woman who gave birth to a "maide hairy like a Beare," because "her mother earnestly beheld, in the very instant of receiving and conceiving the seed, the image of St. *John* covered with a camells skinne, hanging upon the poasts of the bed."[31] These examples appear in a book on "monsters and prodigies," appended to Paré's book on generation, and are both to be found in the chapter that deals with "monsters who take their cause and shape by imagination."

Monsters and prodigies could appear for other reasons as well: because of a lack or plenitude of seed (the generative material thought to be provided by both the mother and the father); to glorify God or to punish some societal evil; or as warnings or prognostications that destruction might befall a particular community.[32] Yet the monsters owed to the imagination were perhaps the most unsettling in the early modern period, because they revealed two seemingly occult kinds of actions. First, these monsters demonstrated that the immaterial imagination could have serious material effects, and second, such monsters opened up the possibility (and the threat) of a *maternal* imagination. What the examples above emphasize is how the incredible force exerted by the mother's imagination might be transposed onto the child: the noblewoman "strongly and continually" holds in her mind the picture of the Ethiopian, while the mother of the hairy girl is said to have "earnestly beheld" a picture of St John.

Such stories speak to a pervasive anxiety surrounding the pregnant woman's mental state, for apparently good reason. Midwifery manuals of the period are full of proscriptions for the pregnant woman. Some of them are familiar to us even today (don't let her carry heavy things, temper her diet, don't make her ride on horseback), but others are quite peculiar: "Let her avoid Fastings, Thirst, Watching, Mourning, Sadness, Anger, and all other Petrurbations [*sic*] of the Mind," advises one manual. "Let none present any Strange and unwholesome Thing to her, not so much as name it, lest she should desire it, and not be able to get it, and so either cause her to Miscarry, or the Child to have some Deformity on that Account."[33] Fasting, excessive thirst, and insomnia sound reasonable enough for a pregnant woman to avoid. But extreme emotions, and perturbations of the mind, or a longing for a "Strange and unwholesome Thing," point to an especial concern in the period with the expectant mother's mental state and the power her mind could exert upon the developing fetus.

Another manual, by Jakob Rüff, cautions that physical changes could occur in the fetus because the mother experiences "terrours," and thus she should be kept away from "sudden sights of Hares, Swine, or other Cattell," explaining that "this sudden terrour troubl[es] and mov[es] the conceived seed." That deformities could result from how the mother's mental state acted on the fetus is of particular concern: she should not be surprised or frightened, lest her mind, so disturbed, trouble her child. Rüff describes examples of the sorts of problems one might encounter, and the list gives us a sense of the mechanism behind the transformation: "Againe, through longing & terrors, many are borne, which have divers spots and markes imprinted on the body, to wit, of Hares, of Mice, of divers colours, of a bunch or cluster of grapes, of flames of fire, and other things."[34] The "spots and markes imprinted on the body" may take the shape of the things that had actually frightened the mother (hares, mice, fire) but, as my earlier examples suggest, could also derive from things that the mother had longed for but could not have (fruit which may be out of season or otherwise difficult to come by). The very desire for an absent thing could cause her imagination to "imprint" its image onto the fetus's body.

Because the anxiety concerns both the force of the imagination and its specifically feminine vehicle, these examples strongly imply a correspondence between the brain and the womb, as though the former's formidable capabilities had somehow been transferred onto the latter. And given the tenacity of the doctrine of signatures, as Michel Foucault has shown, as a means of producing knowledge, physicians and midwifery manual writers were quick to point out the connection themselves: in describing a cure for barrenness, one author explains that his remedy "is very good to strengthen both the Womb and the Head, which are commonly afflicted together by Sympathy."[35] What was the nature of this sympathy, then? For writers dealing with pregnancy and delivery, the connection superseded mere likeness, and indeed literalized or materialized the metaphor. Indeed, at times, it seemed almost as though the womb could think for itself. Another author surveying explanations for barrenness held that

> the most frequent reason why this orifice [i.e., the cervix] opens not in this Act [of copulation], to receive the seed, is the insensibility of some Women, who take no pleasure in the Venereal Act; but when they have an appetite, the Womb being covetous of the Seed opens it self to receive it.[36]

In this instance, the mental state of the woman determines whether the womb will open to receive seed or not; if a woman possesses "an appetite" the womb acquires a corresponding mind of its own, becoming "covetous" and acting to satiate its appetite, whereas women who "take no pleasure" in intercourse find that their wombs shut themselves up in accord. (For men, by contrast, barrenness was attributed to excessive intercourse – but midwifery manuals list little else by way of explanation, and especially nothing so blatantly psychological.)

Paré's own explanation for the mechanism by which the imagination affects fetal development derives from Classical authority. The ancients, he writes,

> understanding the force of the imagination to be so powerfull in us, as for the most part, it may alter the body of them that imagine, they soon perswaded themselves that the faculty which formeth the infant may be led and governed by the firme and strong cogitation of the Parents begetting them … or by the mother conceiving them, and so that which is strongly conceived in the mind, imprints the force into the infant conceived in the wombe.[37]

While conceding that both parents may impart some mental mark on the developing child, Paré's examples, as with those cited above, place the responsibility solely on the pregnant mother. The "powerfull" imagination is seen as causing a physical change in the mother ("alter[ing] the body of them that imagine"), which is in turn forcefully "imprint[ed]" upon the child. Paré does note that some people speculate that a child cannot be harmed in this manner after about seven weeks past conception – at that point the proper external form has been so far paired with the physical matter of the fetus that nothing can really alter it. However, his emphatic insistence on imagination's power, the "firme and strong cogitation" of the parents, leaves no doubt as to his own anxiety concerning what the consequences of imagination might be. Dismissing such speculations – "whether it be true, or no, is not here to be enquired of" – Paré continues, "truly I think it best to keep the woman, all the time she goeth with childe, from the sight of such shapes and figures" that may invoke the frightening power of her imagination.[38]

Paré's proposed mechanism (which, unfortunately, he does not elaborate upon) invokes the idea of printing – the term Paré's English translator uses is *imprint* – with the implied suggestion that the strength of the intellectual conception corresponds to that with which the conceived object is received by the fetus. No doubt derived from Aristotelian theories of biological conception, where the form provided by the father stamped itself onto the matter of the mother, this sense of imprinting is crucial in that it allows us to see more clearly the links between writing and conception, both biological and intellectual. (We might note that Paré's translator uses *conceived* thrice in the last sentence above, thereby strengthening the association between brain and womb.) As the previous chapter discussed, writing in this period was doubly linked both to handwriting and to printing, and thus using the term *imprint* in such proximity to the *conception* suggests both the creative act of handwriting as well as the act of reproduction more closely allied to the workings of the printing press. As we shall see in more detail shortly, Cavendish's metaphor of "patterning out" as a kind of stamping and drawing upon sensitive matter, in order to explain how sensory perception creates material thoughts in the brain, echoes this dual sense of originality and copying.

A midwifery manual written by Jane Sharpe (one of the only female midwife manual authors in the period) speculates at greater length upon how the maternal imagination gets transferred onto the fetus:

> [T]he spirits and humours are disturbed by the passions of the mind, and so the forming faculty is hindered and overcome with too great plenty of humours that flow to the matrix, or the spirits are called off and gone another way. But the imagination is so strong in some persons with child, that they produce real effects that can

> proceed from nothing else … How the imagination can work such wonders is hard to say, but there must be some strength of mind that can convey the species from the external senses to the formative faculty, for by this means there is a consent between the faculties superior and inferior.[39]

The biological formation of the fetus was reasonably well understood towards the end of the seventeenth century, thanks in some measure to Harvey's pioneering work in *De generatione*, a work of comparative anatomy which studies developing chick embryos as analogies for embryos of humans and other mammals. But for the broader culture, the terminology used to discuss generation still relied on older anatomical systems, as is suggested by Sharpe's reference to humoural psychology, that is, to the balance of the four humours (blood, phlegm, yellow bile/choler, and black bile/melancholy) as well as their more schematic relatives, the forming faculties or spirits. "Spirits and humours," "disturbed by the passions of the mind," affect the faculty that shapes and molds the matter in the womb (or matrix). If these humours are perturbed from their proper operations by some trouble of the mind – if, say, the mother longs for some fruit she cannot have, or if she is frightened by fire – they will, in turn, disturb the formative faculty and so have a physical effect on the fetus.

But Sharpe also posits a more direct line of connection that bypasses the humoural entirely: "the imagination is so strong in some persons," she claims, that "they produce real effects that can proceed from nothing else." In these individuals, any spiritual intermediaries are rendered unnecessary, and while the actual physical mechanism may remain obscure, it becomes necessary to posit "some strength of mind" to account for how sensory impressions get transferred to the physical "formative faculty." This belief in imagination's power is essential to these accounts of maternal imagination. The fear, again, is that *anything* the mother encounters, *via any of her senses*, is liable to provoke her imagination, which may act on its own or via the spirits and humours of the body. The seemingly inconsequential act of *thinking* too much about an image can cause a direct and corresponding alteration in the matter of the womb, and this alteration may well be direct: the *fear of fire* cannot translate into anything other than a very literal *fire birthmark* on the infant's skin.[40]

The crucial elements of maternal imagination that interest us here involve the fear of the woman's too-powerful mental faculties, a concern about the impossibility of explaining the actual link between brain and womb, and the emphasis on the powers of conception (both intellectual and biological) residing in the woman. Women may have been regarded as "physical mistakes, imperfect men," but the power they potentially wield, even if unconsciously, over their own bodies and the bodies of their developing children, remains problematic – and potentially monstrous.[41]

Yet, to abstract a little, the problem of maternal imagination can be recognized as simply a more specific iteration of what will become understood as a basic problem underlying Cartesian dualism: that is, its failure to provide an adequate account of how its two substances, mind and matter, interact. A prominent critic of such dualism, Margaret Cavendish turns to questioning the precise mechanism through which, in such a system, the immaterial mind could affect the material body. As these instances taken from Paré and other contemporary midwifery manuals show, no empirically observable explanation was readily available to account for their interaction, and, as we have already seen, this lack tormented Harvey, too.

Insisting instead on the shared materiality of biological and intellectual conception, Cavendish offers an alternative account that responds to these anxieties occasioned by the maternal imagination by appropriating them towards other ends. As we have seen earlier, for her the relationship between the brain and its thoughts is generally as of container and things contained. But how do these thoughts operate? From what are they derived? It is from her related notions of *patterning* and *figuring* that an actual procedure emerges. Patterning and figuring both involve a kind of mimetic representation, constituting together a mode of copying that echoes an underlying connection to Pare's depiction of maternal imagination as an imprint. Not only do printing, imprinting, patterning, figuring, copying, all describe modes of reproduction, but the emphasis for Cavendish is particularly writerly – albeit the process is very different from the writing-on-the-brain that Hamlet must engage in.

Cavendish was always highly sensitive to the idea of her works being in print. In an epistle to book II of *The World's Olio*, for instance, she relates the act of writing to printing, even as she pointedly distinguishes between these two ways of reproducing herself, the movement from the one to the other turning out to be fraught indeed.

> The Reason why I print most of what I write, is, because I observe, that not only the weak Writings of men get Applause in the World, but the infinite weak Translators of other Works; thus there are many simple Books take the World by the Ears; but I perceive it is not the wit, or worth of what is written, that begets a delight to the

> Readers, and a Fame to the Writers; but it must fit the Genius of the Age[.] (*WO* 93)

Reproducing the self in writing and through print are very different endeavours, and there is no direct correlation between the written children of one's fancy and their printed copies, the children that the world sees. Not only do readers misjudge the worth of the original but, far from truly representing the original, the copies themselves are always liable to be simulacra instead, since the "infinite weak translators" insert themselves between the writing and the printing. At the same time, without the supplementary form of reproduction that printing provides, the self begot in writing is condemned to ephemerality, for it is through the "delight" of readers that fame accrues to the writer, ensuring her continuance – though even this is subject to the fickle tastes of the time, "the Genius of the Age." Her ambivalence regarding her *paper children* suggests a different attitude towards writerly production than was evinced in how male writers made use of the brain as womb. Even as Cavendish is deeply invested in print as the medium for the broader and more lasting circulation of ideas, and for the transmission of wit and fancy to the reader (hopefully begetting "a delight" in the process), she remains attuned to the vagaries of the material processes through which such paper bodies are disseminated.

Thus, her discussion of the printing of written work brings to the fore another operant, one which the male-birth model tended to occlude. As Cavendish continues in this epistle, she complains that she lives in a "Carping age," where critics ostracize her for poor grammar, meter, and spelling. "As for the Orthography," she writes, "the Printer should have rectified that; for I think it is against Nature for a Woman to spell right, for my part I confess I cannot" (*WO* 93). In such cases, she expects the printer who set the received manuscripts in type to have intervened as a corrective editor, to mend at least the misspellings. In other matters, however, the printer may be a dangerous corruptor of meaning: "as for the Rimes and Numbers, although it is like I have erred in many, yet not so much as by negligence of those that were to oversee it; for by the false printing, they have not only done my Book wrong in that, but in many places the very Sense is altered," and thus her book "is lamed by an ill Midwife and a Nurse, the Printer and Overseer" (*WO* 93). Even as Cavendish frames the printer's role in language similar to the masculine use of the metaphor of conception, she attributes a far more active role to the "midwife" and "nurse" in shaping her texts – apparently generally for the worse, to the detriment of both them and her fame. (It is in this epistle that she also describes wit, "wilde and fantastical," as originating in the brain.)

The movement from written to printed work, then, is not merely duplicative, but should ideally involve amendment or refinement, without, however, interfering with the original work of wit. The aim must be to copy as precisely and correctly as possible the original, but along the way rectify small deformities by correcting the naturally feminine bad habit of poor spelling. This account of relationship between the two paper bodies of her children, the written original and the printed copy, echoes Cavendish's theories about matter, both sensory and reasoning, and in particular the process of "patterning out" discussed above. Patterning works in self-knowing, self-moving matter to account for action and reaction in ways quite different from those prevailing in standard mechanistic theories. Central to this operation is what Eileen O'Neill terms "sympathetic affinity."[42] To illustrate this idea, O'Neill makes use of Cavendish's own example of a hand throwing a ball. From a mechanistic perspective, the motion of the hand transfers momentum to the ball, and there is no further connection between the two entities. For Cavendish, however, "[t]he hand is the exemplar and efficient moral cause, which occasions the ball to 'pattern out' *n* degrees of motion … Rather than the transference of motion, there has been a replication or imitation that does not violate the conversation of motion in the system of nature."[43] The hand and the ball, both part of self-knowing and self-animated matter, share a direct relationship in which the ball copies the motion suggested to it by the hand, taking its imprint as it were. A similar mechanism occurs in perception, where the sensory organ receives the pattern of the thing it senses.

In short, patterning out thus involves a form of copying that echoes the process by which writing becomes printed matter. Cavendish explains that "all perceptions are made by figuring … all are made by imitation or patterning" (OEP 170). Arguing in the *Philosophical Letters* against John Baptista Van Helmont's assertion that ideas are immaterial, she elaborates on this claim, explicitly establishing a crucial association between the metaphor (and practice) of printing and the products of mental activity (taken in a broad sense):

> Nature is a corporeal substance, and without a substance Motion cannot be, and without Motion opposition cannot be made, nor any action in nature, whether Prints, Seals, Stamps, Productions, Generations, Thoughts, Conceptions, Imaginations, Passions, Appetites, or the like[.] (PL 242)[44]

Unsurprisingly, nature, for Cavendish, encompasses all things, effacing the dualistic distinction between material and immaterial: thoughts are, as everything in nature, material. But of note here is the fact that her list of quasi-mental entities, such as thoughts and appetites, is preceded by a set of items that seem at first pass categorically different: prints, seals, and stamps. In a later letter to an apparently confused imaginary interlocutress, she elaborates further on her logic: "By *Prints* I understand the figures of the objects which are patterned or copied out by the sensitive and rational corporeal figurative Motions," such that "by Prints I understand Patterns, and by printing patterning" (*PL* 539–40). She is careful to point out that this is a metaphorical kind of printing, which she explains using the example of a seal and wax. "[W]hen a Seal is printed in Wax," she writes, "the Seal gives not any thing to the Wax, but is onely an object patterned out by the figurative motions of the Wax in the action of printing or sealing." So it is that a tree may pattern itself out, or pass its image, onto the eye, but "verbal or vocal sounds … [also] print words and set notes in the Air" (*PL* 540).

Sounds printing words in the air is a fascinating – and rather charming – idea, but Cavendish would have us remember that these acts of the body's senses receiving "patterns" from external objects equally encompass the operations of the mind. The key difference is that the patterning experienced by sensory organs is passive and involuntary, whereas rationality engages in what she terms "voluntary actions of figuring" (*OEP* 170). The mind does not have the "external" objects that the senses do, so it must create for itself things with which it figures, "[a]s for example, imaginations, fancies, conceptions, passions, and the like; are made by the rational, corporeal, figurative motions, without taking any copies of foreign objects." Through this, the mind becomes a self-generating printing machine, making figures "of [its] own accord, without any imitation, patterns, or copies of foreign parts" (*OEP* 170). Of course, voluntary rational figuring can also intersect with involuntary sensory patterning, as in dreams, "when the sensitive motions make voluntary figures on the inside of the sensitive organs, the rational take patterns of them" (*OEP* 171).[45]

But, as with the translation of the written manuscript to the printed book, patterning and figuring may well go awry. As Kourken Michaelian explains, "[a]lthough patterning out is a sort of copying, it is an imperfect copying: when the figurative motions of a thing pattern out those of another thing, the former does not come to instantiate the latter … there is always a difference between the copy and the original of which it is a copy."[46] Cavendish notes that the sensory organs "do not always make perceptions of exterior objects, but many times make figures by rote; as is manifest in madmen, and such as are in high fevers and the like distempers, which see or hear, taste or smell such or such objects when none are present" (OEP 189). In other words, the imagination may fall prey to delusions and as such misrepresent the real world, just as Cavendish's manuscripts are at the mercy of a "Printer or Overseer" who may have over- or under-corrected at the printing house, turning the comely original into a monstrous copy.

Throughout, the brain is the active agent in which both sensory perceptions and rational perceptions take place. In *Poems and Fancies*, Cavendish paints the brain as an organizing principle for the waves of information with which it must contend:

> [T]he senses bring all the materials into the brain, and then the brain cuts and divides them, and gives them quite other forms, then the senses many times presented them; for of one object the brain makes thousands of several figures, and these figures are those things which are called, imagination, conception, opinion, understanding, and knowledge, which are the Children of the brain. (PF 20)

Cavendish's "Children of the brain" are very different from the children that emerge from a masculine process of intellectual conception. Rather, these children emerge after the brain copies out all the sensory "materials" brought to it, which then become the various operations of the mind itself. The brain generates and contains its own children, and when they are "put into action," they become the "arts and sciences, and every one of these have [*sic*] a particular and proper motion, function, or trade" (PF 20–1). The imagination (which she allies with the conception) "builds, squares, inlayes, grinds, moulds, and fashions all," while "understanding distinguishes the several parcels [marked out by the brain], and puts them in right places." Ultimately, "when the brain works upon her own materials, and at home, it is called poetry and invention, but when the brain receives and works journey-work, which is not of its own materials, then it is called learning, and imitation" (PF 20–1). Here, then, we circle back to the variety of metaphors in *The World's Olio*, for this formulation seems to encompass them all: the brain contains infinities of copies, figured out across the different mental activities. Poetry and invention occur when the brain utilizes its own powers, while learning derives from the far less exciting action of the brain's casting-out to external entities, copying what comes from elsewhere. Far from acceding to contemporaneous fears surrounding the maternal imagination, then, Cavendish embraces them, while remaining aware of the dangers of conception. She productively revalorizes the maternal imagination, inventively turning this alleged source of monstrosity into the very condition of both poetic production and its eternalization, the proliferation of the brain's children, first as original and then as copy.

The Blazing World

> If you wonder, that I join a work of fancy to my serious philosophical contemplations; think not that it is out of a disparagement to philosophy; or out of an opinion, as if this noble study were but a fiction of the mind; for though philosophers may err in searching and enquiring after the causes of natural effects, and many times embrace falshoods for truths; yet this doth not prove, that the ground of philosophy is merely fiction, but the error proceeds from the different motions of reason, which cause different opinions in different parts, in some are more irregular than in others; for reason being dividable, because material, cannot move in all parts alike[.] (*BW* 124)

Let me conclude this chapter by turning briefly to the best known of Cavendish's children, *The Blazing World*, which was appended to her *Observations upon Experimental Philosophy* (echoing the way Sir Francis Bacon's unfinished prose utopia, *The New Atlantis*, appeared alongside accounts of the scientific observations and experiments gathered in *Sylva Sylvarum*). No doubt, the argument developed above may seem to find its most direct instantiation in the bizarre conclusion of Cavendish's utopian fantasy, where a female community emerges out of a transmigration of brains, externalizing Cavendish's wish – cited at the beginning of this chapter – that "her Idea may live in many brains." But I want to focus here instead on how the distinction developed in the preceding section – between fancy's originals and the preservation of learning or scientific knowledge through printed copies – lays the groundwork for her attempts to marry fiction and philosophy. Cavendish explicitly reflects upon this issue in the prefatory letter to the reader by defending the legitimacy of conjoining "a work of fancy," that is, *The Blazing World*, with the "serious philosophical contemplations" of the *Observations*.

As with its diction, the logic of the passage quoted above is labyrinthine, but we can break her defense into a number of discrete, closely associated steps. Cavendish begins by positing the standard opposition between *philosophy* and *fancy* (or *fiction*). She is concerned that her reader will mistake her intentions in *joining* these two genres and will erroneously conflate her mature philosophy with the fanciful prose of *The Blazing World*, thereby eliding the distinction between philosophy and fiction and, more dangerously, "disparag[ing]" philosophy by treating it *as* a sort of fiction. On the surface, then, Cavendish's opposition of fancy to "serious philosophical contemplations" comes across as a preemptive strike against anyone who would not read her philosophy as philosophy.

Her next step is temporarily to equate fiction – and specifically "fiction of the mind" – with error: it is precisely because philosophy *can* err in its search for causes that one may be inclined to see "this noble study" as a form of fiction. Though she undoes it almost in the next breath, the *Blazing World* echoes this presumption, especially in the scene where the Empress upbraids the bear-men astronomers for preferring telescopy over "natural eyes" (BW 141). She disparages telescopes as "false informers," which, "instead of discovering the truth," "delude" the sense of the bear-men (BW 141). Countering this criticism, the bear-men explain that they "take more delight in artificial delusions, than in natural truths" (BW 142). In Cavendish's time, the "art" of the telescopes was in fact associated with the human ability to discern through instruments philosophical truths regarding nature that were not immediately available to the unaided eye. But she herself always refused to grant such mechanical intermediaries legitimacy, seeing them instead as distortions that deceived the eye's natural ability to discern the truth. This much the bear-men grant, defending instead the "delight" to be found in these delusions. Any kind of Art – and by extension fiction – seems thereby to be aligned with error and deception, while philosophy stands as a form of unmediated observation that leads to the truth.

Yet the bear-men, too, undo this opposition in ways that echo the logic of the epistle to the reader. They continue: "[W]ere there nothing but truth, and no falsehood, there would be no occasion for to dispute" (BW 142). For it turns out that whatever the ideal of philosophy might be, in practice error is unavoidable, indeed even indispensable for philosophy to progress towards a true understanding of things. Disputation, according to the bear-men, leads to wisdom, and wisdom leads to truth. The "To the reader" forwards a similar sentiment: while error undoubtedly exists in philosophy, its presence does not automatically render philosophy equivalent to fiction, because it functions to enable reason's ability to dispute with itself by allowing "different opinions in different parts." Here she concedes that philosophy may "embrace falshoods for truths," and in this concession fiction moves farther away from error. Error, it transpires, is not to be associated with any particular genre of writing, be it fiction or philosophy, but results rather from the very "material" of reason, that fallible operation of the soul.

Being in itself heterogeneous and imperfect, working differently in different domains, reason may not "hit" the truth of nature:

> and since there is but one truth in nature, all those that hit not this truth, do err, some more, some less; for though some may come nearer the mark than others, which makes their opinions seem more probable and rational than others; yet as long as they swerve from this only truth, they are in the wrong: nevertheless, all do ground their opinions upon reason; that is, upon rational probabilities, at least, they think they do. (*BW* 123)

The errors of philosophy arise from distortions internal to reason's material operations, and reason itself is thereby drawn closer to the mechanical instruments it has invented, that is, to the telescopes that the Empress condemns. As an imperfect tool it may or may not attain the truth – and insofar as it misses its target, an inch is as good as a mile, even if the nearness to the truth may delude one into thinking otherwise. Reason offers at best a set of "rational probabilities" that guide philosophy's journey towards truth – but these being only probabilities, philosophy is always prone to "swerve" from the "one truth in nature" at which it aims. We can fully trust neither reason nor our belief in reason, for even those who claim to "ground their opinions upon reason" may be mistaken ("at least, they think they do").

In this regard, fiction's relationship to truth and error turns out to be very different.

> But *fictions* are an issue of man's fancy, framed in his own mind, according as he pleases, without regard, whether the thing he fancies, be really existent without his mind or not; so that reason searches the depth of nature, and enquires after the true causes of natural effects; but fancy creates of its own accord whatsoever it pleases, and delights in its own work. The end of reason, is truth; the end of fancy, is fiction. (*BW* 123)

Because fiction is the "issue of man's fancy," it need not correspond to a truth that exists "without his own mind." Cavendish's argument here echoes the famous claim proffered by Sir Philip Sidney in the *Defense of Poetry*: "Now for the Poet, he nothing affirmeth, and therefore never lieth: for as I take it, to lie, is to affirme that to bee true, which is false."[47] Just so for Cavendish's account of philosophy, for which reason, being an instrument of truth-finding, is necessarily engaged with both truth and error in a manner fiction does not have to be. Fiction is, in a sense, the "truth" of fancy – that is, the tangible expression of what it creates – but because a thing created by fancy need not be true, neither need it err. In effect, this Sidneyan account of fiction overcomes Cavendish's original set of oppositions: philosophy, good because true; fiction, worrying because necessarily false. Instead, we have: philosophy, aiming to be true but potentially erring; fiction, neither true nor false because a product of fancy.

The verbs through which Cavendish frames the difference between reason and fancy are equally telling. Reason *searches* and *enquires*, verbs that do not possess any particular emotional charge; by contrast, fancy *creates ... whatsoever it pleases* and *delights* in its products, suggesting both fancy's self-generative potential and the *pleasure* it takes in its act of creation. Here, too, the echoes of Sidney are palpable. Arguing for philosophy's dependence upon poetry, he writes: "Wher[e] the Philosophers as they scorne to delight, so must they be content little to move…, which Plato and Poetius [Boethius] well knew and therefore made mistresse Philosophie borrow the masking raiment of Poesie."[48] Indeed, for Cavendish, rather than being bound to the search for truth, fancy can delight in its very disconnection from truth, for that very disconnection lies outside of the truth/error opposition. Fancy, even when it is dangerously unconstrained, is tasked with creating anew rather than unearthing the ancient. It is telling that the Empress allows the bear-men their telescopes, but only conditionally: they must keep their academic and philosophical disputes cordoned off, away from the governing state. In this instance, the boundary between fiction and philosophy is elided and the creative power of debate as a means to uncover the "one truth" is confined within the walls of academia. By contrast, like her creator, the Empress has an unrestricted capacity to fashion and fabricate, but always in a positive direction. Concerned now with the governance of the Blazing World, she simply rids herself of the truth/error tangle the philosopher-bear-men have gotten caught in.

As we have seen, this favourable view of the kind of intellectual conception proper to fancy characterizes Cavendish's fictional pursuits, but distinguishing fancy from reason does not mean giving up fancy's claims to rationality:

> but mistake me not, when I distinguish *fancy* from *reason*; I mean not as if fancy were not made by the rational parts of matter; but by *reason* I understand a rational search and enquiry into the causes of natural effects; and by *fancy* a voluntary creation or production of the mind, both being effects, or rather actions of the rational parts of matter; of which, as that is a more profitable and useful study than this, so it is also more laborious and difficult, and requires sometimes the help of fancy, to recreate the mind, and withdraw it from its more serious contemplations. (*BW* 123–4)

Having first contrasted fancy/fiction with philosophy through its differing relationships to reason and the attendant truth/error opposition, Cavendish in characteristic fashion now collapses the boundary between them. As we have just seen, reason can lead to error, while fancy can open the way for new truths. Consequently, she now reveals that, like reason, fancy too belongs to the "rational parts of matter" (i.e., the actions of the soul upon the body), both being "effects, or rather actions" of this matter – though they activate different aspects of the soul's actions upon the body. Rather than being opposed, then, they may as such actively work together when the mind is engaged: philosophy being "a more profitable and useful study … [as] also more laborious and difficult" than fancy's creative production, reason can be enlightened by fancy, which will not only "withdraw it from its more serious contemplations," but also help "recreate the mind."

Cavendish's term here, *recreate*, must be taken in both senses. On the one hand, she intends the act of making anew or making over or making again, through which fancy breathes new life into reason's endeavours. It produces new possibilities and potentialities upon which philosophical reason can operate. On the other hand, fancy also offers *recreation* in the sense of providing pleasure and diversion for the exhausted reason. Thus fancy, and by extension fiction, are shown to be not only useful to but even necessary for the philosophical venture. Following the logic of the Derridean supplement, fancy is both accessory or ancillary to reason but at the same time necessary for its operation. Reason and fancy, and philosophy and fiction, are thus represented as being mutually beneficial to one another, and perhaps even more: reason seems to require fancy for its recreation, but fancy can exist complete in and of itself.

It is upon these grounds that Cavendish finally justifies the connection she has forged between these two *areas of study* and their practical exemplars, the *Observations* and the *Blazing World*:

> And this is the reason, why I added this piece of fancy to my philosophical observations, and joined them as two worlds at the ends of their poles; both for my own sake, to divert my studious thoughts, which I employed in the contemplation thereof, and to delight the reader with variety, which is always pleasing. (BW 124)

Her reference to the "joining" of her works "as two worlds at the ends of their poles" is gentle self-mockery, for this joining of worlds is precisely what enables the plot of the *Blazing World* to advance. One may of necessity hold fiction and philosophy apart, and they may always be distinct enough from one another as to suggest "variety," but they will (at least for Cavendish) touch somewhere, and it is that point of contact that interests us in considering Cavendish's self-fashioning. Her own reason had required recreation – but so, too, will the reader require such recreation. Philosophy creates "studious thoughts" which need the diversion supplied by fancy (though, again, the reverse appears not to be the case).

The Blazing World thus emerges as fictional proving-ground for Cavendish's theoretical philosophy, as well as a swerve of divertissement from gruelling philosophical labour:

> But lest my fancy should stray too much, I chose such a fiction as would be agreeable to the subject treated of in the former parts; it is a description of a *new world*, not such as *Lucian*'s, or the *Frenchman*'s world in the moon, but a world of my own creating, which I call the *Blazing World*: the first part whereof is *romancical*, the second philosophical, and the third is merely *fancy*, or (as I may call it) *fantastical*, which if it add any satisfaction to you, I shall account myself a happy *creatoress*; if not, I must be content to live a melancholy life in my own world. (BW 124)

Fancy, being unconstrained by a necessary relationship to reality and to the truth/error dichotomy, does on occasion require some formal restrictions – thus Cavendish has not composed a copy or an imitation of worlds asserted by others, but rather brought forth a tripartite self-created world that is itself bound to her philosophical meanderings as the two worlds joined at the ends of their poles. Framed by *romance* and *fantasy* – and thus informed by and informing them – Cavendish's *philosophy* serves as the second part of the *Blazing World*.

Cavendish's celebration of her self-fashioning, or rather of her self-generation, extends even to how she responds to her critics: enjoy this, she says, and I'll be a "happy *creatoress*," but enjoy it not, and I will still inhabit my own world (even if I am melancholy). Unlike Sidney in this regard, for whom the reader's delight was crucial for poetry to carry out its didactic intent, her world need not conform to her readers' expectations in order to exist; it depends upon neither their approbation nor their censure.

> I cannot call it a poor world, if poverty be only want of gold, silver, and jewels; for there is more gold in it than all the chemists ever did, and (as I verily believe) will ever be able to make. As for the rocks of diamonds, I wish with all my soul they might be shared amongst my noble female friends, and upon that condition, I would willingly quit my part; and of the gold I should only desire so much as might suffice to repair my noble lord and husband's losses. (*BW* 124)

The *Blazing World*, to be sure, contains just so many riches and wonders as its creatoress imbues it with – certainly more so than in the regular world, where "chemists" must labour under scientific philosophy to produce gold. In the Blazing World, though, the gold in a sense creates itself.

> Translation between these two worlds proves, alas, no easy task: for I am not covetous, but as ambitious as ever any of my sex was, is, or can be; which makes, that though I cannot be *Henry* the Fifth, or *Charles* the Second, yet I endeavour to be *Margaret* the *First*; and although I have neither power, time nor occasion to conquer the world as *Alexander* and *Caesar* did; yet rather than not to be mistress of one, since Fortune and the Fates would give me none, I have made a world of my own: for which no body, I hope, will blame me, since it is in every one's power to do the like. (*BW* 124)

With the comparisons to these monarchs, Cavendish reaches the core of her philosophy of creation, explicitly revealing its connection to her gender and its engendering capacities, which are here opposed to masculine lines of succession, their projects of conquest contrasted with the power (and limits) of female self-making. That Henry is the Fifth implies four other, presumably interchangeable, Henrys before him, while Alexander and Caesar stand for the usual Classical models of masculine command, conquest, and occupation. Margaret, though, shall be the First, none before her – though through this she creates the potential for others to come after her. Indeed, she invites the reader to do as she has done and make another different world, extending an invitation that she will echo even more forcefully in the epilogue. Masculine history is here portrayed as exclusionary and dominating, whereas feminine imagination unfolds in a curiously ahistorical and highly inclusive space.

To conclude: the letters to her reader accomplish at least two things at once. First, on the surface, they operate as exercises in definition: here is what you are about to read, here is its relationship both to other things I have written and to the various genres in which I have written. Next, they exist as explicit moments of self-fashioning, sometimes self-deprecatory but more often self-aggrandizing, in which Cavendish explores the capacities bestowed upon her by her writing. Through her acts of creation she becomes Margaret the First, Empress of the Philosophical World, and most broadly *creatoress*, a gendered term that marks her out as the prime mover of her own ideas. It may be such moments of self-definition that readers saw as monstrous, for in these Cavendish unapologetically takes the rei(g)ns and makes them hers. Her generative acts have, as we have seen, discomfited her audience precisely because she is so candid about her imaginative procreations. Even if her ambition in and of itself is not problematic, the aims of her ambition were clearly seen to be: she wishes to usurp masculine power by projecting a new, often weird feminine creative ability.

Yet, even as she insists that her worlds do not depend upon whether readers approve of them or not, the letters often also extend an invitation, and in this way, Cavendish seeks subtly to control her readers' responses to her fictive progeny. Do not blame me for my own ambition, she says, for you may certainly be ambitious yourself, and of course you can imagine a new world for yourself. Of course, it might be easier for you to be my subjects, but if your feeble imaginations can't bear that thought, you certainly can create your own – just stay the hell out of my world. The reader would be banished for being an "unjust usurpe[r]" rather than a patient willing subject (*BW* 225). And yet usurping what the surrounding culture claims as male prerogative is precisely what Cavendish seeks to do. After all, the only way one can exist in her Blazing Philosophical World is by serving the Empress. That this warning comes in the "Epilogue" casts a different light on the *Blazing World* we have just read, as well as upon its invitation to us to imagine other worlds that are ours: we didn't just read it, we entered into it, and the only way we could have done so was as the Empress's subject. We didn't read – we were ruled.

CHAPTER FIVE

The Bookish Brain: Moxon, Willis, and the Transformation of Flap Anatomy

"But yet the body is his book"

Through one remarkable anatomical text and its travels, this chapter traces a convoluted path: from the body as book to the brain as book. Let us commence this journey by studying the image before us – surely a bizarre one, by any account. Dominating a bleak landscape stands a man, one foot atop a skull, a mottoed colchium flower blossoming between his legs – and around him floats an array of partially dissected body parts and vascular systems. The man appears content enough, and even raises his left hand as if to beckon us to peer closer – though, if we do, we notice that his torso seems to have been sliced in at least three places. Curled above the crook of his arm, level with his head, is a cartouche, upon which a sleeping putto reclines between bubbles and a balanced hourglass. Making explicit and playing upon this traditional *vanitas* theme, a verse from Psalm 144 proclaims "*HOMO vanitati similis factus est. dies ei[u]s sicut umbra, praeterunt*" ("Man is like to vanity; his days are a shadow that passeth away," KJV 144:4). Perhaps this admonition applies even to us, the viewers being warned that spending too much of our precious and limited time staring at the man and his accompanying viscera itself constitutes a form of navel-gazing, both literal and metaphorical.

The warning is differently iterated in the scroll beneath the colchium flower, which explains "*Ut colchium florescit mardicum sic & HOMO graminis instar, puris*" ("As the withered colchium blooms: thus, too, Man, in the likeness of grass, rots [and is resurrected]").[1] Lyle Massey points out that the colchium flower was popularly used as a method for treating gout, a disease linked to excess and wealth, but also to "scholarly rumination." Thus the putto's warning in the cartouche above is echoed here, for

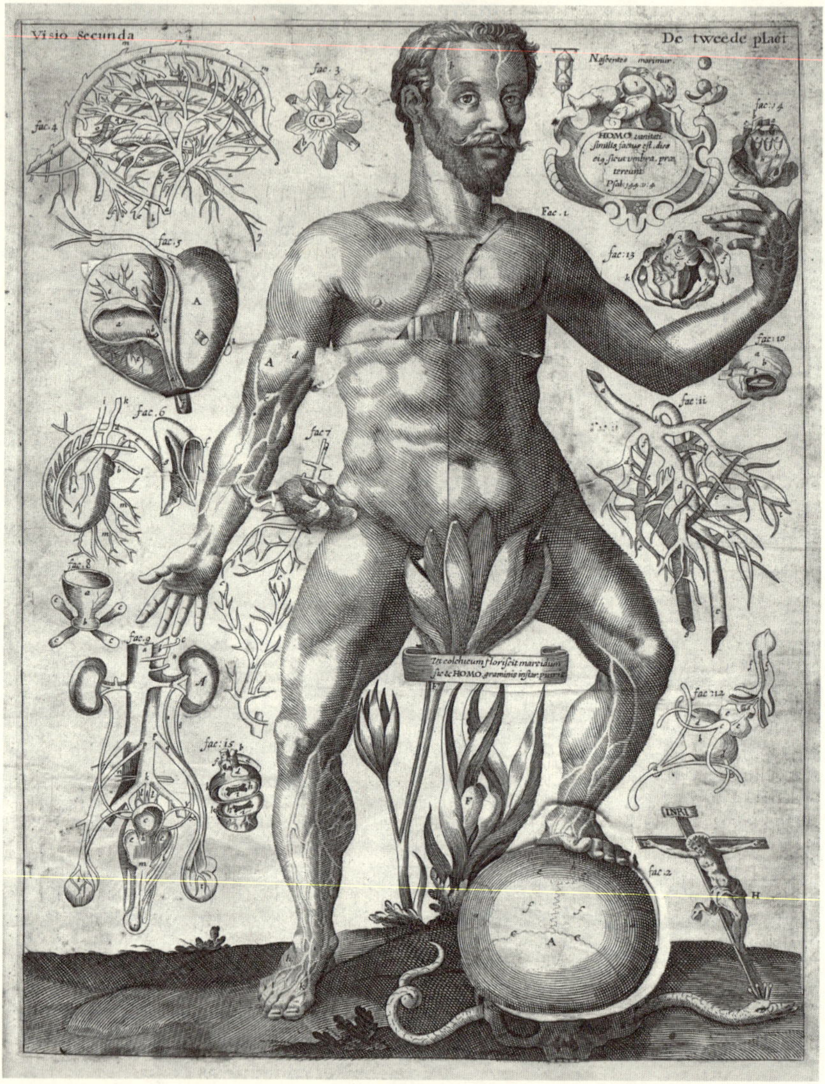

5.1 Remmelin's "Adam" of 1670. Johann Remmelin, *An exact survey of the Microcosmus, or Little World…*, trans. John Ireton (London: Joseph Moxon, 1670), Visio Secunda. The Wellcome Library, London. EPB G. O/S F.2277.

if we gaze overlong on this miracle of creation we may well become susceptible to the very disease this coy covering is meant to heal.[2] The clock continues to tick: the colchium withers as mankind shall, as grass does – and as indeed the freshly brought forth organs around him would too, were they not eternalized in print.

This is a page – the second in a set of three so-called *visios*, that is, visions, apparitions, or likenesses – drawn from Johann Remmelin's *Catoptrum microcosmicum*, a flap anatomy originally published in Ulm in 1613; and subsequently "Englished" by one John Ireton (after being "Perused and Corrected, by several rare Anatomists"), so that it would appear in England for the first time in 1670 from Joseph Moxon's printing house under the altered title *An exact survey of the Microcosmus*. The vagaries introduced by its geographical translation will occupy us later in this chapter, but the *visios* in this original text themselves mark a turning-point in a particular kind of anatomical representation – and specifically, as we shall see, in how the brain would be depicted on the printed page. In the previous chapter we were concerned with how thought, and its relationship to both mind and brain, could be represented in philosophical and literary texts. Margaret Cavendish proposed that the brain itself was "thinking matter," thus effectively collapsing any clear distinction between the physical and the mental. But how does such thinking matter – so active, so vital, constantly creative, a womb of activity and the very knot binding soul and body – become active on the page? How does the early modern period textually capture the workings of what anatomist Thomas Geminus called the "wonderfull and marveylous" but also "very straunge" brain?[3] Whereas earlier in this book we examined metaphors linking the activities of the brain and soul to writerly actions and literary production, this chapter will ultimately focus on the means of making and transmission, following the line of a foundational metaphor that allowed those earlier metaphors to be expressed as they were. In short, the brain in this case will become a *book*, to be opened, leafed through, and read; in the process it will advance both anatomical knowledge and the printing press itself.

But before we arrive at that point, we need to look more closely at how the body becomes a book, and to spend a little more time with the man in Figure 5.1. He bears a striking resemblance to Andreas Vesalius (Figure 5.2), and perhaps this explains why the organs and vessels surrounding him have been largely pilfered from that master anatomist's 1543 *De humani corporis fabrica*. The *Fabrica* would still have held sway in the medical community in 1613, when Remmelin's man made his first appearance in Ulm, and the *Fabrica* (perhaps astonishingly) continued to exert

5.2 Andreas Vesalius performing an anatomy. Andreas Vesalius, *De humani corporis fabrica* (Basel, 1543), title page detail. Yale University, Harvey Cushing / John Hay Whitney Medical Library, New Haven, CT.

its influence over anatomical representation when the man appeared in England more than half a century later.[4] The clearest and probably most ironic Vesalian remnant hovers just next to the man's right ear, as if whispering a particularly salacious secret to him: the *rete mirabile*. As we noted in chapter 1, this dubious organ was granted existence and pictured by Vesalius in his *Tabulae anatomicae sex* of 1538, and still appeared in the *Fabrica* five years later. But in the later text Vesalius printed a retraction alongside the image, denying that the wonderful net existed and insisting instead that the functions this Galenic artifact was thought to perform – i.e., subliming the vital spirits into the animal spirits and sending them to the brain so that the soul could work upon them – were in fact carried out in the brain's ventricles and choroid plexus.[5]

Adjacent to this revived representation of the *rete* in the "visio secunda" is an elegant drawing of what the brain's arterial and venous systems were presumed to look like in the *Fabrica* and its numerous descendants. Below this, running down along the man's right arm, is a disproportionately large image of the liver. He seems to thrust his right arm, with its pronounced veins, through a melee of gallbladder, various gastric vessels, and disembodied heart, and the outward-turned fingertips of his right hand just barely graze the top of a depiction of the "bladders" for "seed and urine" connected to the "yard" (the name shared, as we observed in chapter 3, by both the male generative organ and its putative visual cousin in the brain, the pineal gland). The left knee parts the bladder with its ligaments from the veins of the liver, which extend to curl up beneath his elbow. Incongruously, a dissection of the womb nestles against his left forearm, while his fingers curve around two views of the heart.

And yet, this profusion of anatomical oddities does not capture the strangest and most distinctive element of the image, which is signalled by the numerous cuts across the man's body. These incisions turn the apparently two-dimensional page into a three-dimensional one, for each visible scar marks the existence of a series of flaps hidden beneath the picture plane: openings that lead us into the body's interior, secret doors that reveal everything normally hidden under the skin – and not as disembodied partial objects floating on the image's surface, but as a series of connected slices of what transpires beneath, each taking us deeper into the body's recesses. And at the same time they force us always to hold in mind the presence of the enveloping body that holds all the sheaves together, relating them to one another as *its* organs. Folding back the colchium petals, for example, reveals the man's external genitalia, as well as the fact that the man's torso has been sliced transversely across the bottom of his abdomen, perpendicular to the already-visible longitudinal section running from his pelvis up to his breastbone, and parallel to the two semi-circular cuts around each of his pectorals. Each of these cuts in turn opens up to display progressive sections of the membranes, muscles, organs, venous and arterial systems, and bones within the man's torso. Some of these layers have been bound together along one side, so that the viewer may turn them along one seam as pages within a book – hence the metaphor with which this chapter is concerned. Some organs "float," that is to say, they can be moved about slightly to reveal other details otherwise concealed: notably, the heart has been threaded through the diaphragm and some of the cartilage surrounding the trachea. On the underside of the pectoral flaps, images of the lungs and ribcage have been pasted. And once all the viscera have been folded up or tucked away, the man's kidneys and spinal column are exposed.

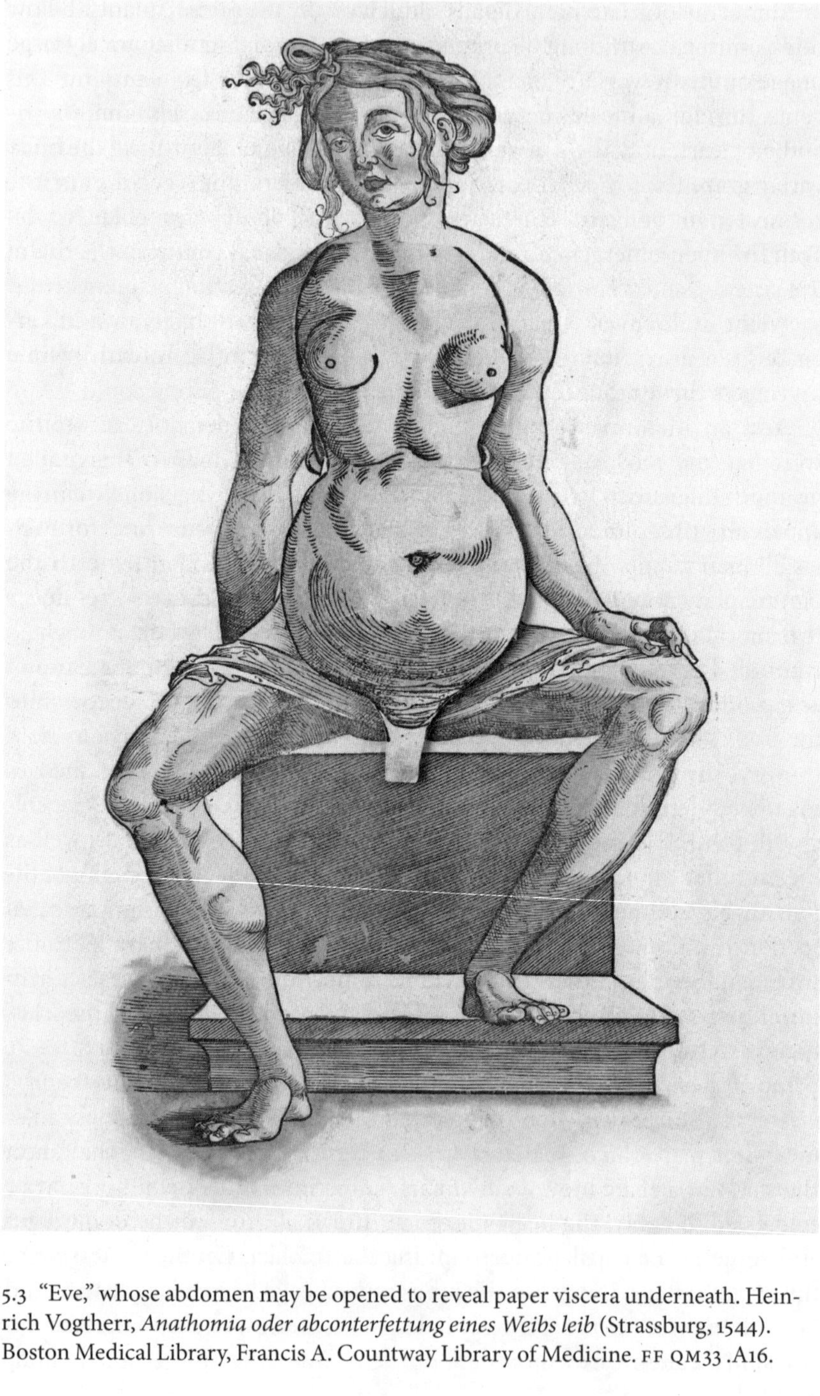

5.3 "Eve," whose abdomen may be opened to reveal paper viscera underneath. Heinrich Vogtherr, *Anathomia oder abconterfettung eines Weibs leib* (Strassburg, 1544). Boston Medical Library, Francis A. Countway Library of Medicine. FF QM33 .A16.

This array of anatomical details hidden under the picture plane and the body's surface testifies to the presence within Remmelin's allegorical image of a relatively new genre of anatomical illustration: the flap anatomy. This genre calls for a rethinking of the purpose of anatomical representation – and, not least, of how through such texts the idea of anatomizing the body was disseminated to a broader public, one for whom purchasing, say, the massive eight-volume *Fabrica* was neither feasible nor desirable. A brief sketch of their emergence and historical role is thus warranted to arrive at a clearer sense of Remmelin's contributions and innovations.

Public anatomical "theatres" were common in the sixteenth and seventeenth centuries, both in England and on the Continent, with Leiden serving as the most successful arena. Members of the public could pay to witness an anatomical lecture, and certainly the students in attendance were likewise part of the audience. Yet legal statutes limited the time of year in which such anatomies could be performed (i.e., largely during winter months, given the absence of preservatives). The number of bodies allotted to any one city was limited, too: anatomy was reserved as an "extra" punishment for criminals, whose bodies would then have to be transported to a city other than the one in which they lived – and in which they presumably had relatives who might object to a public dissection.[6] Some other form of information dissemination was therefore necessary for more remote students and scholars, either to supplement or to replace the "live" anatomy. Doubtless anatomical textbooks such as the *Fabrica* presented one solution to this problem – but these were often expensive and the object of study of specialized scholars.

For other kinds of readers, a special class of medical illustrations, the flap anatomy, emerged in both Germany and Italy during the late 1530s in the form of so-called fugitive sheets, that is, single sheets of paper intended for sale on their own or to be bound with other manuals, astronomical treatises, medical recipes, and so forth.[7] The flap anatomy's history has been well documented, and it is a testament to early modern printing ingenuity that such a variety of these texts still survive today.[8] These sheets were apparently intended for use by barber-surgeons, anatomy students, or an educated lay public, and almost exclusively they show the torso of a seated man or woman (sometimes both), which could be folded up to reveal the body's interior. The reader/viewer could lift up a flap or series of flaps from the midsection of each figure to display the different layers of muscles, organs, and arterial, venous, and nervous systems, as well as the skeleton below. Often, the pictures were accompanied by rudimentary notes about the anatomy being presented, along with brief moral poems encouraging the reader to ponder God's creations more seriously.[9] These accompanying textual supplements were generally in the vernacular, reinforcing the likelihood that the intended audience for the flap anatomy was not a specialized one.[10] Figure 5.3, for example, presents the "Eve" figure of Heinrich Vogtherr the Elder, printed in Strasbourg in 1544 – and pirated significantly thereafter.[11]

The value of these paper flap anatomies lay in the fact that they could be cheaply produced and thus cheaply sold, in a sense democratizing medical knowledge, even as they placed that knowledge in an already-interpreted moral framework. For a student, by contrast, a more complex flap anatomy (such as that in Vesalius's *Epitome*) might make a useful aid to direct study. But, by and large, such fugitive sheets were intended less to supplant direct medical knowledge acquired by palpating and dissecting bodies than to draw attention to the body and its workings in the first place.[12] Indeed, these fugitive sheets were often woefully out of touch with the most current understanding of the human body.

And yet, despite their scant relationship to more serious (not to mention more accurate) medical data, flap anatomies embody a key step in the evolution of anatomical representation. Not only do they popularize academic and specialized knowledge, but they also allow the viewer to become an active participant in the act of anatomy: by lifting up flaps and folding back hinges, the viewer becomes a paper anatomist. They are likewise a form of printing innovation – particularly in the case of the *Catoptrum/Survey*, which has over a hundred organ flaps, occasionally up to nine sections beneath one flap, and which was likely sold pre-assembled.[13] Indeed, the appeal of Remmelin's text, too, seems largely to have been because of its unusual layout, for its anatomical information was, even at its first printing, generally out of date.[14] But the original *Catoptrum* nonetheless took up the genre of fugitive sheets in a different and new way: on the one hand, by incorporating them into complex multi-layered allegories, and, on the other, by re-collecting them as coordinated series of anatomical illustrations within a book. As Cali Buckley remarks of this particular feat of paper engineering, "Few surfaces are without surprises."[15]

The import of the technique runs even deeper, for it transforms the way we approach the page. Confronting the *Catoptrum* one is struck, in Suzanne Karr Schmidt's words, by "print's possibilities when it transcends the flat page and enters the third dimension."[16] This is a dazzlingly modern-feeling idea, and part of what I want to suggest here is that the

5.4 *Adam and Eve*. Albrecht Dürer, 1504. The Metropolitan Museum of Art. Art Resource, New York.

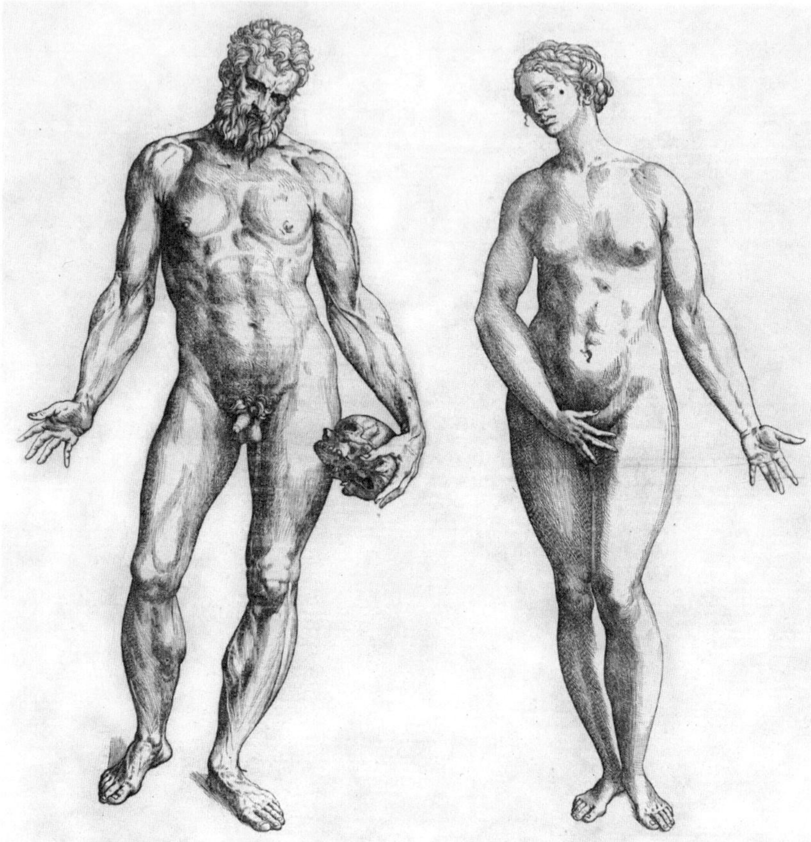

5.5 A Vesalian Adam and Eve. Andreas Vesalius, *De humani corporis fabrica librorum epitome* (Paris, 1560), sig. K_1^v, L_1^r. Yale University, Harvey Cushing / John Hay Whitney Medical Library, New Haven, CT.

innovations introduced by Remmelin's anatomy brought a new kind of legibility to the human body: the body became something of a *mise en abyme* of the structure in which it was housed, a microcosm not simply of the universe but more particularly of the book itself. Through such anatomies, the printed body – and, as I shall later suggest, the brain in particular – was itself rendered as a kind of book, one to be "opened" and "read" in a way that the real body – fragile, temporary, constantly withering – could never be. Such a desire for legibility undeniably underpins modern imaging techniques such as the CAT scan and the MRI.

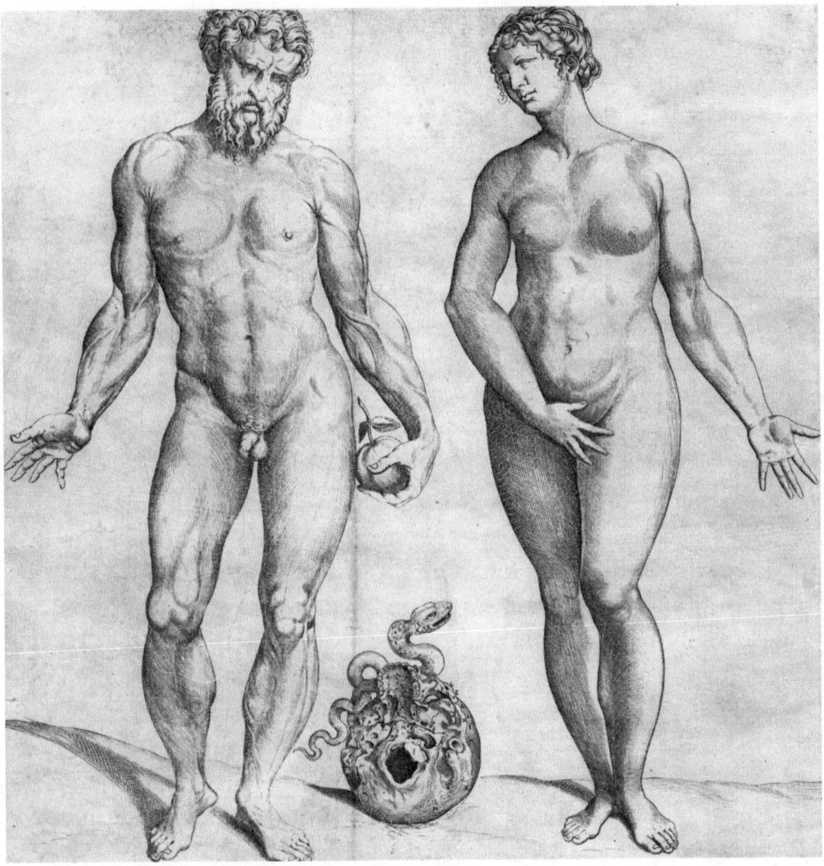

5.6 Adam and Eve following Vesalius; note that the skull Adam holds in Figure 5.5 has been replaced by an apple, while the skull now rests upturned on the ground between the figures. Thomas Geminus, *Compendiosa totius anatomie delineatio* (London, 1545), sig. B_{ii}^{v}. Yale University, Harvey Cushing / John Hay Whitney Medical Library, New Haven, CT.

For the fantasy of the early modern anatomist is that the living body may be opened to reveal its secrets: laid bare, skin undone, the viscera tidy and the veins taut, webs of nerves flourishing their intricate connections – and revealing above all else the soul shining in the brain. The anatomist desires to be able to read the body as a book, aiming to understand its organization and thereby locate the soul. The ideal would be to witness the soul in action in the brain, but since vivisection was (for obvious reasons) condemned, the next best alternative was to key or to map a thorough understanding of the living, thinking body to the bare anatomized body. If one could theorize exactly where in the body the soul interacted with physical flesh, one might then be able to reason out how the soul controlled various bodily and mental processes, and, even more, to grasp how soul and body interacted on the most fundamental level of all. From this perspective, printed anatomies contributed to the desire to locate the soul in the body by providing an opportunity for the viewer to engage with both physical and spiritual reflection – one attraction of the flap anatomy was as an aid for the reader to know himself or herself.

Consequently, fugitive anatomies were invested in showing, as far as possible, the living figure and the anatomy in context, resulting, for example, in the lifelike poses of Remmelin's man and woman. To read the body as a book required being aware of the elements of their – the book's and the body's – production. Demanding not just reading but fingering, touching, pulling-apart, and putting back together, flap anatomies translated for the casual reader the experience of anatomical dissection. But, of course, this work could be accomplished without the usual mess: for, when you read a flap anatomy, you are not so much dissecting a body as you are the print and paper with which it was crafted. You elide, that is, flesh and bone with paper and print, and thereby realize a metaphor: the body is like a book that can be opened to reveal and unlock knowledge. Of course, you are not yourself opening the body; rather, in perusing its replacement, you open up possibilities foreclosed by actual bodies, since the paper body will not resist multiple attempts at puzzling out internal structures. It allows itself to be viewed repeatedly, and the figures it depicts are inviting, friendly, calling out to the reader to open and read them.

As the *visios* from Remmelin make clear, however, the act of reading leads not only to the acquisition of anatomical knowledge in its own right but to an allegory concerning the import and limits of such knowledge. These two elements remain in tension: anatomy as a worldly immanence, its fascination with bodies contending against their transcendental devaluation. For the flap anatomy would ideally also function for the layperson as a *memento mori* and a means of uncovering the gifts of God. It is thus no accident that the fugitive sheets based their male and female figures on Adam and Eve. As Mimi Cazort notes,

> Anatomical representations were deeply affected by Renaissance concern with the human body as the incarnation of ideal beauty. It was natural that artists would use the prototypes of Adam and Eve, perfect in health and immortal before their expulsion from

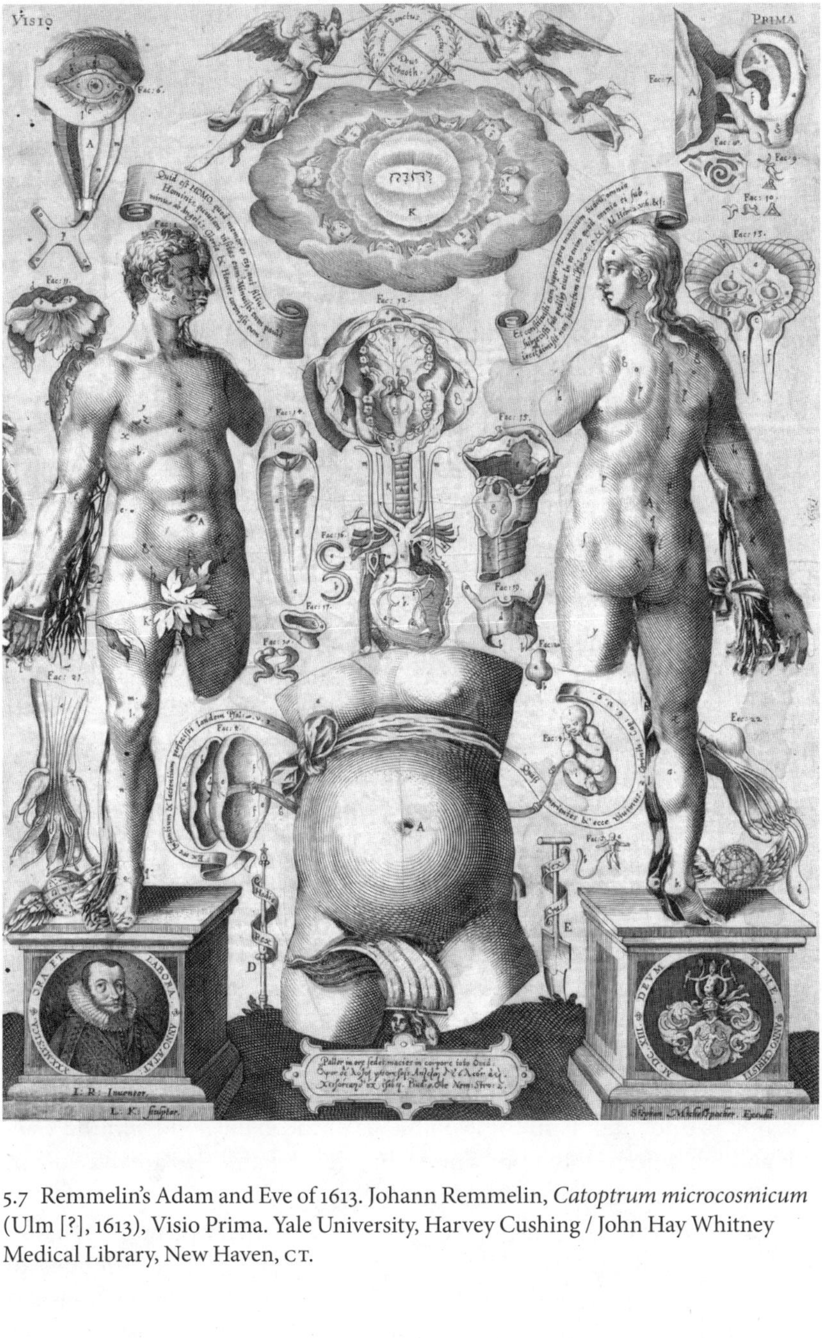

5.7 Remmelin's Adam and Eve of 1613. Johann Remmelin, *Catoptrum microcosmicum* (Ulm [?], 1613), Visio Prima. Yale University, Harvey Cushing / John Hay Whitney Medical Library, New Haven, CT.

the Garden, their pristine physical state reflecting their innocence before the Fall.[17]

Many anatomical full-body depictions could in fact be traced back, Cazort asserts, to a single, influential painting, Albrecht Dürer's 1504 engraving of *Adam and Eve*.[18] Besides its theme, the significance of this painting for subsequent anatomical illustration lay in how it modelled the bodies, exhibiting their musculature and lending them, through shading, posture, and other painterly techniques, a sense of volume. That is to say, in the depiction of the bodies, the two-dimensional picture plane here forcefully laid a claim to three-dimensionality, setting a standard and offering a blueprint for what later anatomies – both two-dimensional and flap – would seek to achieve. Thus, for example, Vesalius and his engraver, Stephen von Calcar, sought to achieve a comparable solidity for the human form by developing existing techniques of engraving and adapting them to anatomical representation.[19] Figure 5.5 shows the "Adam and Eve" figures from the *Epitome* of Vesalius's longer work *De humani corporis fabrica*. Framed by print, these figures seem almost to step from the page. Shadows help create the illusion of contours; Adam's muscles and veins in particular are evident. Then, too, their expressions – Adam deep in reflection, Eve looking sorrowfully to him – project them not simply as anonymous bodies but as the bodies of the progenitors of the human race, an implication given weight by the skull that the male figure holds against his thigh. To know ourselves, these figures suggest, we must turn to our origins.

Another telling example may be found in Thomas Geminus's 1545 *Compendiosa*, a cribbed version of Vesalius's *Fabrica*, which expands pictorially upon its biblical allegory. The *Compendiosa* Adam and Eve retain their poses but are this time set against a landscape, and the pronounced corporeal shadings reinforce their physicality. Instead of a skull, Geminus's Adam clutches an apple, emblem of the Fall. The skull, now on the ground between the two figures, has been turned so that its base faces the viewer; its jaw has been removed, and a serpent has been run through it. The link is evident: it is to these emblems of the expulsion from the Garden and the descent into bodily mortality that Adam's eyes are drawn. Sin gives these figures depth on the printed page.

The *visios* from Remmelin's *Captotrum* find their place in relation to this chain of images. They bring together aesthetic techniques of engraving used in anatomical texts such as Vesalius's with the volumetric possibilities opened up by flap anatomies. This dual lineage is explicit in the allegorical anatomy with which the book opens. The *Captotrum*'s

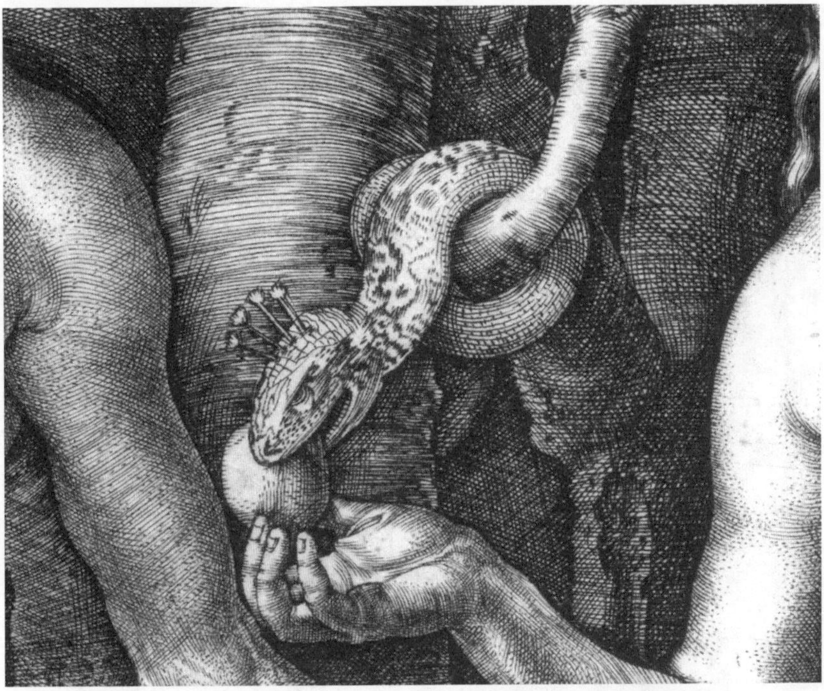

5.8 Dürer's serpent; detail from *Adam and Eve*. Albrecht Dürer, 1504. The Metropolitan Museum of Art. Art Resource, New York.

"Visio prima" is at first glance utterly perplexing. The two "Adam and Eve" figures stand looking at each other atop plinths (the original images on these architectural supports would later be scrubbed in Moxon's English editions of the *Exact Survey* in order to insert dedications to Samuel Pepys), arranged so that one offers a frontal perspective while the other has its back turned to us. The musculature and shading on the bodies make apparent their debt to anatomical illustration in the wake of Vesalius and his contemporaries.[20] At the same time, the human figures flank the flap anatomy of a pregnant torso, which can be opened up to reveal the fetus developing inside. Were we to do so, we would see that, though pregnant, the torso has its hymen intact, reinforcing the mystical suggestion of this as a virgin birth. Above this flap anatomy, a two-dimensional dissection of the throat and head has been tilted back to gaze agape at the tetragrammaton encircled by a cloud whose folds are curiously reminiscent of the convolutions of the cerebral cortex. Mouth open below, eye

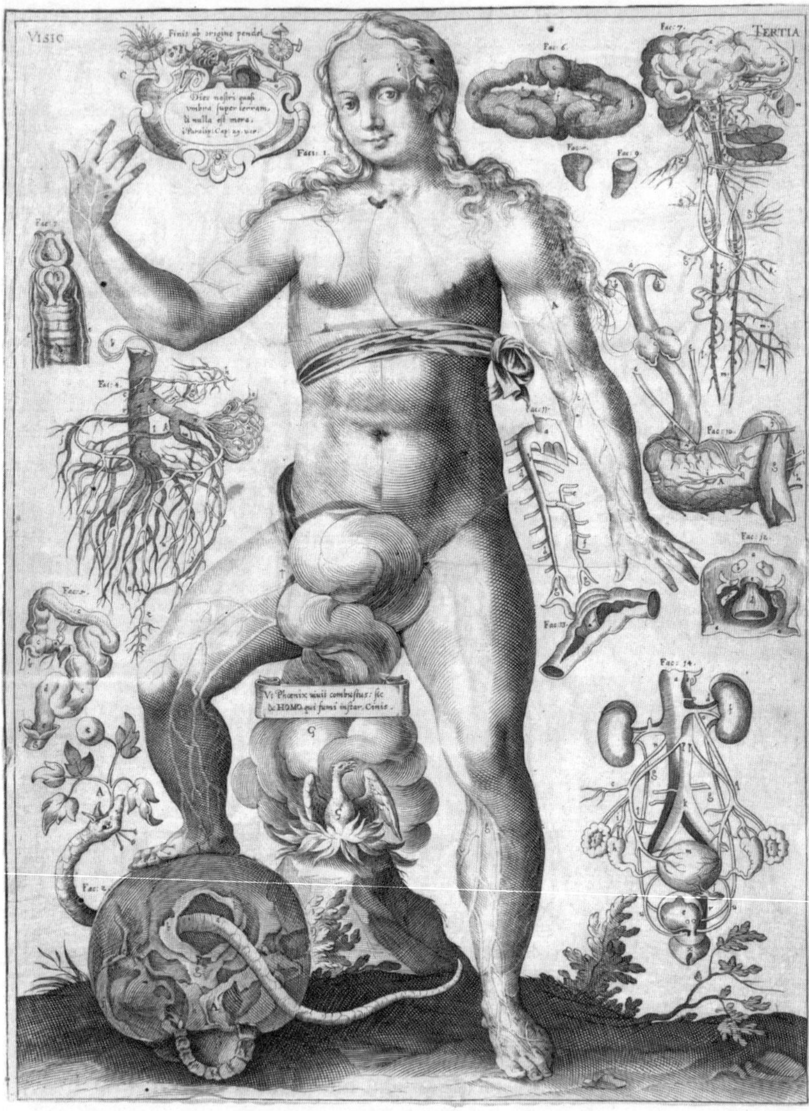

5.9 Remmelin's Eve of 1613. Johann Remmelin, *Catoptrum microcosmicum* (Ulm [?], 1613), Visio Tertia. Yale University, Harvey Cushing / John Hay Whitney Medical Library, New Haven, CT.

and ear on either side: hovering between these sensory organs, the cloud suggests an association with commonsense, as the place within the brain where data from the individual senses come together. Lifting up its flap reveals a succession of figures on smaller flaps underneath: an angel, a man who might be Remmelin himself, a devil figure, and finally an inscription from Syrach/Ecclesiasticus 38.2 proclaiming "A DEO *est omnis* MEDELA" ("For of the most high cometh healing," KJV 1611).[21] This motto hints, perhaps, at a link between the two flap images on this page, likening the physician or anatomist, who possesses a quasi-divine knowledge of the human body, to God Himself, from whom spiritual healing will come.

Yet this connection is only available to the viewer once the flaps have been folded back and "read." The divine book of God and the book of the human body have to be leafed through carefully to uncover the relationship between them: to be able to marvel at the human body as material symbol of God's creation and, at the same time, grasp its transitoriness, its mortification in the light of redemption. As Walter Benjamin writes of the baroque image,

> [i]ts beauty as a symbol evaporates when the light of divine learning falls on it. The false appearance of totality is extinguished. For the *eidos* [essence] disappears, the simile ceases to exist, and the cosmos it contained shrivels up.[22]

But the body nonetheless endures on and in the page, its function as a reminder of mortality balanced by the pleasures of uncovering secrets hidden beneath the flesh. In comparison to its predecessors, Remmelin's images require active interpretation, turning a viewer into a complicit anatomist-reader, with one foot firmly lodged in empirical practice while the other floats on the clouds of the occult. The *Catoptrum* appears to have been popular in Europe, as it went through multiple reprintings and at least one re-engraving in Holland. Despite its seeming fragility, multiple copies have survived to this day, remarkably intact, given the intricate delicacy of such moving parts.

Remmelin's Skull

My brief description barely scratches the surface of the extraordinarily complex baroque allegories of the *Catoptrum microcosmicum*. But I want to focus here on the intricacies of just one prominent image out of the multitude we see on (and within) Remmelin's *visio* of the

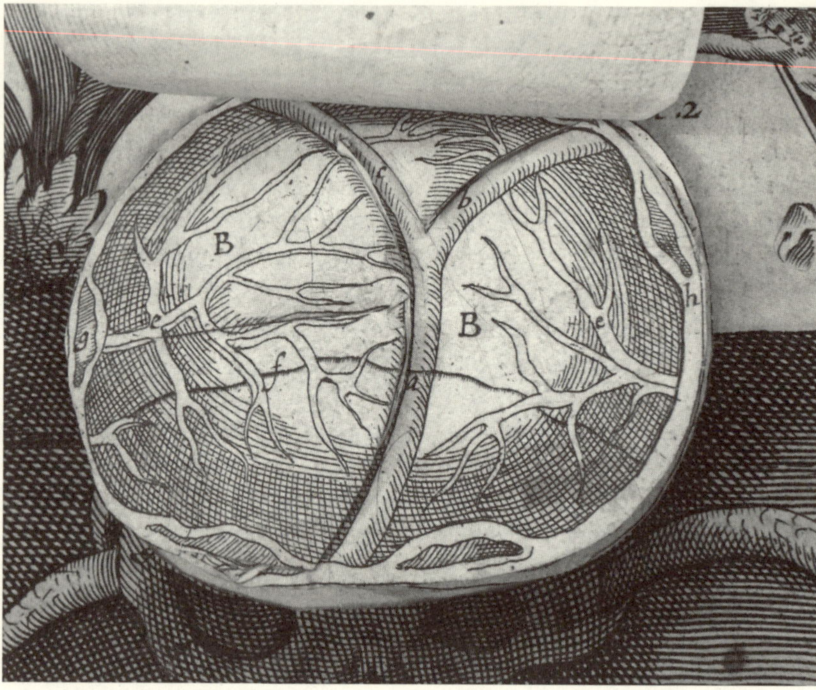

5.10 Detail of Remmelin's Adam, 1670, skull flap lifted. Johann Remmelin, *An exact survey of the Microcosmus, or Little World ...*, trans. John Ireton (London: Joseph Moxon, 1670), Visio Secunda. Author's photo. The Wellcome Library, London. EPB G. o/s F.2277.

man, with whom this chapter began (Figure 5.1): the disproportionately large skull upon which his left foot balances. The skull has been tilted forward to give the viewer a clear picture of the frontal, sagittal, and lamboid sutures. In this position, the face of the skull seems to be peering over the page's edge, perhaps anticipating the turning-up of its own paper hinge. This we shall shortly do, but before we open the skull, let us observe the serpent – surely a descendant of the one haunting Dürer's Garden – threading the pockets of space around the ear sockets, between the zygomatic arch and the mandible of the seemingly empty cranium. The snake sports four plumes. Juxtaposed to the skull is a microchristus (that is, a small personalized anatomical Christ figure on the crucifix), whose business end has thwarted the serpent's continued endeavour to tempt man, leading perhaps to these plumes of blood

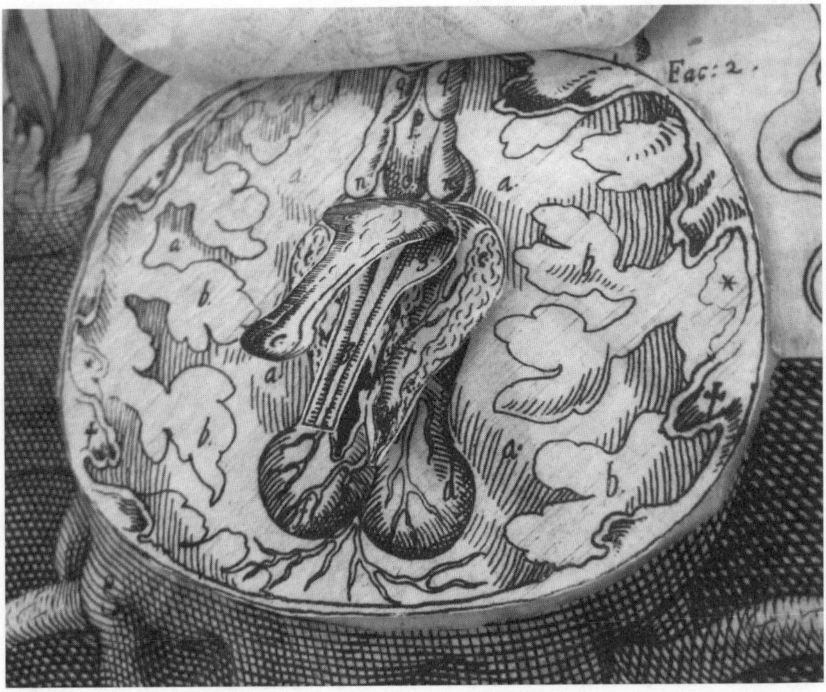

5.11 Detail of Remmelin's Adam, 1670, *dura mater* flap lifted. Johann Remmelin, *An exact survey of the Microcosmus, or Little World …*, trans. John Ireton (London: Joseph Moxon, 1670), Visio Secunda. Author's photo. The Wellcome Library, London. EPB G. o/s F.2277.

rising from its head. Though the microchristus has been labelled "H" it has no corresponding entry in the anatomy's key, but the crucifix's proximity to the skull makes the double valence of the latter as *memento mori* apparent.[23] The skull lies on the hinge between life and death, reminding the viewer of the transitoriness of life as well as signalling the resurrection, the promise of eternal life through death, the overcoming of sin by Christ, and the wonder of God's creation open to the pious reader. When we turn the page to view the next *visio*, depicting the man's female counterpart, we see that the colchium flower has been replaced by smoke from a combusting phoenix, blowing up to cover her genitals. While the snake has now acquired an apple, emblem of the Fall, the phoenix symbolically conveys the same dual sentiment concerning death and the promise of resurrection.

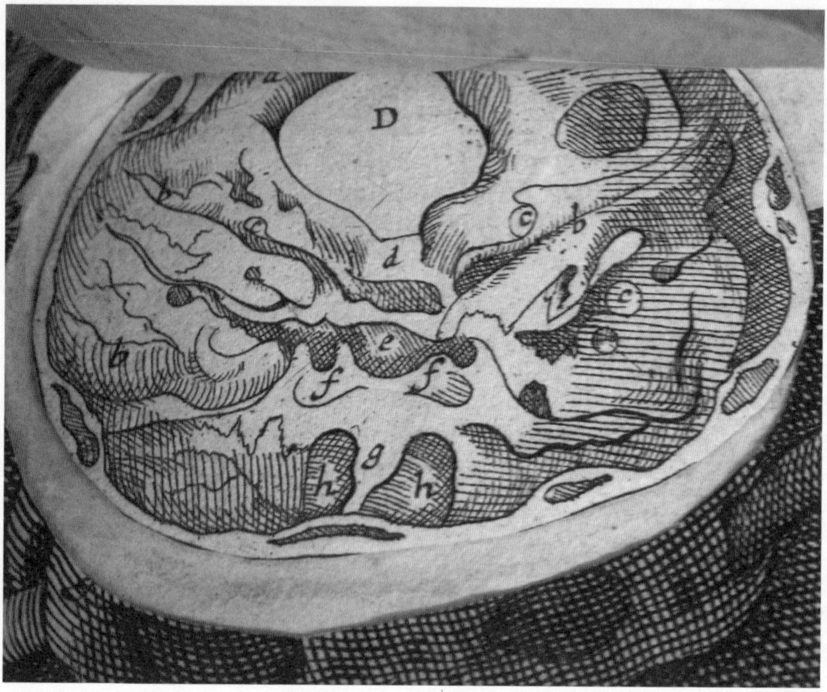

5.12 Detail of Remmelin's Adam, 1670, "yard" flap lifted. Johann Remmelin, *An exact survey of the Microcosmus, or Little World …*, trans. John Ireton (London: Joseph Moxon, 1670), Visio Secunda. Author's photo. The Wellcome Library, London. EPB G. o/s F.2277.

But the skull itself literally offers another kind of hinge between death and life, since it opens up to show the brain's interiority: far from being hollow, this death's head has folded into it all that it would have contained in life. And this feature brings us to another, crucial respect in which Remmelin's *Catoptrum* turns out to be distinctive – and this distinction is preserved in Moxon's editions of the *Exact Survey*. For there was one highly curious elision in flap anatomies from their inception in the early sixteenth century through their disappearance towards the end of the eighteenth century: virtually none of them represented the internal structures of the brain in a manner comparable to how they treated bodies. Apart from three curious and minor exceptions,[24] the *Catoptrum*'s images constitute the period's only fold-out anatomies of the brain; no other anatomy truly offered the brain as a book to be read. As we have seen earlier, dissections of the brain could easily be rendered into print (see, for example, Figures 1.1 and 1.2, which compare two brain anatomies, one by Vesalius and the other a copy by Crooke). And surely a lay reader, having overcome any squeamishness about, say, opening up a female belly to view its internal architecture in paper form, would not be daunted by the far more abstract renderings appropriate to the brain's contents. The skull that our biblical progenitors stand upon in the two *visios* from the *Catoptrum* may have dropped from the hand of Vesalius's Adam – just as the snakes threading the skulls have wended their way from Geminus's representation into Remmelin's. But here the skull under Adam's feet is no mere empty, lifeless reminder of mortality, which only finds its revivification at the resurrection. Instead, this skull teems with knowledge; it opens up to reveal a living brain – and suggests that the knowledge gleaned from construing the body, and especially the brain, as a book can overcome the work of the devil.

Speaking of baroque allegory, Benjamin remarks perceptively on the contrast between "the uniform verses of the emblem-books, the 'vanitas vanitatum vanitas,' and the fashionable bustle with which they appeared, on each others heels." If an object becomes, he continues,

> allegorical under the gaze of melancholy, if melancholy causes life to flow out of it and it remains dead, but eternally secure, then it is exposed to the allegorist, it is unconditionally in his power. That is to say … such significance as it has, it acquires from the allegorist … In his hands the object becomes something different; through it he speaks of something different and for him it becomes a key to the realm of hidden knowledge; and he reveres it as an emblem of this.[25]

In John Webster's *The Duchess of Malfi*, the eponymous protagonist proclaims that she knows that

> [D]eath had ten thousand several doors
> For men to take their exits; and 'tis found
> They go on such strange geometrical hinges,
> You may open them both ways. (4.2.218–21)[26]

The flaps of the skull in Remmelin's *visio* express just such a strange geometry: closed, the skull and microchristus together function as standard vanitas emblems, reminding the viewer of sin, death, and resurrection; open, however, they promise something else: the petrified yet preserved

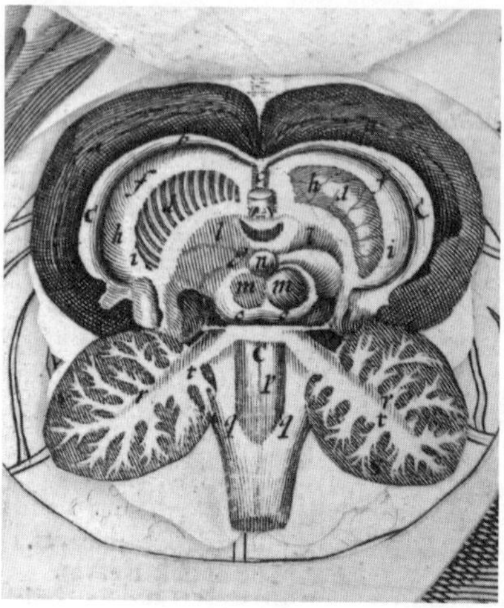

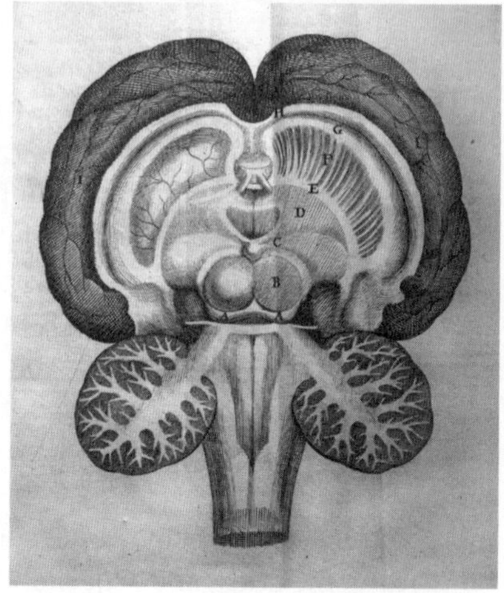

5.13 Left: Detail of Remmelin's Adam, 1702, skull flap lifted. Johann Remmelin, *An exact survey of the Microcosmus, or Little World* …, ed. Clopton Havers (London, 1702), Visio Secunda. Author's photo. The Wellcome Library, London. EPB G. o/s F.350.

Right: The original engraving taken from Thomas Willis, *De anima brutorum* … (Amsterdam, 1672), pull-out opposite sig. L_2^r, Tabula VII, "Exhibet prominentias orbiculares …" Author's photo. Yale University, Harvey Cushing / John Hay Whitney Medical Library, New Haven, CT.

knowledge hidden in the dead, the path to a different kind of heaven, one represented by the early modern brain.

But what kind of knowledge is this? Here, intriguingly, the answers diverge, depending on which edition we focus on. Were we looking at this image in the Ulm *Catoptrum microcosmicum* of 1613 or its corrected edition of 1618[27] – or, for that matter, in the first three English editions of the *Exact survey of the Microcosmus* from 1670, 1675, and 1691 – lifting up the skull flap would reveal a brain largely unrecognizable to us.[28] As D.H.M. Woollam notes, "it was the fresh unpreserved brain with which alone" the early anatomists and many of their early modern acolytes

> were familiar, and the brain preserved in formalin, on which so much of our modern anatomical ideas are based, bears very little resemblance to the brain that Galen and Herophilus knew.[29]

Hence, lifting back the skull on the original page, we would see images of something that looked like a bowl of thick noodles, or perhaps a tree

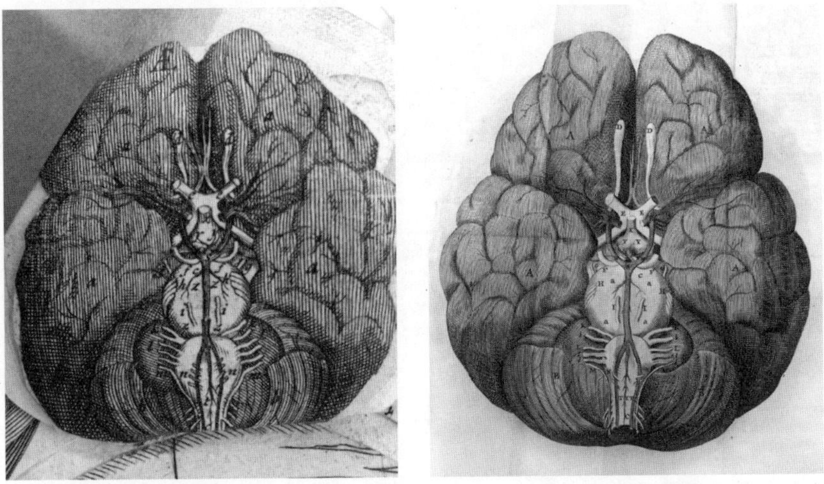

5.14 Left: Detail of Remmelin's Adam, 1702, pasted on verso of first flap. Johann Remmelin, *An exact survey of the Microcosme* …, ed. Clopton Havers (London, 1702): Visio Secunda. Author's photo. The Wellcome Library, London. EPB G. o/s F.350.

Right: The original engraving taken from Thomas Willis, *Cerebri anatome, nervorumque descriptio & usus* (Amsterdam, 1664): figure 1a. Author's photo. Yale University, Harvey Cushing / John Hay Whitney Medical Library, New Haven, CT.

trunk around which are strewn vessels (nerves? arteries? veins?), seemingly at random. This oval draped with spindly branches is the *dura mater* or "hard mother," the brain's first internal covering. Peeling back this full flap then reveals the *pia mater* or "soft mother," surrounded by clouded curls of cortical matter. The next full flap leads to an interior view of the brain's ventricles that bears an uncanny resemblance to the external private parts so prudishly hidden by the colchium flower blooming between the man's legs. Resting in the midst of these is a series of four delicate smaller flaps that allow the viewer to prise up the "arch" between the ventricles of the brain (i.e., the corpus callosum), the choroid plexus (which the tables helpfully remind us is also known as the "thumb" of the brain), the pineal gland ("the yard of the brain"), the "nates" of the brain (testes), and, finally, its "buttocks."

Below these we encounter the last flap, raising which shows the convolutions of the cranium, emptied of the cortical matter. This, then, is knowledge as it was: preserved and transmitted for centuries, from Galen onwards, largely unchanged despite intervening transformations in

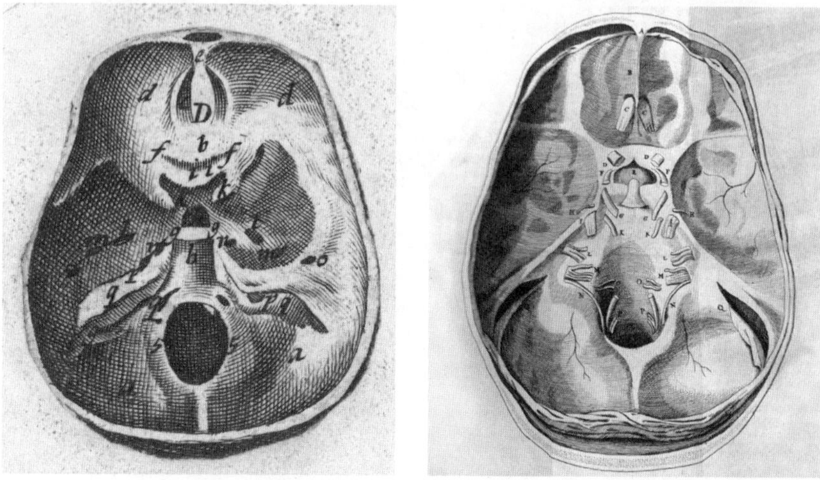

5.15 Left: Detail of Remmelin's Eve, 1702, both cranial flaps lifted. Johann Remmelin, *An exact survey of the Microcosme* ..., ed. Clopton Havers (London, 1702), Visio Tertia. The Wellcome Library, London. EPB G. O/S F.350.

Right: The original engraving taken from Thomas Willis, *Cerebri anatome, nervorumque descriptio & usus* (Amsterdam, 1664), figure 5. Author's photo. Yale University, Harvey Cushing / John Hay Whitney Medical Library, New Haven, CT.

anatomical knowledge, and based in large measure on assumed similitudes between man's external body and the brain's internal structure. As far as its interior is concerned, then, the brain very slenderly resembles itself – at least as we know it today. We would be forgiven for finding it completely unrecognizable.

However, this chapter is concerned with a glaring exception. For, were we to probe within the same skull in 1695 and 1702, we would encounter a very different brain. While some of the flaps would remain the same, faithfully imitating the earlier English editions of the *Exact Survey*, sandwiched between them would be two utterly unexpected images that are simply stunning in their detail and artistry, and which would be instantly recognizable to an anatomy student from the late seventeenth century – and indeed seem familiar to us today as well. These are miniaturized pictures of brains pilfered from foundational texts by Thomas Willis that had been illustrated by Christopher Wren: one from the *Cerebri anatome* and the other a collapse of two images from the *De anima brutorum*.[30]

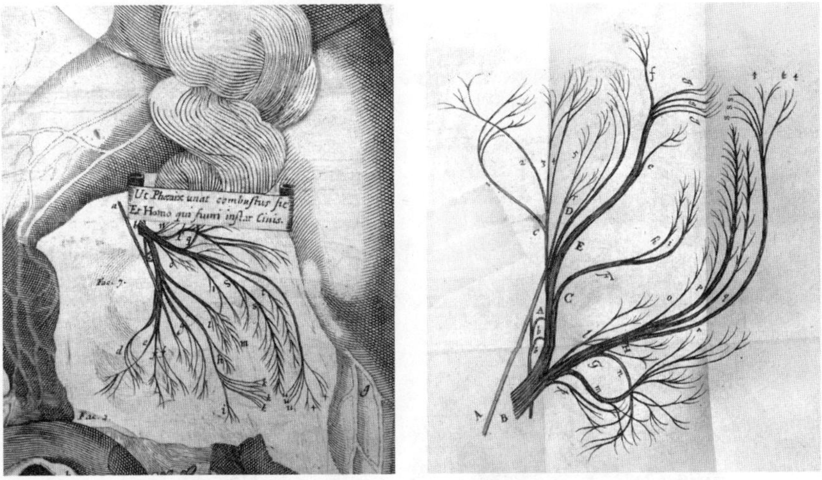

5.16 Left: Detail of Remmelin's Eve, 1702, between legs. Johann Remmelin, *An exact survey of the Microcosme* ..., ed. Clopton Havers (London, 1702), Visio Tertia. Author's photo. The Wellcome Library, London. EPB G. o/s 66323/F

Right: The original engraving taken from Thomas Willis, *Cerebri anatome, nervorumque descriptio & usus* (Amsterdam, 1664), 180. Author's photo. Yale University, Harvey Cushing / John Hay Whitney Medical Library, New Haven, CT.

As noted in the introduction, Willis is commonly considered the first English anatomist to focus on the brain, and the *Cerebri* the first text devoted solely to neuroanatomy. In the 1695 and 1702 editions of the "Englished" Remmelin, the reproduced image from the *Cerebri* (Figure 5.14) is a particularly brazen copy, since it preserves most of the letter labels and even mimics the shading of the surrounding cortical folds. The third *visio* in the series, depicting the man's female counterpart, likewise flaunts illustrations of the base of the skull and the fifth and sixth cranial nerves, taken without acknowledgment from the same Willis text.

While direct copies, the *Exact Survey* images are not, however, absolutely identical to those in the source text. The main difference is that the imitations have been shrunken (albeit with their proportions preserved) in the transposition. Since Willis's texts were originally published in quarto and octavo, whereas the Moxon edition of the English Remmelin is in folio (albeit with miniaturized organs with respect to Willis), there were three different page/image sizes that had to be coordinated, which

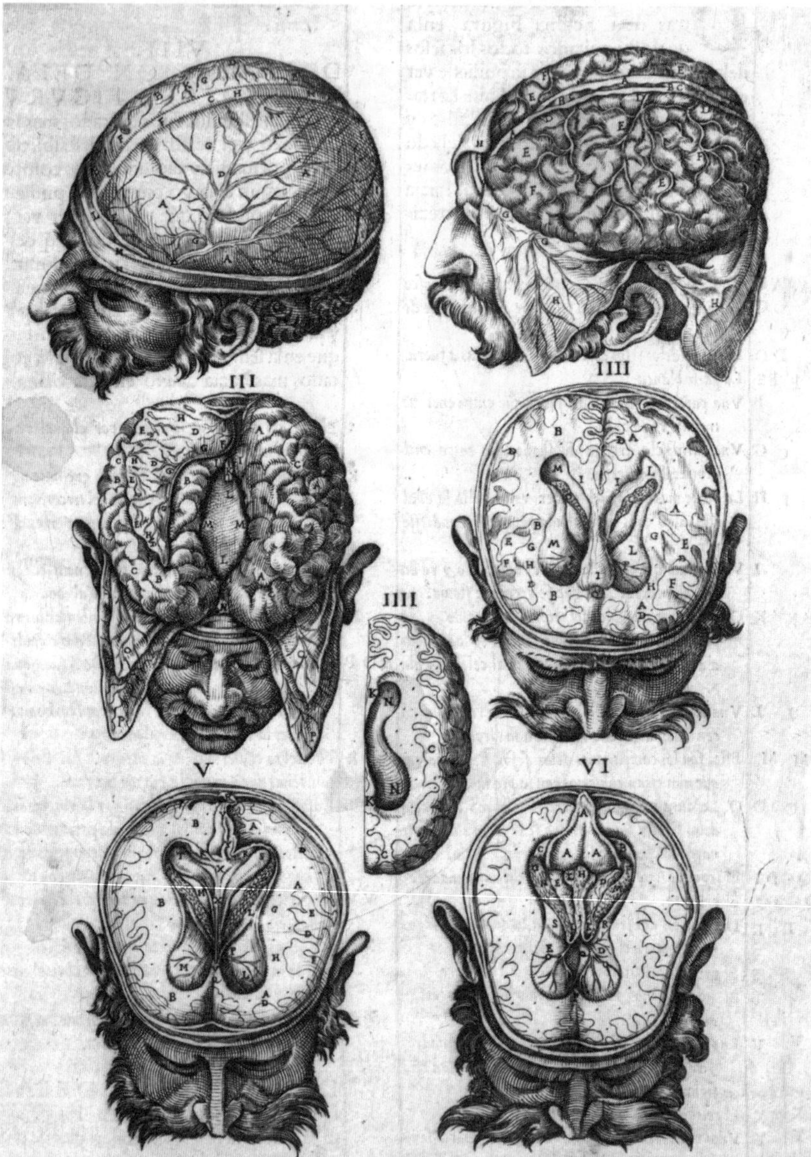

5.17 A progressive cranial anatomy. Juan Valverde, *Historia de la composicion del cuerpo humano* (Rome, 1556), 114. Yale University, Harvey Cushing / John Hay Whitney Medical Library, New Haven, CT.

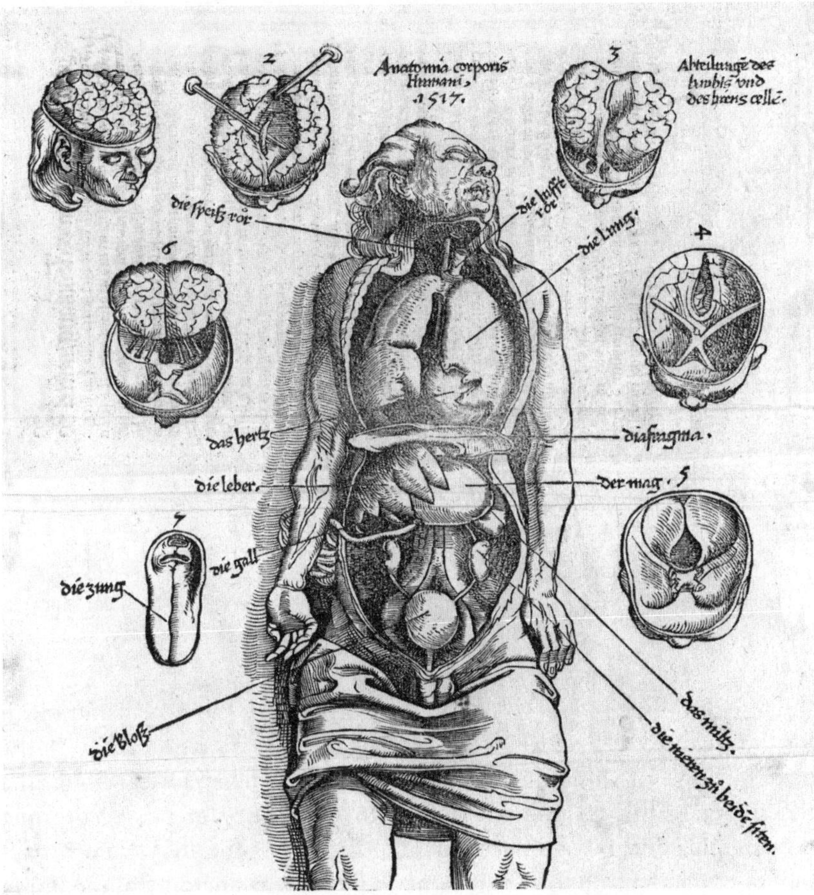

5.18 Brain dissection and anatomy. Hans von Gerstorff, *Feldbuch der Wundartzney* (Strassburg, 1517), table 2. Yale University, Harvey Cushing / John Hay Whitney Medical Library, New Haven, CT.

meant that the original depictions of these iconic brains circulating in England at the close of the seventeenth century would have had to have been painstakingly re-engraved. The smallest of these reproductions has remained hidden until now: no one else writing about the history of anatomy (and about English editions of Remmelin, more specifically) seems to have noticed the discrepancy in what the flaps contained, and so we are privileged to witness them now, as it were for the first time.

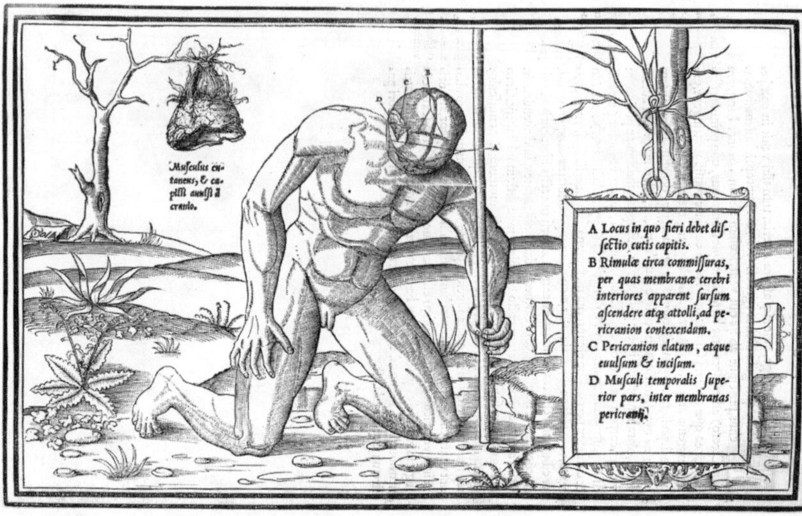

5.19 First cranial section. Charles Estienne, *De dissectione partium corporis* (Paris, 1545), sig. D_i^{v-r}. Yale University, Harvey Cushing / John Hay Whitney Medical Library, New Haven, CT.

How did these brains get there? And what does their inclusion mean for the anatomy as a whole? After making the case more forcefully, in the next section, for thinking about the brain as a metaphorical book, the remainder of this chapter will return to these questions by recounting the complex and intriguing publishing history of the 1695 *Survey*, paying especial attention to how it came to update the anatomy of the brain. I shall suggest that this edition (the fourth to appear in England) transformed and, arguably, transcended its original in this particular respect at least, thereby laying the claim to be not simply an anatomy, in the sense of a passive repository of inherited knowledge, but a scientific and artistic instrument in its own right, a body-book that presented the reader with the most current knowledge of this miraculous organ. As we shall see, understanding how and why it did so requires examining the circumstances of the *Survey*'s publication, and its peculiar status as the sole anatomical text to come out of the publishing house of Joseph Moxon, a celebrated maker of maps, globes, and mathematical paper instruments. It was under the direction of his son James and the physician Clopton Havers that the new edition was produced, and as a consequence the 1695 *Survey* takes up a unique position at the conjunction between Moxon's printing establishment and the burgeoning Royal Society, two

5.20 A later cranial section, with the *dura* and *pia mater* removed. Charles Estienne, *De dissectione partium corporis* (Paris, 1545), sig. Q_{ii}^{r}. Yale University, Harvey Cushing / John Hay Whitney Medical Library, New Haven, CT.

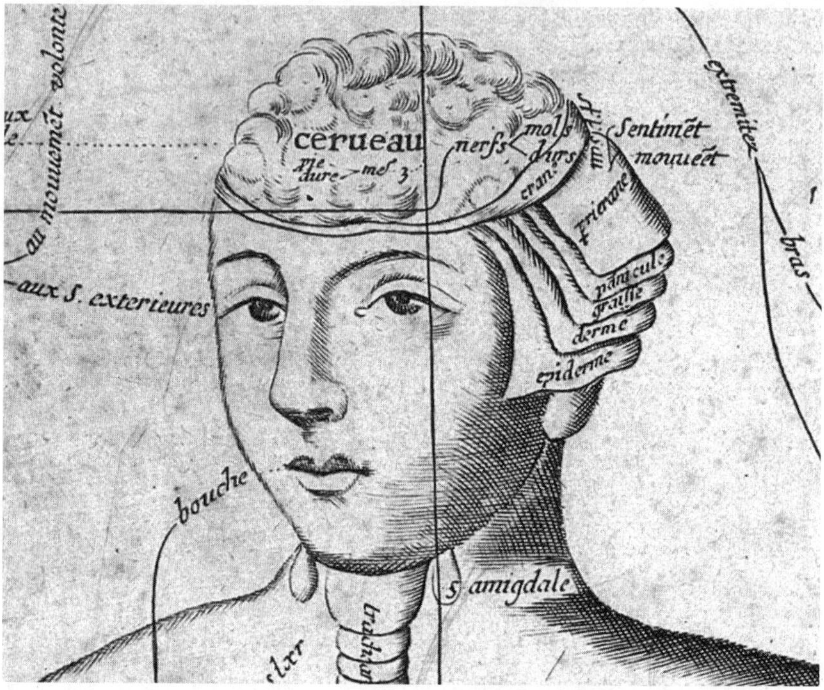

5.21 A cranial anatomy with epidermis unfolded. I. Dumoulin, after Jean Blanchin, *Abrégé de l'anatomye* (Paris, 1679), detail. Wellcome Library, London. Iconographic Collection 647800i.

intertwined institutions that aimed – at least rhetorically if not always in fact[31] – to extend knowledge of the natural world as well as of human nature. Weaving together the strands of publishing and scientific history involved in shaping the metaphorical relationship between brain and book, we shall see how the "slices" of the brain turn into the leaves of a book, which the reader must then turn in order to read.

Leaving the Brain

Before delving into Moxon and the *Survey*, however, I want to make the case more forcefully for thinking of the brain in particular as a metaphorical book. This requires tracing printed images of the brain that may have inspired Remmelin in his endeavour. In many ways, even before the emergence of flap anatomies, this organ – perhaps more so than any other in the early modern period and certainly to a significant degree in

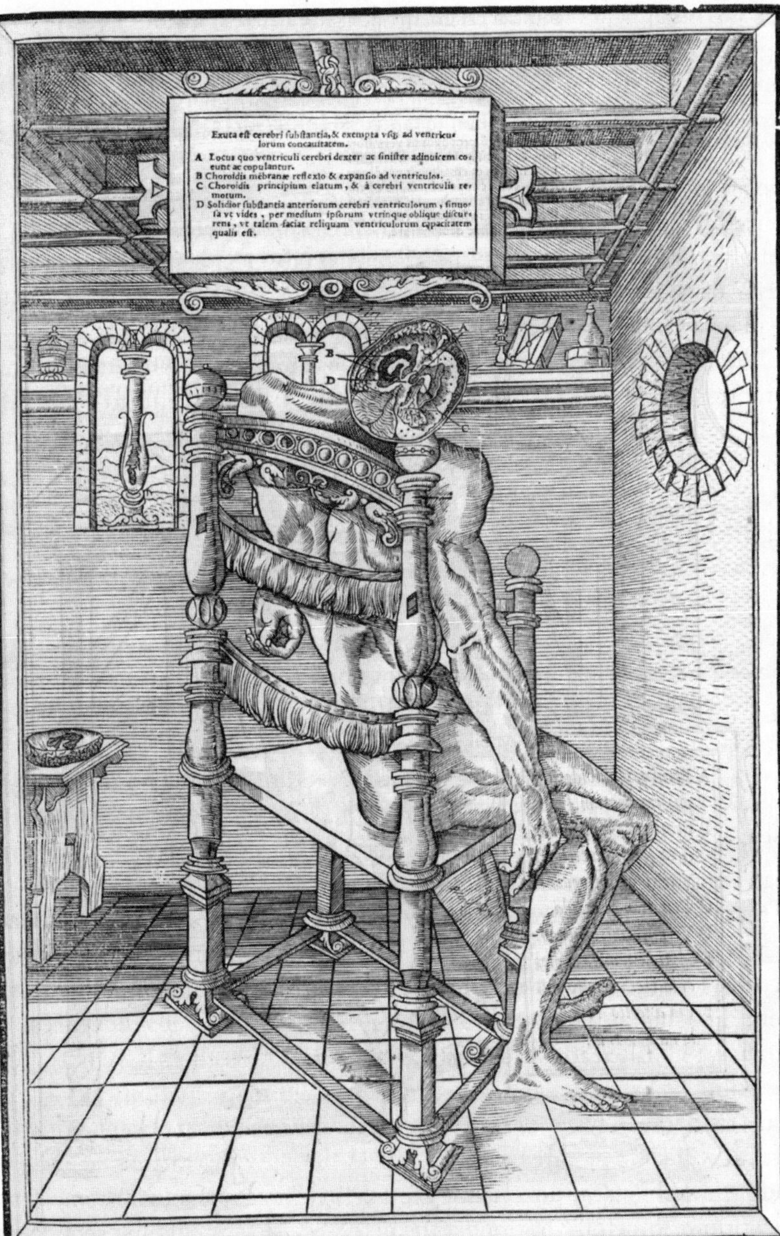

5.22 The fourth cranial section; note the bookshelf. Charles Estienne, *De dissectione partium corporis* (Paris, 1545), sig. Q_{iii}^{r}. Yale University, Harvey Cushing / John Hay Whitney Medical Library, New Haven, CT.

our time – appears to offer itself up most readily to topographical layering. While other organs such as the eye and the heart were certainly also shown in "levels" of dissection, these were generally disembodied.[32] The closest correlate to the brain in progressive dissection was, as chapters 3 and 4 demonstrate, the womb, in representations of which women would appear happily to fold back different abdominal and uterine layers.[33] And so too with the brain, for in the sixteenth and seventeenth centuries the brain was often represented in standard anatomical textbooks through a series of successive slices, sometimes with the rest of the (usually male) head providing (disconcerting) context.

Though we have already seen how Vesalius – and Crooke after him – represented brain dissection (see chapter 1), it is useful to revisit the technique here. Figure 5.17 is a dissection series from yet another Vesalian plagiarist, Juan Valverde.[34] Here, successive layers of the cranium have been peeled back to expose the matter beneath. To "read" the image one begins, as usual, in the upper left-hand corner, where the skull has been removed to show the *dura*, and then moves to the upper right-hand corner, where the *dura* has been removed to expose the cortex, and then to the left-hand image in the middle, which opens the cortex, and so on.[35] The rest of the man's head is at least partially visible at all times, providing the viewer a ready means of bodily orientation. Figure 5.18 gives us an even clearer sense of the body from which the brain has been separated. In von Gerstorff's rendering, the brain dissection starts again in the upper left-hand corner and progresses first horizontally and then vertically – that is, after the manner in which a contemporary Englishman or European would read words on a page.

Mapping the movement of the eyes onto the temporal order of dissections was, of course, not limited to the case of brain dissections. But the fact that the brain and the book are connected is made especially prominent in Charles Estienne's arresting *De dissectione partium corporis humani*, published in Paris in 1545. As Carlino and others have observed, the utility of Estienne's images extends beyond mere anatomical instruction – rather, in a series of stylized, vividly delineated, and even faintly (or forcefully) erotic pictures, we encompass the activities of a *living, lived-in* body (though Singer and Rabin grouse that this is "the ugliest anatomical work we know").[36] No longer the disembodied systems and sections of the more prominent textual antecedents – stomach treated of in one chapter, bones in another, generative organs (worst of all) segmented off by themselves – this work aims to present the paradox that the talented anatomist and artist is tasked with imagining, if not to producing outright:

5.23 Joseph Moxon, pictured with a selection of his published works. Joseph Moxon, *Mathematicks Made Easie …* (London, 1692), author portrait opposite title page. Yale University, Harvey Cushing / John Hay Whitney Medical Library, New Haven, CT.

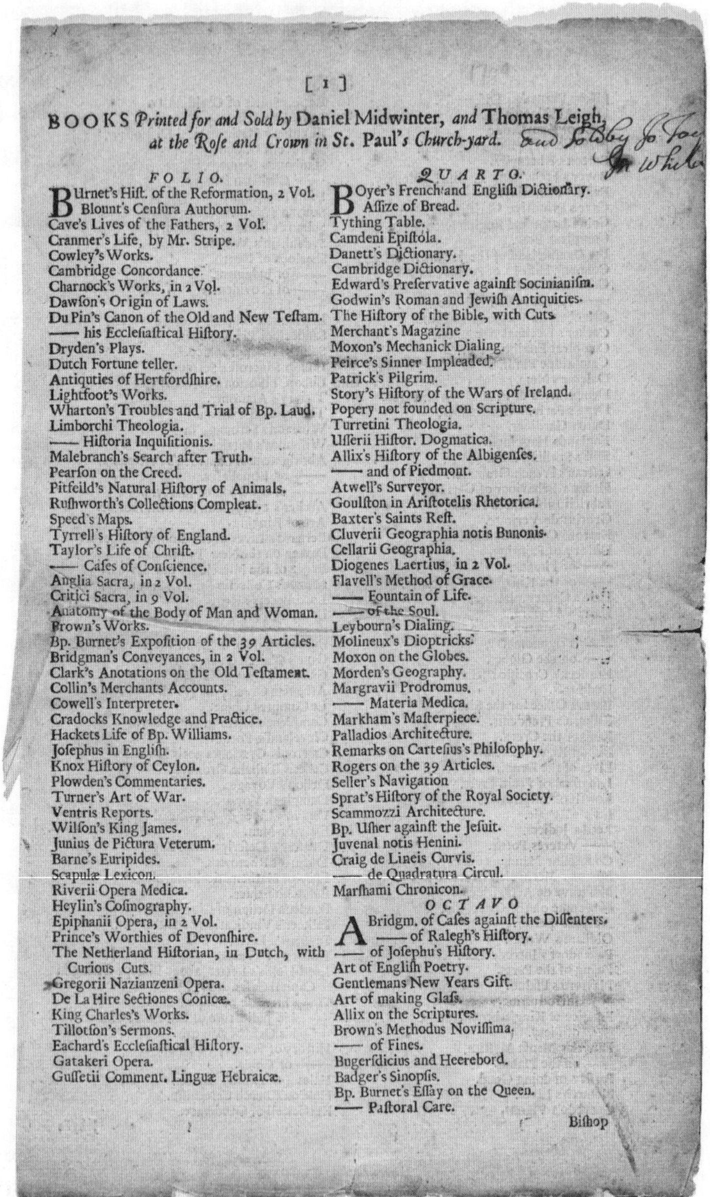

[1]

BOOKS *Printed for and Sold by* Daniel Midwinter, *and* Thomas Leigh, *at the Rose and Crown in* St. Paul's *Church-yard.*

FOLIO.

Burnet's Hist. of the Reformation, 2 Vol.
Blount's Censura Authorum.
Cave's Lives of the Fathers, 2 Vol.
Cranmer's Life, by Mr. Stripe.
Cowley's Works.
Cambridge Concordance.
Charnock's Works, in 2 Vol.
Dawson's Origin of Laws.
Du Pin's Canon of the Old and New Testam.
—— his Ecclesiastical History.
Dryden's Plays.
Dutch Fortune teller.
Antiquties of Hertfordshire.
Lightfoot's Works.
Wharton's Troubles and Trial of Bp. Laud.
Limborchi Theologia.
—— Historia Inquisitionis.
Malebranch's Search after Truth.
Pearson on the Creed.
Pitfeild's Natural History of Animals.
Rushworth's Collections Compleat.
Speed's Maps.
Tyrrell's History of England.
Taylor's Life of Christ.
—— Cases of Conscience.
Anglia Sacra, in 2 Vol.
Critici Sacra, in 9 Vol.
Anatomy of the Body of Man and Woman.
Brown's Works.
Bp. Burnet's Exposition of the 39 Articles.
Bridgman's Conveyances, in 2 Vol.
Clark's Anotations on the Old Testament.
Collin's Merchants Accounts.
Cowell's Interpreter.
Cradocks Knowledge and Practice.
Hackets Life of Bp. Williams.
Josephus in English.
Knox History of Ceylon.
Plowden's Commentaries.
Turner's Art of War.
Ventris Reports.
Wilson's King James.
Junius de Pictura Veterum.
Barne's Euripides.
Scapulæ Lexicon.
Riverii Opera Medica.
Heylin's Cosmography.
Epiphanii Opera, in 2 Vol.
Prince's Worthies of Devonshire.
The Netherland Historian, in Dutch, with Curious Cuts.
Gregorii Nazianzeni Opera.
De La Hire Sectiones Conicæ.
King Charles's Works.
Tillotson's Sermons.
Eachard's Ecclesiastical History.
Gatakeri Opera.
Gussetii Comment. Linguæ Hebraicæ.

QUARTO.

Boyer's French and English Dictionary.
Assize of Bread.
Tything Table.
Camdeni Epistola.
Danett's Dictionary.
Cambridge Dictionary.
Edward's Preservative against Socinianism.
Godwin's Roman and Jewish Antiquities.
The History of the Bible, with Cuts.
Merchant's Magazine
Moxon's Mechanick Dialing.
Peirce's Sinner Impleaded.
Patrick's Pilgrim.
Story's History of the Wars of Ireland.
Popery not founded on Scripture.
Turretini Theologia.
Usserii Histor. Dogmatica.
Allix's History of the Albigenses.
—— and of Piedmont.
Atwell's Surveyor.
Goulston in Aristotelis Rhetorica.
Baxter's Saints Rest.
Cluverii Geographia notis Bunonis.
Cellarii Geographia.
Diogenes Laertius, in 2 Vol.
Flavell's Method of Grace.
—— Fountain of Life.
—— of the Soul.
Leybourn's Dialing.
Molineux's Dioptricks.
Moxon on the Globes.
Morden's Geography.
Margravii Prodromus.
—— Materia Medica.
Markham's Masterpiece.
Palladios Architecture.
Remarks on Cartesius's Philosophy.
Rogers on the 39 Articles.
Seller's Navigation
Sprat's History of the Royal Society.
Scammozzi Architecture.
Bp. Usher against the Jesuit.
Juvenal notis Henini.
Craig de Lineis Curvis.
—— de Quadratura Circul.
Marshami Chronicon.

OCTAVO

Abridgm. of Cases against the Dissenters.
—— of Ralegh's History.
—— of Josephu's History.
Art of English Poetry.
Gentlemans New Years Gift.
Art of making Glass.
Allix on the Scriptures.
Brown's Methodus Novissima.
—— of Fines.
Bugersdicius and Heerebord.
Badger's Sinopsis.
Bp. Burnet's Essay on the Queen.
—— Pastoral Care.

Bishop

5.24 The 1704 catalogue of Daniel Midwinter and Thomas Leigh, who purchased the Moxons' stock. In the *folio* column, Remmelin's *Survey* is listed as "Anatomy of the Body of Man and Woman." (London, 1704), sig. *$_1$^{r}. Bodleian Library, Oxford University.

that is, Estienne's figures demonstrate vivisections. His are corpses actively engaged with their landscape, resting a hand on a descriptive cartouche while holding another arm aloft to illustrate the composition of the major abdominal muscles, or wearily prying back a leaf of skin to expose the glands around the throat.[37] What is more, Estienne arranges several of his figures in sequence, so that turning the page of the book turns the page of the body as well. Nowhere is this more striking than in the series of brain dissections.

There are eight cranial sections in all, and in each case the man is looking down so that the top of his head is presented in a vertical plane to the viewer. There is something inherently unsettling about this graphic and contexualized dissection, not least because the male figure (I hesitate to call him a corpse because he seems so active) appears to be readily assisting the invisible anatomist. In some of the plates there are distant observers; others take place in the middle of nowhere. What we never see, in any of the sections of the skull and brain, is a hint of the man's facial features.

In the first image, he holds a pole to steady himself as he kneels in a landscape sparsely decorated with alien blooms. In the background something is suspended from a tree branch; closer inspection reveals it to be the man's scalp. This is remarkably different from the sections Vesalius had published less than a decade prior. Naturally, several of the Vesalian images echo the lifelike poses seen in Estienne's work. Yet the hard anatomical work in the *Fabrica* takes place in the pages where organs and systems and bones have been prised out and arrayed carefully so that they do no more than hint at the life that once existed. For Vesalius and his direct descendants, adherence to natural form only served the empirical anatomist so far; a realistic disembodied head was a concession to the idea that the body being dissected needed to be recognized as having once been human. We are given traces of identity: the shape of the man's ear and the line of his nose, the fact that he was mustachioed, how his forehead and eyes would have looked – these are all details that Estienne, by contrast, withholds from the viewer of his cranial anatomies, even as he pours a richer context of both natural and bodily surroundings into the image.[38] For instead he provides us not just with the anatomy itself but also with the necessary trappings and leavings that would have attended the event.

Thus, in Figure 5.19, it is as if Estienne has followed the man as he removed the top of his skull, to hang it on a neighbouring branch as a literal skull-cap. The man then bends over to show the *dura mater* and vessels covering the top of the brain. In the next image, he is seated, holding the explanatory cartouche, *dura mater* likewise discarded on the bench beside him, to bring the folds and gyri of the cortex into view, and to reveal some of the layers of the *pia mater* gently peeling themselves from the interhemispheric fissure like pages in a book or the unrolling of a scroll. The unfurling develops a telling contrast between the brain as a textual space, covered in as yet indecipherable scribblings, and its secret hidden depths, opening out to the viewer-reader, promising (but forever holding back) an explanation of the wonders concealed therein.

Even in two-dimensional anatomical images, the unrolling or opening-up of flaps was not an uncommon way of depicting the successive layering of skin and membranes. Vesalius employed this technique, and so did his copiers after him, and the technique survived well into the seventeenth century. An odd Parisian fugitive anatomy from the late seventeenth century, which seems to derive some of its surrounding dissections from Remmelin's text, presents, in the words of its cataloguer, "a very eccentric representation of the meninges (folded back from the brain like pages in a sample-book)."[39] The plate, showing bodies in several poses and surrounded by various body parts in different stages of dissection, is especially notable for the way in which the meninges have been *folded* away from the brain. The "pages" have been labelled: *epiderme, derme, graisse, panicule, prienme,* &c.; and their foldings-back from the central line take both Vesalius and Estienne to an extreme. This method very likely represents an attempt on the part of the artist to preserve the complexity of the layers separating the brain of the cadaver from the inquisitive eye of the anatomist, and yet, as with Estienne's depictions, the figures here are so lively that we are encouraged to think of them as living, interacting entities, rather than as lifeless empirical proving-grounds for anatomical knowledge. With the layers arrayed as they are, cascading neatly over the figure's temple and ear, one might easily be able to put them back together again, risking no damage to the *cerveau* beneath beyond the fact of its having been seen.

The fourth of Estienne's brain sections emphasizes the connection between brain and book even more explicitly. Here, unlike most of the other figures in this series, the man is found inside a room, propped up in a chair, his back facing us. In the corner of the room, a wooden bench holds a portion of the lobes of the upper cortex that has been sliced off again.[40] The lines of perspective, though somewhat inexact, would seem to point the viewer toward the brain, but the eyes of the viewer – as well as the eyes the dissected body presumably still possesses – are drawn to the upper left-hand corner and especially to a set of items on a shelf.[41]

Metonymically, the head has been tilted towards a small book, placed between a candle and a flask. This connection may be coincidental, but given the weight of other additions to the various landscapes – gibbets, anatomical instruments, people, brain sections – the presence of the book points to and underscores a connection between brain and book, between the layers of the one and the leaves of the other.

As each successive section of the brain reveals itself, as we are drawn further into the living labyrinth of the soul, the chance of glimpsing that animating spark appears to present itself ever more tantalizingly. To put a Derridean spin on the book metaphor: "Exhibiting, baring, stripping down, unveiling … this is an old routine: the metaphor of truth, which is as much as to say the metaphor of metaphor, the truth of truth, the truth of metaphor."[42] With each turn of the page of Estienne's book, another layer is added to or removed from the part of the body in question. Each set of images, then, tells a story: first, the arraying of the skeleton with nerves and sinews, veins and arteries, organs, muscles, and finally skin; then the undoing of the torso, muscle by muscle, intestines there then gone, kidneys and liver and stomach and reproductive system shown in turn; then ribs, lungs, trachea. Finally, we reach the brain, sectioned slice by slice, before concluding with the woman's womb and reproductive organs. Build a body, watch it form, then disassemble it part by part – this is how clockmakers were said to learn their art. Each layer, then, corresponds to a page of the book, such that the book itself becomes a body, one or several, pressed together. Slicing the bound pages unlocks the work carried out by the anatomist's scalpel, allowing the reader to pry into the secret corners of the body, taking (perhaps) as much (erotic?) joy in witnessing as Estienne's figures do in participating.

Moxon's Choice

To begin at the end: Joseph Moxon appears to have stopped publishing by 1684,[43] and his son James would eventually take over the press and business. A year after his father's death, James Moxon began to reissue a number of his father's works, starting in 1692 with a new edition of *Mathematicks Made Easie…* This volume bears an engraving of the elder Moxon, and above him we see a shelf indicative of the books that he had published. From top/right to bottom/left, we read: "Mathema. Dictionary. / Mechanick Exercises. / Use of the Globes. / Practi. Perspective. / Copernican Spheres / W. Cor. of Errors / the English Globe. / Water Works. / Tutor to Astrology / Christologia. / Vignola. / Euclidi Curiosi / Anatomy / Mechan

Dyalling / Mathema Jewel. / 3. Or. of P. Letters / Sa. Geogrphy [*sic*] / U. of the Quadrant. / U. of the Astro. Cards / U. of Ne. Bones." Taken as a set, these titles encapsulate Joseph Moxon's legacy as a draughtsman and innovative printer invested in novel scientific and mathematical developments.

But among these, the *Anatomy* sticks out. Although Joseph Moxon (and, later, his son) printed broadly scientific and mathematical texts, neither had previously printed – and nor would they afterwards publish – any anatomies apart from the English edition of Remmelin's *Catoptrum microcosmicum*. It is thus almost certain that the *Anatomy* above Joseph's head recalls the fact of his having (repeatedly) issued "Englished" versions of Remmelin's famous book. When James disposed of a quantity of his father's leftover stock in 1698, the *Exact Survey* is the only anatomical or medical piece listed in the stock summary.[44] A 1704 catalogue of books printed for and sold by Daniel Midwinter and Thomas Leigh – who apparently took over some part of James Moxon's inventory[45] – includes in its list of folio volumes an "Anatomy of the Body of Man and Woman," referring to the *Survey*. Only one other anatomy title is listed in the catalogue, Daniel Tauvry's *A New Rational Anatomy* (1701). Not only does this book appear after Joseph Moxon's death, but it is less an anatomy in the familiar mold of Vesalius, Crooke, or Remmelin than an application of mechanistic philosophy to anatomical study.[46] (Notably, Thomas Willis's *Receipts for the Cure of All Distempers* [1701] is also included in this later list, and this may not have been coincidence, given the important and unexpected way Willis's neuroanatomical figures appeared in the 1695 *Exact Survey*.[47]) Moreover, the *Exact Survey* is the only medical or anatomical work in Midwinter and Leigh's inventory to be published in folio – other folio items are maps, histories, and volumes of collected works. Consequently, its repeated publication by the Moxons signals a curious departure in terms of size and subject matter.

To say nothing of its elaborate design: after all, printing the *Catoptrum* for an English audience was a serious undertaking, especially in comparison to other texts from this printing house. While anatomical books and flap anatomies were reasonably popular at the time, the decision to produce so complicated and visually arresting a piece of paper engineering was unusual for someone who was not remotely a publisher of medical books. Thus, for Joseph Moxon to print this set of anatomical illustrations is itself something of a wonder. So what could have compelled the printers to publish the *Exact Survey* over and over again, and the public to purchase it? Why would this anatomical curiosity appeal to a map-maker and tradesman in scientific books? And once we have addressed these questions, we need to return to another, which has been posed earlier: What led to the exceptional 1695 edition of the *Exact Survey*, published after Joseph's death, in which the original text underwent major revision – including, as we have seen, the extraordinary replacement of its images of the brain?

Joseph Moxon's professional biography, taken together with the list of titles above his commemorative portrait, provides a clue in that it establishes the context in relation to which his choice of the *Catoptrum* needs to be understood. To fit a complex anatomical text among books on astrology, geometry, geography, mathematics, and mechanics frames that work very differently from how it might have been situated among its flap forebears. Had we found the *Exact Survey* catalogued as it generally is today, for example, with Vogtherr's fugitive sheets, alongside Vesalian texts, and next to practical manuals of physick and chirurgery, we could continue to read this ornate flap anatomy as a peculiar document in the history of medical thought. But it reads very differently when seen amid the oeuvre of the Moxon printing house. Certainly, the 1695 *Exact Survey* post-dates James Moxon's reissue of *Mathematicks Made Easie* ..., but the image of his father below the set of exemplary titles for which he was responsible arguably forces us to rethink the nature of Joseph Moxon's original interest in Remmelin's text back in 1670 as well.

There is little by way of information regarding when and how Joseph acquired plates from the *Catoptrum*; generally the anatomy is left out of accounts of his publishing history,[48] as well as that of his son.[49] Given that he spent a portion of his childhood in Holland and returned there for two years in 1652, it seems likely that Joseph purchased the Dutch strikes of the *Catoptrum* during his travels abroad. By the time he left for Holland as an adult, Moxon had spent the Civil War years with his father – also called James – publishing Protestant/Puritan religious pamphlets,[50] and then, around 1650, he turned his attention to map-making.[51] 1655 saw his first solo venture, a translation of Barozzi's *Vignola*, for which he is, as Carey S. Bliss enthuses, "the translator, the engraver, the printer, and finally the publisher and bookseller!"[52] In 1661–62, his growing reputation as a map- and globe-maker earned him the title of Hydrographer to the King; in his petition for this position he claims he possesses "the desire ... to benefit his Native Country with the most exact and perfect Waggoner [that is, a navigational treatise] in the English Tongue that is yet extant in any Language whatsoever."[53] Moxon's insistence on the exactitude and perfection of his work, not only in English but in "any Language whatsoever," is characteristic of the image he sought to

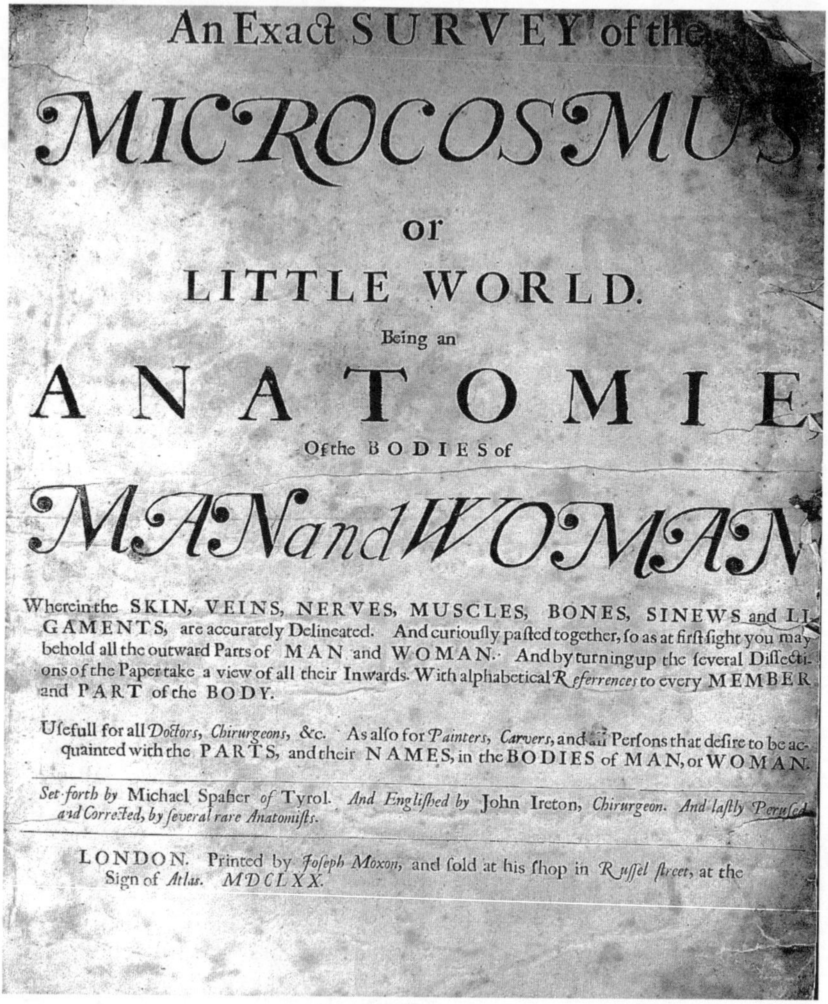

An Exact SURVEY of the

MICROCOSMUS

or

LITTLE WORLD.

Being an

ANATOMIE

Of the BODIES of

MAN and WOMAN.

Wherein the SKIN, VEINS, NERVES, MUSCLES, BONES, SINEWS and LIGAMENTS, are accurately Delineated. And curiouſly paſted together, ſo as at firſt ſight you may behold all the outward Parts of MAN and WOMAN. And by turning up the ſeveral Diſſections of the Paper take a view of all their Inwards. With alphabetical *References* to every MEMBER and PART of the BODY.

Uſefull for all *Doctors, Chirurgeons,* &c. As alſo for *Painters, Carvers,* and all Perſons that deſire to be acquainted with the PARTS, and their NAMES, in the BODIES of MAN, or WOMAN.

Set forth by Michael Spaher *of* Tyrol. *And Engliſhed by* John Ireton, *Chirurgeon. And laſtly Peruſed and Corrected, by ſeveral rare Anatomiſts.*

LONDON. Printed by *Joſeph Moxon,* and ſold at his ſhop in *Ruſſel ſtreet,* at the Sign of *Atlas.* *MDCLXX.*

5.25 The title page of Remmelin's 1670 *Survey*, listing his friend Michelspacher (sometimes given as "Michael Spaher") as author and John Ireton as translator. Johann Remmelin, *An exact survey of the Microcosmus, or Little World ...*, trans. John Ireton (London: Joseph Moxon, 1670). The Wellcome Library, London. EPB G. O/S F.2277.

cultivate, namely, in Adrian Johns's words, that of the Vitruvian "master printer."[54] The signatories on the back of his petition claim to be "most of us Professors [i.e., practitioners] of Mathematicks" who are eager for Moxon's globes, as these tools will be "very advantagious to Navigation and other Mathematicall studies."[55] As Graham Jagger points out, the support of such luminaries demonstrates Moxon's ability to surround himself with contacts to his advantage, as the signatories are mostly Oxford Royalists.[56] By this time, then, Joseph Moxon had positioned himself as a particular sort of craftsman, one who was elite in the knowledge of his trade. While his publishing output slowed initially after his accession to the position of Royal Hydrographer – Jagger finds nothing of note between 1660 and 1665 – over the twenty years of his tenure Moxon would be very nearly prolific, putting out almost forty volumes, either "as printer, publisher, translator, or author."[57]

During this time, Joseph became especially well-known for his maps and globes, attracting the likes of Samuel Pepys, who records visiting Moxon's shop to commission globes.[58] Globes, not unlike the *Catoptrum*, were complex objects to produce. The particular type of globe Moxon specialized in was the pocket globe, which consisted of a globe of the earth encased in a separate hollow sphere, the inside of which showed a celestial map. It seems likely that he was the first to market, if not to manufacture, this variety of globe.[59] The utility of such a globe, as D.U. Bryden explains, lies in the fact that

> lining the case with celestial gores allows the sky to be depicted as seen from Earth, that is on the apparently concave surface of the bowl of the heavens. The conventional celestial globe [i.e., a standard convex globe] requires observers to imagine themselves outside the celestial sphere. The asterisms of the stars provide a "god's eye-view," rather than a "man's eye-view," somewhat confusingly for the learner and for making comparisons with a two-dimensional star chart. Moxon, however, merely pasted onto the concave surface gores suitable for a conventional convex celestial globe.[60]

Thus while Moxon succeeded in bringing an important innovation to England, he did not perfect the pocket maps; as such, they tended to be curiosities seemingly intended for students and educated laypeople.[61] In this regard, the case of the pocket globe was not unlike that of the visually fascinating but – despite its title – often scientifically inaccurate *Exact Survey* of 1670. Moxon did, however, partner with Roger Palmer, Earl of Castlemain, to publish the latter's *Essay on the Use of the Celestial and Terrestrial Globe* in 1679, and to produce the so-called "English" globe, an immobile version much-beloved of the mathematicians of Moxon's day. In a letter of testimony signed by Isaac Newton and Henry More among other Cambridge greats, the academics praise the English globe for making possible "Spherical Trigonometry" in a fashion that "Vulgar Globes" do not, and "exhibit[ing] to sight the Solution of many Questions at once."[62]

Moxon's map-making prowess, along with his connections to people such as Robert Hooke, earned him the interest of the Royal Society;[63] in 1664 the Society's treasurer was ordered "to pay for a globe ... which Dr. Wren should choose at Mr. Moxon's."[64] Though this is not such an unusual crossing of paths given the intellectual pursuits of both men, it does seem almost fated in retrospect, for 1664 is, of course, the year that Willis's *Cerebri anatome* appeared, with its elegant engravings of the brain by Wren. More than thirty years would pass before Wren's illustrations appeared in miniature in the *Survey*, but this meeting marks at least a virtual inception of the relationship between the Moxons and Willis's brains.

Also of note from the perspective of print innovation was Moxon's 1659 *Tutor to Astronomy and Geography*, which contained, among other things, a clear description of how to read a globe, how to use a quadrant and an astrolabe, and "Ancient Stories of the several Stars and Constellations. Shewing the Poetical Reasons why such Various Figures are placed in Heaven. Collected from Dr. Hood."[65] The text deploys multiple fonts, in-line images of varying sizes, and sophisticated mathematical and astrological symbols. Strikingly modern in its layout, alongside its wealth of technical information it seems to offer a style guide for the potential that print would come to have. And in this respect, it calls to mind the innovations in print that Vesalius and his heirs introduced to the anatomical textbooks in the complex interaction of text and image on the page. Moreover, the inclusion of a "poetical" rationale for cosmic order reminds us of a parallel in the vitality of anatomical etymologies stretching back to Homer, as well as of George Puttenham's enthusiastic declaration that because poets were "the first that entended to the observation of nature and her works" and also "the first observers of all naturall causes & effects in things generable corruptible," they were likewise "the first Astronomers and Philosophists and Metaphysicks."[66] Such an integrative approach, both in connecting together varying intellectual systems and in connecting word and image, seems to have primed Moxon to take on the task of printing the *Survey*.

In 1677, Moxon began to issue installments of his *Mechanick Exercises*, the work for which he is perhaps best known, and in 1684 he published the entirety of the piece in a single volume. The *Mechanick Exercises* constitutes a vital piece of the history of printing and publishing – a trade the secrets of which, as Bliss notes, had not seen the light of day in print – in revealing the essential operations of what Moxon calls "Handy-Works":

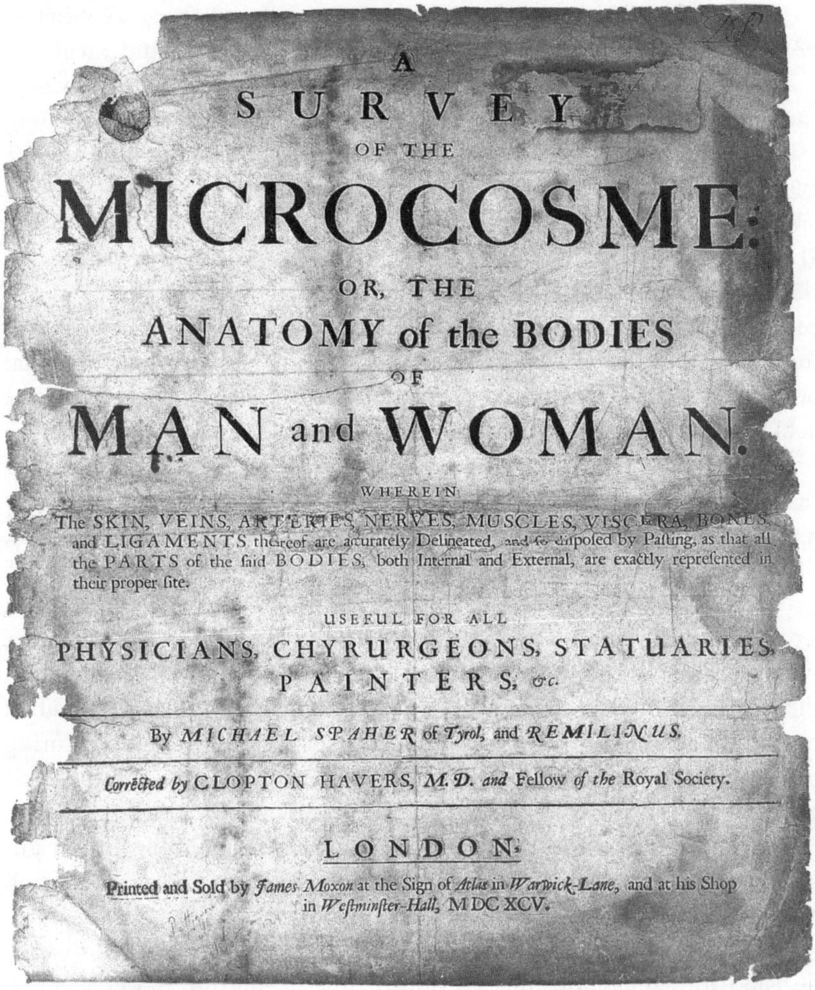
A

SURVEY

OF THE

MICROCOSME:

OR, THE

ANATOMY of the BODIES

OF

MAN and WOMAN.

WHEREIN

The SKIN, VEINS, ARTERIES, NERVES, MUSCLES, VISCERA, BONES, and LIGAMENTS thereof are accurately Delineated, and so disposed by Pasting, as that all the PARTS of the said BODIES, both Internal and External, are exactly represented in their proper site.

USEFUL FOR ALL

PHYSICIANS, CHYRURGEONS, STATUARIES, PAINTERS, &c.

By *MICHAEL SPAHER* of *Tyrol*, and *REMILINUS*.

Corrected by CLOPTON HAVERS, *M. D. and* Fellow *of the* Royal Society.

LONDON:

Printed and Sold by *James Moxon* at the Sign of *Atlas* in *Warwick-Lane*, and at his Shop in *Westminster-Hall*, MDCXCV.

5.26 The title page of Remmelin's 1695 *Survey*, listing (again) Michelspacher and Remmelin as authors, James Moxon as the publisher, and Clopton Havers as the "corrector." Johann Remmelin, *An exact survey of the Microcosme…* (London, 1695). The Wellcome Library, London. EPB G. O/S 66323/F.

That *Geometry, Astronomy, Perspective, Musick, Navigation, Architecture, &c.* are all excellent Sciences, all that know but their very Names will confess: Yet to what purpose would *Geometry* serve, were it not to contrive rules for *Handy-Works*? Or how could *Astronomy* be known to any perfection, but by Instruments made by

> Hand? What *Perspective* should we have to delight our Sight? What *Musick* to ravish our Ears? What *Navigation* to Guard and Enrich our Country? Or what *Architecture* to defend us from the Inconveniencies of different Weather, without *Manual Operations*? Or how waste and useless would many of the Productions of this and other Countries be, were it not for *Manufactures*.[67]

Though it appears to have been relatively unacknowledged in its day, the clarity of Moxon's prose is such that, as Bliss testifies, "it has been and will continue to be a handbook for anyone attempting to learn and understand the work of the handpress printer … Every later printers' manual or guide from that day until the close of the nineteenth century owes a debt, whether admitted or not, to Moxon's work."[68] Not only was it the first book published in England in the form of a serial, but it also served as Moxon's formal introduction to the Royal Society.[69] Diarist John Evelyn (whose wife Mary, recall, wrote witheringly of Margaret Cavendish), then serving as "commissioner for the rebuilding of St. Paul's Cathedral" – a project in which Christopher Wren was intimately involved – wrote a letter of introduction on Moxon's behalf in 1678 to Sir Joseph Williamson, then president of the Royal Society.[70] Moxon presented Williamson with the first six numbers of *Mechanick Exercises*,[71] and not long after, albeit not without some controversy, Moxon was elected as the first (and, during the seventeenth century, apparently the only) tradesman to join the Society.[72] During his time in the Society, he became better acquainted with Hooke, who often mentions Moxon in his diaries – including an intriguing entry in 1677 (which Bliss cites with glee): "Wild, Aubery, Merret, Moxon, &c., here to see comet but missed it. Drank 2 bottles claret."[73]

It is against this background that we need to consider Joseph Moxon's decision to begin publishing editions of the *Catoptrum* under the title *An Exact Survey of the Microcosme: Or, the Anatomy of the Bodies of Man and Woman*. Under his name, three English editions were published, in 1670, 1675, and 1691, and little was altered of their material. The images are taken from the Dutch strikes, as is evident from the designations "De Eerste plaet," "De tweede plaet," and "De Dende plaet" on their top corners. From an anatomical perspective, Moxon altered next to nothing of the original, leaving most of the ornamental scrolls, globes, cartouches, and religious symbols untouched. The only significant alterations were the removal of some surrounding verses of Scripture and Latin poems, the elimination of the flaps beneath the name of God in the *Visio Prima*, and a scrubbing of the plinths upon which the man and woman stand in that

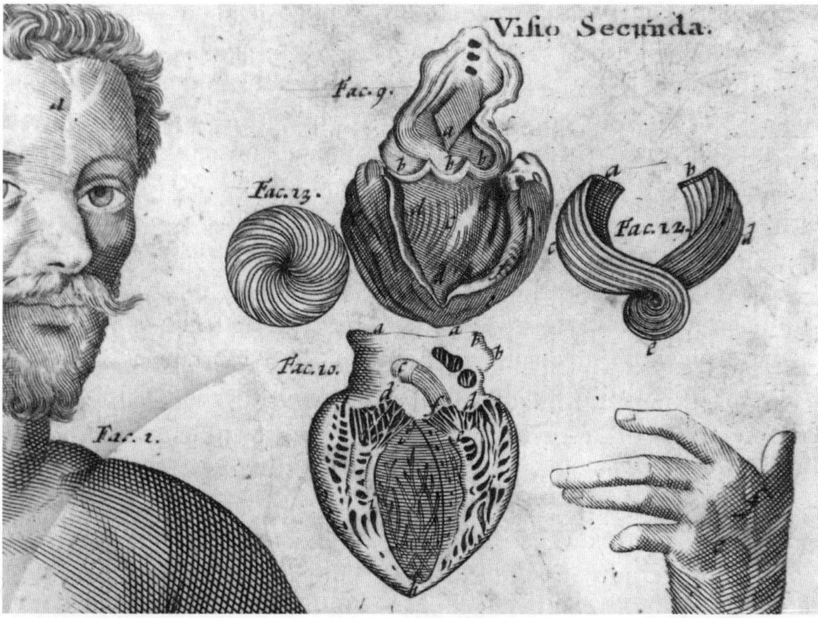

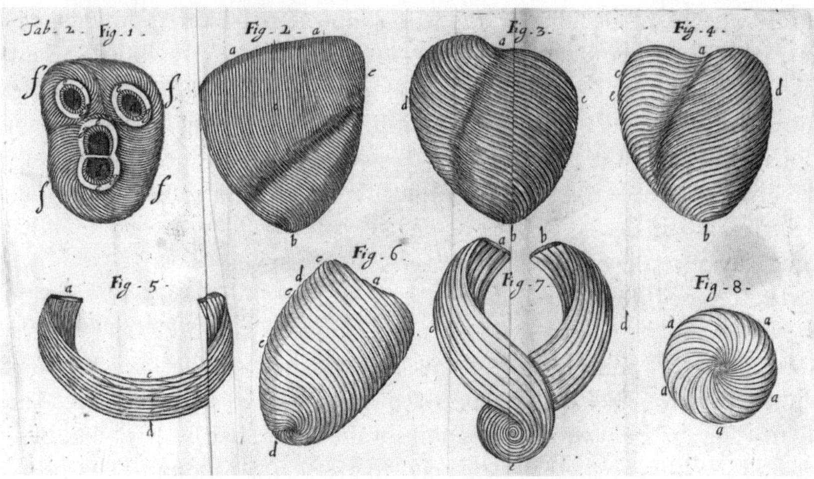

5.27 Top: A detail from Remmelin's 1702 *Survey*, showing a dissection of cardiac muscles. Johann Remmelin, *An exact survey of the Microcosme …* (London, 1702), Visio Secunda. The Wellcome Library, London. EPB G. O/S F.350.

Bottom: A plate showing a cardiac muscle dissection by Richard Lower. Richard Lower, *Tractatus de corde* (Amsterdam, 1669), Tabula 2. Yale University, Harvey Cushing / John Hay Whitney Medical Library, New Haven, CT.

image. The text was ostensibly "Englished" by one John Ireton, who has proven difficult to trace.[74] The tables, too, largely follow (but abridge) the *Catoptrum*'s labelling and explanations.

But given Moxon's interests and publishing history, the *Exact Survey* was nonetheless a very different text from Remmelin's. For, as its new title already suggests, the text that Moxon produced needs to be viewed not (or, at least, not only) as a work of anatomy, but as a work more in line with the cosmographies, histories, maps, and theories of mechanical engineering that populate his catalogues. Reading the Englished *Catoptrum* in this way alters its purpose from the title onwards: a *survey* of the human body recalls not so much an anatomical endeavour as a map-making one, and it is perhaps no accident that that word stands out in a larger, capitalized font on the 1670 title page. At stake is not simply a laying bare of the body's interior.[75] Rather, the body thereby becomes yet another tool among several for charting the natural universe and man's place in it – not primarily as a casually informative *memento mori*, but as instead an exhaustive catalogue of modern information about the human body: no longer a visual aid for lay anatomists *manqué* but a scientific tool comparable to the other materials that Moxon wrote and published about. From this perspective, the pocket globe peculiar to his workshop offered a three-dimensional cartographical counterpart to the *Exact Survey*. And indeed these shared, as we have seen, a remarkable disregard for inaccuracies, despite formal assertions of exactitude and utility. Just as pasting earlier conventional gores onto the underside of the pocket globe confused rather than enlightened the user, so too did the anatomical information of the *Catoptrum* remain unaltered, despite being over half a century out of date. What counted above all was how the enterprise was framed.

Hence, the title page points out that it is in this *Exact Survey* "Wherein the SKIN, VEINS, NERVES, MUSCLES, BONES, SINEWS and LIGAMENTS, are accurately Delineated. And curiously pasted together, so as at first sight you may behold all the outward Parts of MAN and WOMAN. And by turning up the several Dissections of the Paper take a view of all their Inwards. With alphabetical *References* to every MEMBER and PART of the BODY." Intriguingly, what the printer lists is not a promise that the body's organs will be displayed, but that the body's *systems* will be delineated. Everything has been pushed into a category, even the parts themselves, which exist not as individual entities (viz., the heart, the lungs, the stomach, &c.) but as a set of *parts*. That these parts will be *delineated* points to their status as representational figures, the use of the verb suggesting the connection to drawing and tracing, on charts or maps in particular.[76]

The title page then goes on to explain that the bodies so represented will be "useful for all Doctors, Chirurgeons, &c. As also for Painters, Carvers, and all persons that desire to be acquainted with the PARTS and their NAMES." As for the first two categories, despite the fact that, anatomically at least, the bodies are not entirely accurate, if we think of the pictures as akin to rough road maps of the systems of the human body, their utility for practicing physicians and surgeons seems at least possible. For painters and "carvers" the value of the bodies is perhaps more obvious, as their lifelike positions and the variety in the decorative surrounding material lends them some element of artistic appeal. The claim to utility – whether actually realized or not – is accentuated by the few alterations that have been made. As noted above, some of the images have been scrubbed, the most metaphysical of the flaps eliminated, and the marginal poems and verses dropped. Thus where Remmelin's *Catoptrum* aimed to be a mirror in which the viewer might see and know himself, the first three editions of the *Exact Survey* already move away from the aim of sheer contemplation, rhetorically to emphasize use-value instead. That the title page plays up the labelling of parts and stresses the fact of providing "alphabetical references" to these is further evidence of such a shift. The (specious) emphasis on exactness, on the correctness of its representations, further draws this text into the ambit of Moxon's other works: the point of his rebranded anatomy is that it will be of navigatory use to its reader, so the correspondence of parts must needs be "exact and perfect." In short, it would be the anatomical equivalent of the "Waggoner" he had promised the king.

The New Book of the Brain

In spite of the title page's claims, Moxon, Ireton, and the "several rare Anatomists" who had ostensibly "perused" and "corrected" the first three editions of the *Exact Survey* had not in fact done much by way of updating the anatomical information. With the final 1695 edition, published under the aegis of James Moxon after his father's death, this would change dramatically. The task of revision would fall to Clopton Havers, who seems to have been peculiarly well-connected in London medical circles. The title page of the 1695 edition is reproduced in Figure 5.26, and the differences are, if not earth-shattering, at least noteworthy. The font has become a bit more workmanlike, albeit the promises are nearly the same, though the claim to exactitude has been retracted: all of the parts of the body have still been "disposed by Pasting," such that said parts and bodies, "both Internal and External, are exactly represented in their proper size." It is somewhat peculiar for the book to claim it will portray its figures "exactly represented in their proper size," i.e., that they will be represented according to scale, which is not a common (or indeed a practical) claim for an anatomy to make, and it is of course entirely untrue in this case especially: one has only to glance at the floating disembodied eye and ear on *Visio Prima* to find fault with the notion. But such quibbles aside, the alterations are indeed substantial and consequential, even beyond the insertion of Willis's pictures of the brain, which is of most interest to us here.

Havers's own main anatomical work, *Osteologia Nova, or some New Observations of the Bones*, was published in London in 1691, and in it he describes what are known today as the Haversian canals, narrow tubes running through compact (or cortical) bone. His modern champion Jessie Dobson terms him a "pioneer in osteogeny," observing that his book "occupies a permanent place in literature since it … [is] filled with bold speculations based … on careful reasoning from the results of ingeniously contrived experiment and accurate observation."[77] Such an adventurous spirit would have served Havers well in his task of updating the *Exact Survey*, and it is clear that attention to detail helped effect the transformation we are considering here.[78]

After studying at St Catharine's College in Cambridge, Havers was admitted as an Extra-licentiate at the College of Physicians in London, which allowed him to practise medicine outside London and permitted him to apply to practise in Cambridge or Oxford.[79] He was awarded an MD at the University of Utrecht in 1685 and elected to the Royal Society in 1686 (when Pepys was presiding); at this point he became licensed to practise medicine in London.[80] At this point, too, Havers began to focus his own research on the human skeleton. During the late 1680s, Havers presented papers to the Royal Society on bone and marrow morphology and diseases of the cartilage. These papers became the foundation for his *Osteologia Nova*, which ran to three editions in Latin, and was generally well thought of in England.[81] On the strength of this work, he was selected by the Company of Barbers and Surgeons to deliver the Gale Anatomy Lectures in 1698.[82] By all accounts (which alas are not many), he was well respected. Preaching at his funeral on 29 April 1702, Lilly Butler eulogized him thus: "To give a just Character of him, is to describe a *perfect and upright Man*, one of the brightest Examples in his Sphere, of all the Graces and Vertues of our most holy Religion." Of Havers's work, Butler says that "so happily successful was he in these Studies, that I find the learned of the Faculty, both natives and Foreigners, most highly applauding those early and singular Fruits of them publish'd in his *Osteologia*."[83]

Envisioning exactly how Havers and the Moxons might have come to know one another requires some creative speculation. Joseph Moxon was ejected from the Royal Society in 1682, years before Havers was elected, but the Society still seems the most likely point of contact between the two men. It may also have been possible that the contact was arranged by Havers's publisher for his *Osteologia Nova*, Samuel Smith, who operated out of St. Paul's Church-Yard, an energetic stone's throw from Warwick Lane, where James Moxon was working by 1691. One can imagine that Havers might have been eager to see some of his work reach an educated lay public and that his careful eye might have noted immediately the potential for updating presented by the *Exact Survey*. Indeed, as William Munk observed in an 1878 roll of the Royal College of Physicians, Havers was "a minute and very accomplished anatomist."[84]

At any rate, however Havers encountered the Moxons, he was the choice for preparing the "corrected" edition of 1695. To this end, Havers drew upon the work of several prominent anatomists, and, in so doing, he sheds light upon the international web of connections and influences that was growing ever more intricately convoluted at the end of the seventeenth century. Russell notes of the 1695 *Survey* that it was, of all the extant versions of the *Catoptrum*, "the most significant from the anatomical point of view," given the figures that were replaced.[85] To begin with, Havers dispensed entirely with the scrolls and god-cloud on *Visio Prima*; the cartouche, colchium inscription, microchristus, and serpent on *Visio Secunda*; and the cartouche, phoenix and mountain, and serpent on *Visio Tertia*. (That he retained the colchium flower and smoke-cloud for the man and woman respectively is probably due to their practical function as screens for the genitalia.) The more significant departures concern, of course, the actual anatomies. Russell points out that most of the more Vesalian dissections have been removed in favour of engravings from works by Richard Lower, Reinier de Graaf, and Steven Blankaart.[86] De Graaf was best known for his studies of the pancreas and the female reproductive system – as he remarked in a letter to Franciscus Sylvius, "I preferred most to dissect the pancreas and the genital parts since I found there constantly new things, unremarked by anatomists before me."[87] In addition, de Graaf studied the reproductive habits of rabbits very closely indeed; this in turn led him to discover the significance of the ovarian follicles.[88] De Graaf's 1668 *De virorum organis generationi inservientibus, de clysteribus et de usu siphonis in anatomia* and his 1672 *De mulierum organis* are likely candidates for Havers to have read. Blankaart was a specialist in pediatric medicine and anatomy in Amsterdam, and was among the first to have issued a medical periodical in Holland (the *Collectanea*).[89] He tended, as G.A. Lindeboom points out, to compile medical information rather than to write it himself, though his work, especially the *Tractatus novus de circulatione sanguinis per fibras, nec non de valvulis in iis repertis* (Amsterdam, 1676), had some impact.

The most crucial influence on Havers, however, and the most important for our concerns here, was Richard Lower, first a student and later a colleague of Thomas Willis. There is little doubt that Lower's famous work of post-Harveian cardiology, the 1669 *Tractatus de corde*, was a source of inspiration for Havers as he amended the *Exact Survey*. Lower had treated Charles II for uremia, before his own Protestant, anti-Popish beliefs caused him to fall from court grace.[90] Havers doubtless met Richard Lower at the Royal Society, or he would at the very least have been familiar with Lower's critical work on blood transfusion. In fact, Lower gave demonstrations of what are thought to have been the first animal-to-animal and the first animal-to-human blood transfusions in England.[91] Lower was also a crucial collaborator in Willis's neuroanatomical work – though, as K.J. Franklin vigorously argues, Lower was the more careful anatomist, and in this Willis was "so much his [i.e., Lower's] scientific inferior."[92] Willis certainly thanks Lower generously in the introduction to the *Cerebri anatome*:

> Richard Lower, a doctor of outstanding learning and an anatomist of supreme skill. The sharpness of his scalpel and of his intellect … enabled me to investigate better both the structure and functions of bodies, whose secrets were previously concealed. [Soon] the cerebrum and its appendages seemed clearly revealed and thoroughly explored … [then, in] the dissection of the [cranial] nerves, the really wonderful dexterity of this worker and his untiring perseverance were conspicuous in the extreme.[93]

Given that Havers had borrowed some images from Lower already, and given the latter's close association with Willis's work on the *Cerebri anatome*, it seems highly likely therefore that Lower was the source of the new miniaturized images of the brain that replaced the originals from Remmelin's *Catoptrum*.

By substituting the older, obsolete flaps with Willis's and Wren's meticulous depictions of the brain – images that marked, even if he did not realize it, the threshold of a new discipline – Havers transformed an out-of-date but spectacular enterprise, re-imagining a text that itself pushed the very boundaries of the printed page and that urged us to rethink what books could do and how bodies could be represented. Demystifying Remmelin's mystical vision, Havers nonetheless sought, with the aid of Moxon's publishing house, to preserve the text's material so as to turn it into a surveying instrument, and to bring it into line with modern medical knowledge.

The 1695 *Survey*, as the *Catoptrum* before it, is a complex feat of paper engineering, but its images lift the original text out of the ashes of obsolete medical information and into a curiously modern documentation of the human body. For a textual anatomy to be informative, it has to render the body mappable. By the end of the seventeenth century, innovations in printing had already produced ever more sophisticated keyings of text to image, but the flap anatomy as a genre had not exactly kept pace with medical advances, and this held, as we have seen, for even such a visually striking work as the *Catoptrum*. Correcting it, Havers and James Moxon breathed new life into the figures, completing a trajectory only hinted at in Joseph Moxon's earlier editions by moving the *Survey* out of the realm of anatomy and into a broader network of charting and mapping, and doing so specifically in the context of the trade and making of books. To be sure, the *Catoptrum* had never been a *purely* anatomical text – its myriad lines of scripture and its baroque iconography already differentiated it from standard anatomical depictions of its day. Yet first Ireton (whoever he might have been), hesitantly, and then Havers, drastically, would "scrub" the original engravings of their emblematic functions, suggesting the extent to which the English publishers operated with a sense of its potential as a bearer of newer scientific information. We must, then, think as they did: the *Survey* is a compression of the map of the human microcosm into the pages of a book.

And of no other organ in the revised text is this truer than the brain. One of the few remaining emblematic images on the second *visio*, the skull still bears its traditional association with death and mortality. But lift the skull in the 1695 *Survey* and we are confronted with the birth of a new organ. No longer the repository of an ossified, arcane knowledge, the opened skull reveals instead the emergence of a new book in which to read the soul's progress.

With the publication of the 1695 *Survey*, the flap anatomy has shaken off the dust of its own history and become a complex tool for understanding the body. Accomplishing this requires rethinking anatomical representation – and, specifically, how the brain should be represented – as well as the possibilities of the printed page. As transcendent as the *Catoptrum* and *Survey* are, they never get entirely beyond the bounds of their physicality – they always will be, that is, books. What they contain, on the other hand, and what they allow the reader to observe, in their own very different ways, is the living, and newly rendered readable, contemporary picture of the brain. The closest modern correlates are the MRI and CAT/PET scans, which also force an eerie (if not ghostly) confrontation with what was once a vital and intimate part of another human being. But these scans also remove any (literally) visceral connection, given the distancing inherent in the technology involved, to create a clinical, sanitized, and bloodless scene of dissection. The brain-books of the seventeenth century are likewise bloodless, but they are arguably more human and more immediate: the interaction necessary between reader and anatomy echoes strikingly the interaction between reader and book. In turning each page, one delves more deeply into an ever-sought-after but ever-evasive connection, whether between author and text, or between soul and body.

CODA

The Brain of Dr. Deijman

The surviving section of Rembrandt's 1656 *Anatomy Lesson of Dr. Deijman* stands as an unsettling partial emblem for this book as a whole. To some degree this partiality is enforced, because the image at this book's gateway exists today only as fragment, most of the canvas having been damaged in 1723. And yet as remnant it has the virtue of heightening the focus on its central action, which is also the central theme of *The Subtle Knot*: the anatomical investigation of the brain. One might suggest, too, that this later anatomy functions as a pendant to Rembrandt's earlier, much better known *The Anatomy Lesson of Dr. Tulp*, which Francis Barker's now-classic *The Tremulous Private Body* used as its cover, seeing in it the epitome of its central argument concerning the collapse of personal psychological interiority onto public bodily performance.[1] Anatomically, Tulp's lesson concerns the dissection of an arm, but, as Dolores Mitchell and others have noted, the complex network of artistic and scientific patrons represented as standing behind the anatomical performance stare not at the exfoliated limb, but at the book to which the hand points – that is, at the text that has fixed the cadaver as knowledge-object.[2] The body is thus simultaneously central and secondary to the discussion of its own subjectivity, both in the painting and in the broader cultural literature surrounding it. By contrast, the Deijman fragment, while working within a related set of discursive practices, forces us to focus on the strange, shining brain, and emphasizes the tension or opposition between the cranium's plenitude and the emptiness of the body cavity. Tulp's *Lesson* fixes in print a kind of *techné* regarding the body, evidenced in the clinical detachment of Dr Tulp as well as in how the angling of the corpse distances the viewer from the anatomical exercise. Deijman's *Lesson* echoes instead many of the concerns at the heart of my own critical study: it

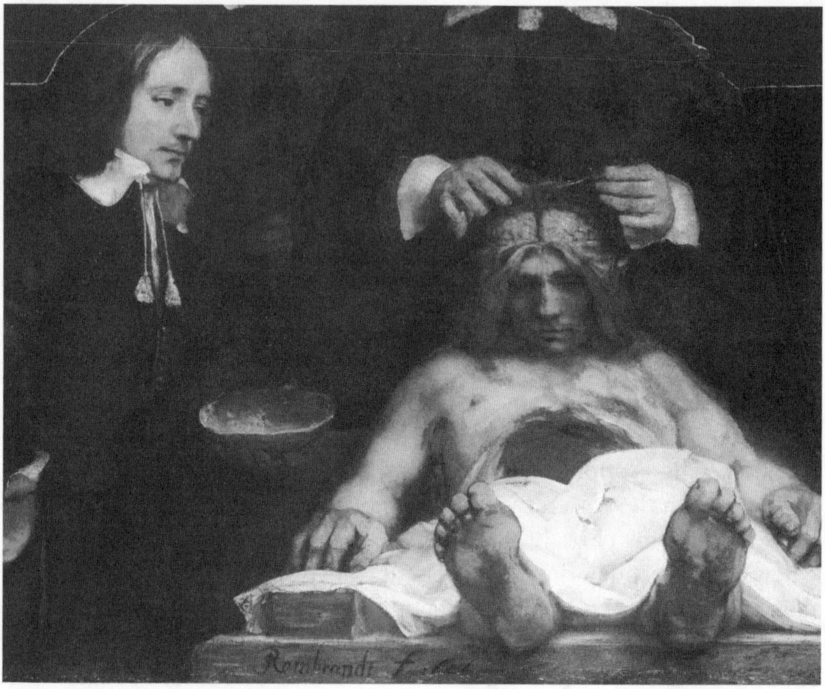

C.1 *The Anatomy Lesson of Dr. Deijman*, Rembrandt van Rijn, 1656. The Amsterdam Museum.

seems to foresee the migration of selfhood from the body's core to the brain, and perhaps concomitantly, to trace a path from the body's physical to its intellectual and moral centre – even as it holds out the promise of rendering the latter's intangible impulses fully material. Much like the metaphors examined in the preceding pages, Rembrandt's second great anatomical depiction emphasizes the passage or the bridge between the immaterial and the material, the figurative and the literal.

This is not to say that Deijman's anatomy forgets the anatomical and theological past out of which the brain emerged as the primary site to investigate whatever it is that makes us human. As Jonathan Sawday has rightly noted, the connection between the cranial dissection here and that of Vesalius (and Crooke after him) cannot be overstated.[3] The far reach of the Flemish anatomist is apparent not only in the tilt of the cadaver's head and in how the meninges have been unfurled gently downward, but most importantly in the fact of the brain's exposure as an integral part of

C.2 Rembrandt's original sketch (1656) for the *Anatomy Lesson of Dr. Deijman*. The painting was heavily damaged in a fire in 1723; the sketch allows us to imagine the composition of the painting, including the seven other figures not visible in the painting fragment. The Amsterdam Museum.

a (once-living) body. The debt to antecedent theological and aesthetic traditions is equally evident. The face and pose of the cadaver – that of the thief Joris Fonteyn – recall most directly (as critics have noted) Mantegna's *Cristo in scurto* (c. 1478–85), leading, for instance, to a reading of this particular cadaver as expressing the transformative miracle of Christ.[4]

But I am less concerned in this afterword with traversing again what is already well-trodden ground. And so, largely bracketing the painting's socio-historical grounding as well its theological overtones, I wish to speculate instead on what Rembrandt's painting may reveal to an empathetic viewer in the context of the birth of neuroscience. Consequently, I attend here less to the lost whole, which can be (and has been) projected from a surviving sketch by Rembrandt of the painting's entire composition, than to what the remnant makes central: the brain's abundant materiality. Being a fragment, this painting has often been overlooked as part of Rembrandt's corpus, but once viewed in its own right its depiction of the brain nonetheless exerts an almost repulsive fascination upon the viewer. Whereas we nowadays tend to view the organic brain through a media-mediated distance – a comportment that the Tulp anatomy expresses by turning watchers into readers – the brain that Deijman approaches still seems alive with the promise of hidden knowledge both dangerous and sublime.

In the fragment, Deijman stands behind the corpse, his own head outside the artificial frame, his hands manipulating the cadaver's brain, while an assistant stands to the side, cradling a (literal) skullcap.[5] What draws the eye here are the pinks and reds lain against an immediate background of blackish-brown and a broader perspective featuring pale greys and whites. Indeed, it is easy for the glistening brain – perhaps even more so than the bloodless and empty abdominal hole – to elicit disgust in the viewer. While Sawday is correct in claiming that the image recalls Vesalius, there is one crucial distinction: the Vesalian image, because it is an engraving, is necessarily bloodless, whereas Rembrandt's craniotomy gives us unnerving flesh and blood, and living fingers with which to plumb the body's depths.[6] The fragment of Deijman's *Lesson* thus manages to provoke its viewer, extending an invitation to participation in the image and the knowledge-making tradition of which it is an integral part.

Through this, Rembrandt lends visual reality to an impulse that arguably underlies all the different metaphors we have explored thus far in this book, a desire to render substantial the workings of the human mind, to make tangible thinking and the life that thought enables: whether we seek with Donne and the English anatomists to identify the "subtle knot" that "makes us man"; or, with *Hamlet*, to uncover the brain's hidden workings by materially disturbing the organ; or to translate, as Harvey does, the labour of thought into the pangs of actual birth; or to exult, as does Cavendish, in the brain's mechanical capacity to generate metaphors concerning itself; or, finally, with Remmelin, to turn the brain into a book, whose secrets can be read. Each of these metaphors offers a partial window onto the brain and how it functions, much as the six blind men describe the elephant through whichever part is closest to each of them.

The invitation to the viewer, his or her inclusion in the anatomical scene, is borne out by the painting itself, and, curiously, the fragment emphasizes the position of the viewer in a way that the complete painting might not have accomplished. As noted above, critics have identified in the painting an allegorical connection to Christ, seeing in its oddly foreshortened perspective upon the corpse an echo of earlier representations of Christ's body. Yet it seems equally possible that the foreshortening also reflects anatomical practicality: the upper part of the body has been propped up to facilitate Deijman's dissection of the brain. The propping-up has the added effect of thrusting the cadaver's feet forward and over the edge of the table – and the table itself here performs a kind of double duty, standing both as table and also as a partial wooden frame for the central scene of anatomy. Rembrandt's name and the date of the painting's composition appear along the table's edge, marking a collapse between the depicted world of the anatomical theatre and the real world of which the viewer is a part. The feet further mark this collapse, pushing as they do into the viewer's space. The feet form two points of a triangle of which the brain is the apex, and the assistant's line of sight confirms and reinforces the centrality of the exposed organ. With this set of minor details – wooden table/frame, painterly inscription, lowly feet pointing to capital brain – Rembrandt's painting directs the gaze of the viewer by integrating him or her into the anatomical scene.[7]

Benjamin Binstock's remarks regarding the earlier *Tulp's Lesson* are relevant here as well. For him, Rembrandt "introduces an unprecedented, dramatic narrative action and physical subordination among the figures. Yet he mediates these elements through chiaroscuro, relating the figures to the viewer through space." The effect is "to displace a rhetorical gesture [i.e., a lecture by Dr Tulp, whether moralizing or anatomical in nature] by a visual demonstration [i.e., of the workings of the human body], or speech by sight." Such a displacement has in turn a reflexive quality, for, "[j]ust as the surgeons see themselves in the body of the corpse, the viewers see themselves in Rembrandt's painting."[8] Binstock notes, too, the way in which this reflexive quality operates in the surviving sketch of Deijman's *Lesson*: extrapolating from Rembrandt's rough plan for the whole painting, he observes how "the balustrade of the anatomical theater [has been] curved around to circumscribe our space, so the viewer [is] 'sutured' into the center of the anatomical theater as the third participant performing the dissection."[9]

And yet, we might speculate that, in yielding the viewer a place at the anatomization, in giving visual expression to the impulse to know the brain, Rembrandt testifies equally to the hubris of this desire. Even if the viewer is visually integrated here, the tactile details seem to urge us to do more than look, to reach out and touch the organ, as Deijman's quasi-disembodied hands are doing. No doubt, to give in to that urge would be an artistic breach – we would encounter only dried paint and not the brain itself, and so the illusion of a stilled life would be broken – but the invitation to transgress the limits of the viewer's decorum also alerts us to the breach of anatomical decorum in the painting itself. As Sawday notes, several procedural proprieties have been violated here: the anatomist Deijman "should have moved to the thorax, and demonstrated the heart and lungs to the audience" after removing the abdominal viscera; "[i]nstead, the anatomist has removed the top of the skull, which has been given to the young assistant whose attention is fixed on [Deijman]'s 'reflection' (or turning inside out) of the dural membrane which surrounds the brain itself."[10] These breaches reveal for Sawday a broader link to Cartesian anatomy, specifically the search for the pineal gland: Rembrandt eschews established anatomical procedure precisely in order "to show the primacy of the brain in the investigation of what it was that constituted the human being," and doing this involved forcibly turning the viewer's attention to the search for the organ that housed the soul.[11]

However, as we saw in chapter 1, what frustrated early modern anatomists time and time again was the seeming fragility of the brain and the ephemeral nature of the structure that putatively housed the soul, whether pineal gland or *rete mirabile*. Does the excavation of the body's interior lead only to nothingness, to what is sought evading the anatomist's grasp? Surely, this question haunts the upper half of the fragment too, even if the contrasting fullness of the cranium seems to promise a knowledge that the corpse's body has thus far failed to yield. For is it not the case that Deijman and the viewer might always be too late to uncover the soul in the brain? On the one hand, then, Rembrandt's painting realizes for its viewer the fantasy also offered by Remmelin's paper anatomies: a reader or viewer could enter into the frame of the image, literally or figuratively manipulate the figure, and become an anatomist by proxy. On the other, Rembrandt's fragment perhaps also exposes that desire to know *as* a fantasy, the mocking hollowness of the body cavity reverbing (to borrow Kent's felicitous phrase defending Cordelia) the only too real possibility of failure.

The tension or opposition between these two perspectives may be peculiarly early modern, but what remain resolutely modern are the *data* revealed about the brain by Deijman's lesson. Indeed, according to a recent article in the medical journal *Neurosurgery*, the painting's depiction of the brain displays remarkable accuracy when compared with the anatomical knowledge of the period (especially where Vesalian imagery was concerned), even approaching the kind of fidelity that is obtained via fMRI or other modern digital imagery. As Frank IJpma, Norbert Middelkoop, and Thomas van Gulik put it, "Rembrandt's beautiful color reproduction of the brain in the painting corresponds very well with the real appearance of the brain at dissection. Therefore, it is unmistakable that Rembrandt used a real dissection specimen while painting his anatomy lesson."[12] What Rembrandt captures here is in fact scientific discovery in its context, with the means and method on display alongside the information itself.

Let us recall here the connection chapter 5 made between early modern anatomical representations of brain dissections, specifically the flap anatomy and its visual ramifications, and the idea conjured up of the brain as a book, with layers and meninges figured as pages a reader could manipulate. As we have seen in visual creations by Remmelin, Vesalius, Estienne, and others, so here, too, Rembrandt's image of the brain strongly resembles a book: the Doctor of Medicine stands with scalpel poised directly above what would be the book's spine, its two halves on either side the result of unfolding the "leaves" which constitute the brain. As with Hamlet, the brain is itself a book, to be leafed through and read. Whereas the hand in the Tulp *Anatomy Lesson* points outside itself to the externalization of its interiority in the printed word, here the organ is itself its own representation. The question then becomes: what is revealed when we read the brain? What sort of meaning is at stake? As this book has argued, the brain both houses and expresses the soul in this period, and among the things inscribed upon it are all the early modern – and often recognizably modern – anxieties surrounding the soul. In these depictions of the brain, the soul thus becomes metaphorically linked to the meaning of the text: just as textual interpretation is the condition of linguistic meaning, so too, in order to have any hope of comprehending the soul, we must read and interpret its book, the brain. Rembrandt's painting is a living image, in that it does not merely invite the viewer to look, but it also demands that the reader interact with, read, and interpret the unveiled brain.

Whether or not Rembrandt's picture of the brain can be confirmed by the powerful tools of modern digital imagery, as IJpma et al. suggest, what Deijman's *Lesson* – and the paper anatomies of Remmelin, Vesalius, and others – share with modern mechanisms for imaging the brain is precisely this sense of uncovering and unfolding in order to lay bare information. Then as now, that desired information was fundamentally about the inner workings of the human mind, its ability to sense and perceive, its faculty for remembering, its capacity for imagination. Yet where the viewer was integral to this procedure in the early modern period, today the viewer, insofar as he or she can be posited to exist at all, is less viewer than intellectual mediator or interpreter of data. That is, while there is a thematic and visual connection between Remmelin's anatomies and modern fMRI images in that both involve a gradual uncovering of the brain via manipulatable layers, paper anatomies were aimed at a broadly educated audience, whereas the fMRI requires specialized training for a proper reading. The viewer of the image has not been erased so much as transformed, but with this transformation comes a necessary distance, leaving open the question of how, and whether, modern neuroimaging can be recaptured and brought into the longer cultural moment described by this book. Rembrandt's painting preserves the distinctiveness of the early modern period even as it anticipates later techniques for visualizing the brain – but can these two motivations be reconciled?

An answer might well lie in one of the critical concerns addressed in the Introduction, namely grounded or embodied cognition, and to that end I want to conclude with an example in fiction that itself looks both backwards and forwards, if bizarrely. The following quote, from a short story by Thomas Burke titled "The Hands of Mr. Ottermole," recalls the mantra of *tota in toto, et tota in qualibet parte* we saw play out in chapter 1, the claim that the soul might fulfill its duties in a distributed manner, across the body. This notion is itself an echo of the medieval idea that the soul might well flourish in the Augustinian fingernail as in the brain.[13] At the same time, and in a connected manner, this distribution is also the precursor to the sort of embodied cognition I described in the Introduction, in which the brain is acknowledged as the hub of selfhood but is nourished by a complex web of sensations and perceptions stemming from the body's interaction with the world, on both physical and psychological/emotional levels. The following quote captures both these potentialities:

> Everybody knows that we can't control the workings of our minds. Don't they? Ideas come into our minds without asking. But everybody's supposed to be able to control his body. Why? Eh? We get our minds from lord-knows-where – from people who were dead hundreds of years before we were born. Mayn't we get our bodies in the same way? Our faces – our legs – our heads – they aren't completely ours. We don't make 'em. They come to us. And couldn't ideas come into our bodies like ideas come into our minds? Eh? Can't ideas live in nerve and muscle as well as in brain?[14]

The setting for Burke's story is just-post-Victorian London, recalling the frantic search for Jack the Ripper; there has been a series of mysterious stranglings across the city, and an intrepid young reporter is probing various suspects. He has a moment of inspiration – "Came to me all of a sudden. Quarter of an hour ago. And I'd felt that we'd all been blind. It's been staring us in the face." – and approaches the sergeant in charge of investigating the case, one Ottermole – whose hands, as it turns out, are in fact responsible for the murders.[15] Confronted by the reporter, the sergeant responds as above, seeking to rationalize his actions.

Where *do* ideas come from? The usual fantasy has the mind controlling the body, the soul restraining the material desires of the ever-wayward body. The body here wants to kill, but the mind ought to keep it from doing so. Ideas ought to enter into the mind in order to maintain this balance. But for Sergeant Ottermole, a rift has developed in this otherwise orderly co-operation of parts. Ideas, as body parts, are potentially untraceable, and as such they become free and even radical agents, rendering an "othered" mind inside an equally "othered" body. And if neither mind nor body has a distinct origin, then ideas might well "come into our bodies like ideas come into our minds," living "in nerve and muscle as well as in brain."

Burke's narrator doesn't comment further on Ottermole's pronouncement, but we might imagine how Rembrandt, Deijman, Remmelin, and their contemporaries would have responded. For the early modern period, ideas could well travel the body, but their deployment was never arbitrary; rather, they were always coordinated by the soul or mind operating in and through the brain. Nerve and muscle conveyed ideas, but it was the brain that directed them. Ottermole's hands could not have acted independently; in a sense such an idea would have been foreign to a period that both created and benefited from a Cartesian nervous system. It is no mistake that Deijman's hands are touching Fonteyn's brain, as this makes explicit the necessary connection between material thought (in the brain) and implied action (in the hands). Both embodied cognition and the figure of metaphor could repair this rift – the disjunction between thought and flesh – and both must do so, as we have seen, by utilizing first a neurocentric model of the relationship between mind and body. For better or for worse – but always, it is to be hoped, for better – all roads lead inexorably back to the brain.

Notes

Introduction

1 Crooke, *Mikrokosmographia*, sig. C_2^r.
2 Geminus, *Compendiosa*, sig. B_2^r.
3 Puttenham, *The Arte of English Poesie*, 6.
4 Indeed, as Glickstein observes, "Brain tissue is soft, with a gelatinous consistency. If the brain were placed unsupported on a hard surface such as a table, gravity alone would distort it" (*Neuroscience: A Historical Introduction*, 13).
5 Daston and Park, *Wonders and the Order of Nature*, 13–14.
6 Spiller, *Science, Reading, and Renaissance Culture*, 6.
7 Puttenham, *The Arte of English Poesie*, 5.
8 Crooke, *Mikrokosmographia*, sig. Qq_5^r, C_1^v.
9 Arisosto, *Orlando Furioso*, trans. Waldman, 281. I am utilizing a modern English translation of the original Italian text here for illustrative purposes; below I will consider an early modern English translation.
10 Ibid., 419.
11 The original Italian reads: "Poi giunse a quel che par sì averlo a nui, / che mai per esso a Dio voti non ferse; / io dico il senno" (Canto 34, verse 82), the relevant word here being *senno*, which suggests wits or sense.
12 Ariosto, *Orlando Furioso*, trans. Waldman, 420.
13 Ariosto, *Orlando Furioso*, trans. Harrington, sig. Bb_3^r.
14 "Era come un liquor suttile e molle, / atto a esalar, se non si tien ben chiuso" (Canto 34, verse 83).
15 Ariosto, *Orlando Furioso*, trans. Harrington, sig. Bb_3^r.
16 See the OED, s.v. "body, n.," esp. I.1.a, I.1.b, &c., all of which have to do with solid corporeal entities.

17 Astolfo eventually "snuffs up" a portion of the *wit* for himself, and then takes Orlando's missing "moyst and soft" body back to earth, to apply it rhinally to the mad hero, thereby restoring Orlando from *furioso* back to a relative *normale*.

18 Richardson and Steen, "Literature and the Cognitive Revolution," 3.

19 Crane, *Shakespeare's Brain*, 9.

20 For a solid summary of the ongoing debate about the precise number of neurons in the human brain, see Herculano-Houzel, "The Human Brain in Numbers."

21 Arguably, this remains true today as well, even if we gloss over the connection: the systems-oriented description anticipates its functional role.

22 Stevens, "Sacred Heart," 263. See also Gross, *A Hole in the Head*, 25–52.

23 Stevens, "Sacred Heart," 264.

24 See, for example, Skinner et al., eds., *The Cambridge History of Renaissance Philosophy*, 464.

25 Marvell, "A Dialogue between the Soul and Body," in *Andrew Marvell: The Complete Poems*, 103–4.

26 See, for example, Bossy, "Medieval Debates of Body and Soul."

27 Crooke, *Mikrokosmographia*, sig. B_6^v.

28 See, for example, Arikha, *Passions and Tempers*; Lindemann, *Medicine and Society*; and Siraisi, *Medieval and Early Renaissance Medicine*.

29 Donne, *Sermons*, vol. II, 261–2.

30 See, for example, Brazier, *A History of Neurophysiology*; Clarke and Dewhurst, *An Illustrated History of Brain Function*; Glickstein, *Neuroscience: A Historical Introduction*; Poynter, *The History and Philosophy of the Brain and Its Functions*; and Smith et al., *The Animal Spirit Doctrine*.

31 See, for example, Latour, *We Have Never Been Modern*.

32 I.e., that disorder or chaos in a closed system will increase with time; that mass is the amount of matter an object contains, independent of the gravitational force exerted on it; and that acceleration is the change in velocity per unit of time.

33 Snow, *The Two Cultures*, 15.

34 See, for example, Paul B. Armstrong's cogent critique of Pinker, particularly of his 2003 *The Blank Slate*: "[Pinker's] avowedly opinionated chapter 'The Arts' (400–420) is … the least valuable section of his book, and it shows the dangers of a scientist's presuming to pronounce unilaterally about the humanities ('They didn't ask me,' [Pinker] admits, 'but by their own accounts they need all the help they can get' [401])." While Armstrong agrees with Pinker that there are limits to neuroplasticity, he points out that the book's "sweeping generalizations about modern art and contemporary literary criticism, offered

with much less precision, nuance, and rigorous argumentation than he feels necessary to provide when discussing scientific issues, have understandably and unfortunately alienated many in the audience of humanists he needs to persuade" (*How Literature Plays with the Brain*, 183n3). See also Liza Blake, who, in mentioning debates between Pinker and Leon Weiseltier in *The New Republic*, argues that Pinker and Weiseltier "are not debating two positions on the same question, but two different questions." That is, whereas (in Blake's words) Pinker "argued that the humanities could and should base themselves on the sciences," Weiseltier "argued for the separation of the sciences from fundamentally non-scientific questions" ("The Grounds of Literature and Science," 5).

35 Shapin and Schaffer, *Leviathan and the Air-Pump*, 4.

36 This example in particular has garnered so much attention that it is known as the "Proust Phenomenon." See for example Shepherd and Shepherd-Barr, "Proust Effect," in which they attempt to map out the episode's neural basis. See also Fornazzari et al., "Proust and Madeleine," which explores conscious episodic memory using Proust as a starting-point. Chu and Downes, in "Proust Nose Best," test the phenomenon by having subjects rate the quantity of information and its emotional quality of their controlled autobiographical memories. Several other studies take Proust's madeleine as shorthand for a particular memory pathway: Parker et al., "Odour and Proustian Memory"; Pontius, "Overwhelming Remembrance of Things Past"; Slotine and Lohmiller, "Modularity, Evolution, and the Binding Problem"; Saive et al., "A Novel Experimental Approach"; Todd and Anderson, "The Neurogenetics of Remembering." See also Bray, "Forgetting the Madeleine," who makes precisely this point, arguing that in order to read Proust (for example) as a "neuroscientific literary paradigm," we must "forget the madeleine … [and] forget the neat simplicity of the image" (42).

37 Crane and Richardson, "Literary Studies and Cognitive Science," 127.

38 Glenberg et al., "From the Revolution to Embodiment," 579–80. See also Aziz-Zadeh et al., "Congruent Embodied Representations," in which the authors use functional magnetic resonance imaging (fMRI) to examine activity in the premotor area of the cortex when subjects listened to descriptions of actions utilizing the hand, mouth, and foot, or watched short videos of the action described, observing activation of the premotor area in both scenarios, which suggests that the part of the brain that moves the mouth/hand/foot responds to visual or linguistic scenarios involving the mouth/hand/foot. See also Barsalou, who proposes "a perceptual theory of knowledge" in which sensory-motor areas of the brain are involved in cognition and allow for the construction of perceptual symbol schematics for efficiently interacting with

the outside world and with the individual's previously lived experience ("Perceptual Symbol Systems," 577). Finally, a study by Glenberg and Gallese builds on that of Aziz-Zadeh et al. to argue that all aspects of language require the integration of linguistic and sensory-motor areas of the brain. I am grateful to Barbara Luka for her thoughts about extended and grounded cognition.

39 Cohen, "Next Big Thing in English."

40 See, for example, Armstrong, *How Literature Plays with the Brain*; Crane, *Shakespeare's Brain*; and Massey, *Neural Imagination*. A notable exception is Rose, *Neurology of the Arts*, which, in essays by C.U.M. Smith and Julius Rocca, provides a culturally and historically embedded examination of neurological and artistic phenomena.

41 See, too, Claire Preston, who extends this list to include Galileo, Milton, and others who equally traversed in the early modern period what would later become "modern" cultural boundaries. *The Poetics of Scientific Investigation in Seventeenth-Century England*, 18–19.

42 Puttenham, *Arte of English Poesie*, 148–9.

43 Donoghue, *Metaphor*, 64.

44 Lakoff and Johnson, *Metaphors We Live By*, 28–9.

45 Hayles, *Chaos Bound*, 31.

46 See Hayles, *The Cosmic Web*, esp. 15–59.

47 Foucault, *The Order of Things*, 26.

48 Ibid., 27.

49 Ibid., 21.

50 Ibid., 22.

51 Ibid., 27. Foucault is here quoting Crollius, *Traité des signatures*, 33–4.

52 Blumenberg, *Paradigms for a Metaphorology*, 2–3.

53 Ibid., 3; emphases Blumenberg's.

54 Ibid., 5; emphasis Blumenberg's.

55 Preston, *The Poetics of Scientific Investigation*, 10.

56 Paul, "Vorschule der Ästhetik," 184. Cited in Adler and Gross, "Adjusting the Frame," 204; see 203–7 for a brief catalogue of pre–Lakoff and Johnson studies of the import of metaphor for even everyday language.

57 See, for the clearest example, Biagioli, "Stress in the Book of Nature."

58 Orazio Grassi, Jesuit mathematician.

59 It is worth noting that, when Galileo entered the service of the Grand Duke of Tuscany in 1610, he wished to be titled *philosopher* rather than *court mathematician*, which Drake finds peculiar; see Drake, "Introduction," 222 et passim.

60 Galileo, "The Assayer," 237–8.

61 Daston, "Galilean Analogies," 303.

62 Ibid., 309.
63 See chapter 5 for a discussion of the complex publishing history of the *Catoptrum*.

Chapter One

1 Donne, *Songs and Sonnets of John Donne*, ed. Redpath, 220. Unless otherwise specified, all references to Donne's poetry will be to this edition.
2 Popham, *A Looking-glasse for the Soule*, sig. A_{12}^{r-v}. Popham's primary concern is for the sanctity of his readers' souls. He entreats his readers to turn away from sin in order to avoid a last-minute deathbed contrition that may or may not be attended to by God. The "looking-glasse" is a means to exhort readers to examine their souls, making them better know themselves, the better to be holy. For similar – albeit far more serpentine – discussions of the proto-psychological and religious aspects of the soul, see: Böhme, *Forty Questions of the Soul*; and Crashaw, *Complaint or Dialogue*.
3 Woolnor, *The true originall of the soule*, 50–3.
4 Fletcher, *The Historie of the Perfect-Cursed-Blessed Man*, 8–9. No apparent relationship to Phineas Fletcher of *Purple Island* fame.
5 Henry More, *The immortality of the soul*, 67. Roughly, More aims to confirm the immortality of the soul by opposing Thomas Hobbes's theory of matter – often through quasi-mathematical proofs.
6 I say "most fully" here because another prevailing (though not necessarily oppositional) theory was *tota in toto, et tota in qualibet parte*, i.e., "all in all, and all in part": the soul exists wholly in the body and wholly in each part of the body. For an historical discussion of this theory, see Waddington, "All in All." For a rich examination of the role of the soul in shaping modern conceptions of the mind, see Porter, *Flesh in the Age of Reason*. Other structures which could be candidates for the location of the soul included the choroid plexus; see below for further elaboration.
7 Eggert, *Disknowledge*, 4, 158.
8 I recognize the anachronism of invoking a modern scientific discipline, but the word remains the most convenient term for the body of knowledge under consideration here. See Rousseau, *Nervous Acts*, and below, for elaboration on this point.
9 I locate my argument in the seventeenth century, though the difficulties of dating with precision the composition of Donne's poetry should be acknowledged. See Smith, "Preface," 13–16.
10 Linking these schemes was an underlying anxiety surrounding the extent to which the soul could be said to be embodied, and where and how that

embodiment took place. To describe the landscape of the material body, anatomists, too, would often invoke metaphorical relationships, for instance calling the liver a "fountaine" from which blood flowed as water, and the brain's ventricles "Labyrinthes and Meanders." See Banister, *The Historie of Man*, sig. X_3^r, and Crooke, *Mikrokosmographia*, sig. Oo_6^r. Subsequent references to Crooke will be made in the body of the text.

11 Sugg, *Murder after Death*, 149.

12 Puttenham, *The Arte of English Poesie*, 4–6.

13 Ibid., 149. While necessary, metaphors are also dangerous. As Puttenham complains, the figure is open to misuse: "But if for lacke of naturall and proper terme or worde we take another, neither naturall nor proper and do vntruly applie it to the thing which we would seeme to expresse, and without any iust inconuenience, it is not then spoken by this figure *Metaphore* or of *inuersion* as before but by plaine abuse, as he that bad his man go into his library and set him his bowe and arrowes, for in deede there was neuer a booke there to be found, or as one should in reproch say to a poore man, thou raskall knaue, where *raskall* is properly the hunters terme giuen to young deere, leane & out of season, and not to people: or as one said very pretily in this verse" (150).

14 I draw here upon N. Katherine Hayles's seminal work on the importance of metaphor to the interaction between science and modern literature. See, for example, Hayles, *The Cosmic Web* and *Chaos Bound*. See also Bono, "Science, Discourse, and Literature."

15 Sugg, *Murder after Death*, 2.

16 Roberts and Tomlinson, *The Fabric of the Body*, 125 et passim. For further delineation of the *Fabrica*'s importance, see: Brown, "The Renaissance," 28; Harcourt, "Andreas Vesalius," 28; and Hunt, "On the Localisation," 342–3.

17 By the time of the Italian Renaissance, the public performance of anatomy – as it was introduced by Mondino de'Liuzzi in the early 1300s – had solidly in place its own *modus operandi*: a professor would sit in his tall chair, slowly orating the text of a Latin anatomy, while below, a surgeon performed the actual dissection and an "ostensor" pointed out the parts of the body, as they were announced, to the attending students. For a well-paced and informative discussion of this history, see Carlino, *Books of the Body*.

18 Lind, *Jacopo Berengario Da Carpi*, 15. Berengario also appears to have been the more academic of the two men: "Berengario's dependence on authorities other than Mundinus, whom he frequently mentions, especially in connection with his *Commentary*, is also considerable and consistently acknowledged by exact reference to books and chapters; he sometimes differs with his authorities, as with Avicenna on the fourth part of the blood in the middle ventricle of the heart, something Berengario said he could not see

with his own eyes. He quotes Aristotle, Celsus, Galen, and Avicenna, and refers to other medical men, such as Herophilus; he mentions no works of contemporaries. Vesalius, it may be pointed out, refers to no one but Galen and Asclepius in the *Epitome* and then only to criticize" (15).

19 See Persaud, *The Early History of Human Anatomy*, 50–75; and Sugg, *Murder after Death*, 12–86. While my analysis of anatomical history here in some ways parallels Sugg's in *Murder after Death*, my focus here is more specific; I consider these key anatomical moments in the more limited context of neuroscientific history, whereas Sugg sketches the broader anatomical tradition.

20 For a clear account of this process, along with a helpful diagram, see Singer, *Vesalius*, xviii–xx. For Aristotle, one of Galen's key sources, the *spiritus* is "a subtle vapour or exhalation produced from blood and disseminated throughout the body by the arteries and nerves, which were assumed to be hollow. The source of all activity in the living body, *spiritus* was often referred to by Renaissance philosophers as the 'first instrument' of the soul." Cited in Schmitt et al., eds., *The Cambridge History of Renaissance Philosophy*, 469.

21 Rousseau, *Nervous Acts*, 26.

22 See Julius Rocca, *Galen on the Brain*, 250. Bartholomaeus was active from approximately 1220–1240.

23 Galen, *Claudii Galeni opera omnia*, vol. 3, 696–7.

24 Rocca, *Galen on the Brain*, 249. Also see Rocca's discussion of da Carpi's commentary on an older text by Mondino, 251–3.

25 For more on the medieval history of anatomy, see: Carlino, *Books of the Body*; Cunningham, *The Anatomical Renaissance*; and Roberts and Tomlinson, *The Fabric of the Body*.

26 Quoted in Lind, *Jacopo Berengario Da Carpi*, 458–60.

27 Lind, *Jacopo Berengario Da Carpi*, 146–7.

28 Note, too, the number of times da Carpi mentions forms of the word *subtle*; the import of this word will be discussed below.

29 Cited in Heseler, *Andreas Vesalius' First Public Anatomy*, 220–1.

30 Singer, *Vesalius*, 4.

31 Ibid.

32 Roberts and Tomlinson, *The Fabric of the Body*, 128.

33 See Singer, *Vesalius*.

34 The quasi-translation of the *Compendiosa* by Nicholas Udall actually also incorporates the text of Thomas Vicary's 1577 *A Profitable Treatise of the Anatomy of Mans Bodie* alongside images and descriptions from the *Fabrica* and the *Epitome*.

35 See Roberts and Tomlinson, *The Fabric of the Body*, 140. Of the *Compendiosa*, Roberts and Tomlinson observe that "two figures, of nerves and vessels, had

been copied [from the *Fabrica*] at significantly reduced sizes [while the rest had been lifted at full size]; they were on folding sheets in the *Fabrica*. Geminus engraved the figures on copper ('brass'), and they were accompanied by some of the relevant explanations from the *Fabrica* and by the text, slightly modified, of the *Epitome*."

36 Amusingly enough, however, in Geminus's dedication to Edward VI, he continually refers to the engravings as "mine," thus obfuscating the direct lineage between Vesalius's work and his own.

37 The desire for some almost-tangible link between soul and body was not without its attendant infelicities. More, for example, rather bitterly denounces Descartes's pineal gland theory as resting on "nothing but a poor silly contemptible knob" (More, *An antidote*, 59). Sugg notes of More's scorn that "[t]he recurrent dream of uniting those intangible but deeply felt certainties of human emotion with the pulsing fiber and tissue of the body has here turned unmistakably sour. The tantalizing essence so persistently conjured by writers and poets in the earlier seventeenth century required a remaining space of uncertainty and an associated cloud of mystery in which to survive" (Sugg, *Murder after Death*, 131).

38 The eventual rejection of the *rete mirabile* may have been due in part to the demise of the spirit theory, which William Harvey's work on the circulation of blood precipitated. It seems likely that, at least on occasion, a structure acknowledged to exist today, the circle of Willis, may have been mistaken for the *rete mirabile*.

39 Mitchell, "Donne's 'The Extasie,'" 97.

40 See, in particular, Sugg, "Donne and the Uses of Anatomy" and *Murder after Death*; but also Hirsch, "Donne's Atomies and Anatomies"; Lange, "Humourous Grief"; and Willard, "Donne's Anatomy Lesson."

41 Donne, *The Sermons*, II: 261–2.

42 For an intriguing and useful discussion of the problems of cross-disciplinary influence in the case of medical and literary sensibility in the eighteenth century, see Rousseau, *Nervous Acts*, 173.

43 Targoff, *John Donne*, 55.

44 For a catalogue of such arguments, see Graziani, "John Donne's 'The Extasie,'" 121–4.

45 Targoff, *John Donne*, 50. As Targoff notes, "however intimate the relationship may be between himself and his beloved, he seems to find the notion of exchanging spiritual parts unsettling" (64). Targoff's focus is almost wholly non-anatomical (she does not mention the brain or the *rete*, for example), but her analysis and mine are roughly complementary, approaching as they do the same topic of the complex body-soul intercommunication from different

directions. The difficulty derives in part from the paradoxical early modern idea of the soul, as something both completely indivisible and infinitely divisible, both matter and anti-matter, both distinct and indistinct from the human body.

46 *Pace* Grierson, who complains of "The Ecstasy": "There hangs about the poem just a suspicion of the conventional and unreal Platonism of the seventeenth century. In attempting to state and vindicate the relation of soul and body he falls perhaps inevitably into the appearance, at any rate, of the dualism which he is trying to transcend. He places them over against each other as separate entities and the lower bulks unduly." Donne, *The Poems*, ed. Grierson, II: xlvii.

47 Donne, *Sermons*, III: 235–6.

48 Sugg, *Murder after Death*, 149; Targoff, *John Donne*, 57.

49 See, too, Redpath's gloss on the poem's apex: "Just as blood 'labours' … to beget spirits as like souls as it can make them … so, on their side, souls must come down from the purely spiritual to feelings and powers within the range of the senses, otherwise human love cannot freely pass between a man and a woman." Donne, *The Songs and Sonnets*, ed. Redpath, 325.

50 Davies, *Nosce teipsum*, sig. D_3^v.

51 Marvell, *The Complete Poems*, 103–4.

52 Lewalski, *Protestant Poetics*, 352. For Traherne's program of study, see Inge, "Introduction," xx. For prior misattributions by Alexander Grosart and Bertram Dobell, see Margoliouth, "Introduction," xii; see also Bloom, *John Donne*, 69n1. This is a necessarily limited account of Traherne's corpus, as I am focusing here only on a particular instance of Traherne's relationship to metaphor. For one take on the context of Lewalski's reading, see Jordan, "Thomas Traherne and the Art of Meditation," 381n1.

53 Traherne, *Thomas Traherne, Poetry and Prose*, 60–1, ln. 174, 189–215.

54 Lewalski, *Protestant Poetics*, 356.

55 See OED, s.v. "subtle, a."

56 Ibid. A standard modern Latin dictionary gives the sense of *subtilis* as "finely woven, slender, fine": *Cassell's Latin and English Dictionary*, s.v. "subtilis, -e."

57 Davies, *Nosce teipsum*, sig. I_4^r.

58 Ibid., sig. F_4^r.

59 Sugg, *John Donne*, 150–1.

60 Banister, *The Historie of Man*, sig. Dd_4^r.

61 Banister (sig. Dd_4^v) does express some doubt about Galen's account of the exact location of the *rete*: "These [arteries] hath *Galen* described for the netlike folding … *Collumbus* notwithstandyng boldly affirmeth this (if it be any where at all) to be the marveilous nette: for no where (sayth he) shall you finde such foldes, through foldes, and interweavynges of the least, and

innumerable Arteries, els. But *Galen* did describe the marveilous nette whereof he maketh mention, to be above *Sphenoïdes*, where that *Glandule*, which receiveth every excrement begotten in the braynes, beyng to that office by Nature dedicated, is resident. But whosoever shall seeke the same where *Galen* hath described it, he shalbe frustrate of his purpose. For nothyng els shall he finde there, but certaine litle Arteries, ministryng lyfe to these partes, and to the bone *Sphenoïdes*."

62 Geminus, *Compendiosa*, sig. B_2^r.

63 Ali ibn Abbas al-Majusi, d. 994. In early modern English writing, he is known primarily as Haly Abbas.

64 Geminus, *Compendiosa*, sig. B_2^r.

65 Sugg, *John Donne*, 142.

66 This is virtually a direct quote from Galen; Crooke's inclusion of it as his own seems a rather flagrant disavowal of the Vesalian argument concerning the *rete mirabile*.

67 How the soul actually acts on the animal spirits is something we shall have to wait for Descartes to discover, and it is perhaps no accident that Crooke emphasizes the soul's close connection to the pineal gland – though, as Crooke contends, the pineal gland "is not particle of the Braine … but only lyeth upon it on the outside" (sig. Rr_6^v).

68 Significantly, though, it remains unclear whether Crooke's description refers to the *physical* matter of the brain or the *thought* of the brain; the "Labyrinthes and Meanders" could have to do with the structure of the brain as an anatomical object or with the problem facing anatomists and philosophers when they considered the brain as the receptacle of the soul. Crooke here seems to anticipate the modern (what we might call the post-post-Cartesian) elision between mind and body, but his ambiguity also marks an important step in the process of linking soul to body and, more specifically, to brain. By localizing the soul's functions in the brain, Crooke and his colleagues opened the door to the physical study of immaterial processes – and, crucially, pushed the question of where, precisely, the line between the physical and the immaterial lay.

69 OED, s.v. "curiosity," II.9–11.

70 Chapman, *Jesuitism described*, sig. E_2^v.

71 See the (brief) discussion of this passage by Sugg, who calls attention to Chapman's insistence that the *rete* cannot be anatomized. Sugg, *Murder after Death*, 96.

72 Donne, *The Poems*, ln. 501–6.

73 Vesalius did include a rendering of the *rete* in the *Fabrica*, by way of admonitory instruction; it is predictably similar both to the *Tabulae*'s and Crooke's renditions.

74 Forrester, "The Marvellous Network," 207. Forrester also explains (215–17) that we now understand that in animals which do possess it, the *rete* regulates the transfer of heat to the brain, thus controlling the brain's temperature.

75 Vesalius, *Fabrica*, 310; translation in Forrester, "The Marvellous Network," 208.

76 Mid- to late-century anatomists continued their larger projects under the auspices of the new Vesalian tradition – but they continued as well to reject Vesalius's argument against the existence of the *rete mirabile*. We can now see why such a move would prove appealing, given how tantalizingly useful the metaphor was. Alexander Read's 1650 *Manuall of the anatomy of the body of man* calls the *rete* "a membranous twisting framed of innumerable twigs of arteries … representing a net spread abroad" (214). The word "innumerable" points to the sense of something which cannot be fully comprehended even by the most careful of descriptions. Writing in 1663, Thomas and Caspar Bartholin offer the reader greater detail: "The *Rete mirabile* or *wonderful Net* … is so called by reason of its artificial and wonderful structure, for it shews like many Nets heaped together. Now it hath another structure in Calves and Oxen … though we must not say that it is not in Men as *Vesalius* doth, though hard to discern. I remember nevertheless that it hath been wanting" (*Bartholinus Anatomy*, 138). The Bartholins differentiate between what Vesalius claims occurs only in "Calves and Oxen" and what actually occurs in the human body; in fact, though it is difficult to see, and at times it might have "been wanting" in a cadaver, we are encouraged to believe in this structure, even in the face of the anatomical authority from whom the Bartholins have (like Crooke) filched a picture of a brain dissection. In 1680, the natural philosopher Samuel Haworth defies not just Vesalius, but also Descartes, asserting that the "*Rete Mirabile*, or *Plexus Retiformis*, is at the *Basis* of the *Cerebrum*; it consists of Carotid and Cervical Arteries, brought up from the Heart to the Basis of the Brain, and bring in them Blood and Vital Spirits to this Rete, for the first preparation of the Animal Spirits" (116). Haworth returns us less to the structure of the wonderful net than to its important function, as (here) serving the first part of the final transformation of the spirits.

77 Rousseau, *Nervous Acts*, 165–6. As Rousseau proceeds to explain, such a concrete and absolute localization has important implications for the role of the nerves: "For this organ alone, the brain, depends upon the nerves for all its functions. Once the soul was limited to the brain, scientists could debate precisely how the nerves carry out its voluntary and involuntary intentions … If indeed the soul is limited to the brain … then nerves alone can be held responsible for sensory impressions, and consequently for knowledge; and more consequential, the nerves must necessarily be hollow tubes rather than

solid fibres so that the brain's unique secretion, animal spirits, can freely flow through them to the body's vital organs" (166–7).

78 Willis, *The Anatomy of the Brain*, II: 54. Willis favoured either his eponymous arterial circle or the cerebral cortex for the location of the final sublimation of spirits. In the typical human brain, argued Willis, "noble emotions and great impulses and ardors of the spirits are aroused" which would overtax the delicate fibres of the *rete* – which themselves "would hardly turn the millwheel." Willis, *Cerebri anatome*, 53–7; translation in Forrester, "The Marvellous Network," 211.

79 Snow, *The Two Cultures*, 15. If ever the "two cultures" desired to see eye to eye, Snow contends, there must exist a third culture, one dedicated to uniting the two branches, reconciling the intellectual divides, and knocking down the wall.

80 For an especially egregious example of the modern populist tendency see Jonah Lehrer's *Proust Was a Neuroscientist*. Lehrer claims to uncover in the work of nineteenth-century artists the presaging of various components of our contemporary understanding of the brain. Such claims are certainly intriguing, and they represent to my mind a tentative step in the direction of integration, but Lehrer's key fault is that he fails to realize that literary (and by extension artistic) analysis, like neuroscience, is a study in its own right; to make the connections he desires, he must pay more than lip service to his artists and do more than simply juxtapose ideas to prove a point.

81 For a cogent summary of the debate, and an attempt to find common ground between the two disciplines, see Huxley, *Literature and Science*, in particular 39–44 (which include a brief discussion of the lines of Donne's poem that serve as my epigraph).

82 Littlefield and Johnson, "Theorizing the Neuroscientific Turn," 3.

83 Murison, "'The Paradise of Non-Experts,'" 30, 31.

84 Rousseau, *Nervous Acts*, 163.

85 Jonathan Sawday reads the wonderful net as a precursor to the modern Internet and to modern neuronal mapping. Yet his assertion that the *rete* survived in anatomies *only* "well into the sixteenth century" is, as we can see, inaccurate, and his explanation of the reasoning for the *rete*'s persistence – it "was there because it *had* to be there; what other explanation was available to link the body's evident sensory and motive capacities to its known physiology?" – is reductive, given that there were other theories. See Sawday, "Towards the Renaissance Computer," 38–9.

86 Ibid., 38.

87 I am grateful to Arne De Boever of the California Institute of the Arts for suggesting this idea.

88 Donne, *English Poems*, ln. 278–82.
89 Sawday, "Towards the Renaissance Computer," 36–7.

Chapter Two

1 Willis, *Mnemonica; Or, The Art of Memory*, sig. C_6^v–C_7^r. This text is a full translation of his 1618 *Mnemonica; sive Ars Reminiscendi*, which appeared in partial English translation in 1621. See Yates, *The Art of Memory*, 336n40.
2 Shakespeare, *Hamlet*, ed. Jenkins. All references will be to this edition, unless noted in the text.
3 "Known" lesions are generally those detected by fMRI in human subjects who have experienced trauma or necessary corrective surgery, whereas "intentionally created" lesions are performed upon animal subjects. Du Boisgueheneuc et al., in "Functions of the Left Superior Frontal Gyrus," explain the utility of combining such studies to better localize function: "In humans, functional imaging (PET scan and fMRI) cannot show whether one given region of a network is critical (i.e. the cognitive process cannot be implemented without this region) or accessory (i.e. the cognitive process can be partially or fully functional without it). Therefore, lesion studies in humans are particularly useful to complement experimental studies in animals and functional imaging studies" (3324). For a particularly interesting example, see Geraci et al., "Theory of Mind," a theory which involves "representing and reasoning about mental states" (such as making inferences about another person's emotional state based on his or her facial expression), in patients with damage to particular parts of the prefrontal cortex following traumatic injury (978).
4 See, for example, the controversy surrounding the potential segregation of spatial and nonspatial working memory in the prefrontal cortex, as reviewed by Curtis and D'Esposito, "The Effects of Prefrontal Lesions." See also Rorden and Karnath, "Using Human Brain Lesions," in which Rorden and Karnath question the utility of the lesion study given the wide range and increasing precision of fMRI. Literary scholars are probably most familiar with this method via Roman Jakobson's work on metaphor and metonymy in "Two Aspects of Language and Two Types of Aphasic Disturbances."
5 See, for example: Morris, "Maimed Funeral Rites"; and Holleran, "'Hamlet' as a 'Memento Mori' Poem."
6 For a boiled-down explanation of the idea that damage to a system can reveal its inner workings, see Cook, "Staging Nothing," esp. 90–1. Cook's schematic account of V.S. Ramachandran's work on anosognosia (a condition in which a person who has suffered a given problem, such as paralysis or blindness, as

a result of brain damage is unable to acknowledge that the problem exists) describes as well the "lesion" study of humans.

7 See, for example, Marchitello, "Artifactual Knowledge in *Hamlet*."

8 A notable exception is the Player Queen's ominous exhortation that sleep rock the Player King's brain; see 3.2.222. Then, too, upon encountering his mad sister, Laertes wishes that heat would dry up his brains; see 4.5.154.

9 For the hippocampus, see Manns et al., "Semantic Memory"; Squire and Zola, "Episodic Memory"; and Vargha-Khadem et al., "Differential Effects of Early Hippocampal Pathology." For the limbic system, see MacLean, "Some Psychiatric Implications"; and Papez, "A Proposed Mechanism of Emotion."

10 Arceus, *A most excellent and Compendious Method*, sig. Cii^{r-v}.

11 Ibid.

12 Read, *A Treatise of the First Part of Chirurgery*, sig. $B_2{}^{r-v}$.

13 Ibid., sig. $B_3{}^{r}$.

14 Arceus, *A most excellent and Compendious Method*, sig. Cii^{r}.

15 Crucial, too, is the fact that Arceus identifies the *rete mirabile* as a vulnerable entity.

16 Crooke, *Mikrokosmographia*, sig. $C_1{}^{v}$.

17 Ibid., sig. $E_2{}^{v}$–$E_3{}^{r}$.

18 See Parkin, "H.M." Molaison passed away in 2008, having worked with (i.e., been studied by) neuroscientists for fifty years. His brain was sliced and histologically fixed in late 2009 and currently resides at the University of California at San Diego. See Annese et al., "Postmortem Examination of Patient H.M.'s Brain." Work with Molaison was, of course, not without moral controversy; I think often of that poor man who had to be told on a daily basis that his mother had passed away, and who wept upon receiving what was for him continually new news. Despite the humanity with which the scientific community treated him, it seems he was sometimes seen more as a (crucial) data-point than as a person. See for example Luke Dittrich's recent *Patient H.M.: A Story of Memory, Madness, and Family Secrets.*

19 See Scarborough, "Galen and the Gladiators."

20 See Hodgson, "Long-Range Perspectives," 416: "Several centuries before Galen, [the Indian physician] Sushruta … had discovered that damage to one branch of the recurrent laryngeal nerve of a human resulted in a hoarse voice *and* a loss of taste … The Hindus, who would not dissect pigs (as Galen had done), derived their information from war casualties."

21 See Dunn, "Galen (AD 129–200) of Pergamun"; and Tomey et al., "Herophilus, Erasistratus, Aretaetus, and Galen."

22 The human brain has four ventricles which store cerebrospinal fluid; in the Galenic model, these ventricles housed refined, not-quite-physical spirits,

which served the will of the soul. See, for example, Singer, *Vesalius*, 40. Early modern theories about the utility of the ventricles will be explored in chapter 3.

23 See Baig et al., "The Eastern Heart"; Gross, "Galen and the Squealing Pig"; and Scarborough, "Galen Redivivus."

24 Arceus, *A most excellent and Compendious Method*, sig. Fiv^{r-v}.

25 Ibid.

26 Shakespeare, *The Tragical Historie*, sig. C_4^{v}–D_1^{r}. Given that one theory regarding the relation of the First Quarto and Folio texts involves memorial reconstruction, it is certainly ironic to consider that the word "brain" might simply have been forgotten by the person transcribing Q1. For opposing viewpoints in the discussion of the textual history of the play, see Werstine, "The Textual Mystery of *Hamlet*"; and Foakes, *Hamlet versus Lear*, esp. 90–7. For an intriguing third prong in the debate, and an advocation of a palimpsestic reading of the print and performance history of the play, see Johnson, *The Tain of Hamlet*, esp. 33–45.

27 Had Hamlet remembered what he intended to remember, we would have a very different play indeed, as evinced by Q1's leanness and emphasis on action rather than contemplation. As Richard Helgerson laments, in "What Hamlet Remembers," "Readers sometimes forget what a straightforward play this opening [i.e., of Hamlet remembering what he sets out to remember] would have begot" (70–1).

28 Williams, "Hamlet's Tables," 141.

29 Interestingly, though Williams made this observation in 1979, he seems to have been alone in doing so: none of the major editions of *Hamlet* (e.g., Hudson, Dyce, London, Bevington, Pelican, Arden, Oxford, New Cambridge, New Folger, Norton, Riverside, or Guild) note the potential reference.

30 Heywood, *A Woman Killed with Kindness*, 6.126–7.

31 Thompson and Thompson, *Shakespeare: Meaning & Metaphor*, 166.

32 Further examples of the metaphor abound in the footnotes, glosses, and marginalia of the Folger Shakespeare Library's copy of the third variorum Shakespeare, volume 7 (shelfmark PR2752 1821e c. 2 v. 7). Dante, Montaigne, Dekker, and Beza are among the referents; see Shakespeare, *The Plays and Poems of William Shakespeare*.

33 See, for example, Draaisma, *Metaphors of Memory*, 24–7.

34 Laurentius, *A Discourse of the Preservation of the Sight*, 4.

35 Guevara, "'His fine pate full of fine dirt,'" 1.

36 Lewis, "Hamlet, Metaphor, and Memory," esp. 639.

37 Greenblatt, *Hamlet in Purgatory*, 214.

38 Willis, *Mnemonica; Or, The Art of Memory*, sig. L_6^{r}.

39 For a clear account of this process, along with a helpful diagram, see Singer, *Vesalius on the Human Brain*, xviii–xx.

40 See Sutton, *Philosophy and Memory Traces*, 25–49. My reference here is a simplification of Sutton's rich and compelling metaphor.

41 Shakespeare, *Plays*, 249.

42 See, for example, Triplet, *Writing Tables*.

43 Stallybrass et al., "Hamlet's Tables," esp. 415; but see also Lewis's critique of the study, "Hamlet, Metaphor, and Memory," 610–16. A more modern correlate to the table-book certainly haunts the account by Stallybrass et al.: though the authors never mention Freud's "mystic writing-pad," the idea is remarkably similar. See Freud, "A Note Upon the 'Mystic-Pad.'" See, too, Ayers's assertion in "Reading, Writing, and *Hamlet*" that the "trivial fond records" are actually more metaphorical in nature, comprising the "basic structures of scribal education, the compressed summaries of all fields of knowledge which function as the cues needed to retrieve and employ the larger corpus already filed in his mind" (430).

44 Can the memory willfully be erased in such a matter? Lewis ("Hamlet, Metaphor, and Memory," 614 and passim) argues that such "mnemonic erasure" is impossible, while Sutton (*Philosophy and Memory Traces*, 16) suggests that "neurological ethics" in the early modern period allowed philosophers to theorize that we can be taught "to influence or directly act upon our own brains and spirits, to achieve true virtue by the exercise of moral dominion over our own bodily fluids."

45 Brooks, "'Within the Book and Volume of My Brain,'" esp. 28.

46 See Plato, *Phaedrus*.

47 See the following studies for experimental discussion of this fact: Anderson et al., "Working Memory"; Funahashi et al., "Prefrontal Neuronal Activity"; Haarmann and Usher, "Maintenance of Semantic Information"; and Miller et al., "Neural Mechanisms."

48 Knowles ("*Hamlet* and Counter-Humanism," 1064) points out, "Hamlet does not realize that [keeping the Ghost's commandment alone in his brain] is impossible. He cannot replace a mind shaped by rhetoric with unalloyed feeling. Rhetoric provided not just knowledge, but *how* knowledge was assimilated and understood: it provided a cognitive structure which enforced the Western censure of emotion."

49 As Kerrigan ("Hieronimo, Hamlet and Remembrance," 119) sagely observes, "Revenge is stifled by remembrance."

50 Sullivan, *Memory and Forgetting*, 13.

51 "In the Tony Richardson / Nicol Williamson production the 'My tables' is cut, but 'meet it is I set it down' causes Williamson to 'write' on the stone wall

with his dagger. In a similar fashion, Mel Gibson, in the Zeffirelli film, writes enthusiastically on the stones of the parapet with his sword." (Williams, "Hamlet's Pockets," 195.)

52 See Goldberg, *Writing Matter*.

53 Fleming, *Graffiti*, 75–6.

54 See Carruthers, *The Book of Memory*, 52, 303n31; Finger, *Origins of Neuroscience*, 333; and Manzoni, "The Cerebral Ventricles," 131.

55 Crooke, *Mikrokosmographia*, sig. Yy_6^v–Zz_1^r.

56 Ibid.

57 Guevara, "'His fine pate full of fine dirt,'" 1.

58 See Paster, "The Pith and Marrow of Our Attribute," 261.

59 Derrida, "Fors," 67–8, emphases in the original.

60 *The History of King Lear* 4.5.181 / *The Tragedy of King Lear* 4.5.189; *Othello* 3.3.118–19; *Titus Andronicus* TLN 2637, First Folio, London, 1623, though the Q1 (1594) edition has *souls* instead of *brains* in an intriguing and useful substitution; *Antony and Cleopatra* 2.7.94–5; *Romeo and Juliet* 4.3.52–4. Unless otherwise noted, these citations are from Shakespeare, *The Complete Works*, ed. Wells and Taylor.

61 Norton, Paris, and Wonderlich, "'Strange things I have in head,'" 229–302. They do allow that, "[a]lthough we contend that Macbeth's presentation is compatible with a spongiform encephalopathy, no evidence corroborates that this is what Shakespeare intended," though they argue that "Shakespeare showed an uncannily prescient understanding of prion disease transmission via exposure to neural tissue"; the authors speculate that Macbeth's exposure to prion disease occurred during his initial meeting with the witches, "whose necromantic brews contained a variety of human and animal organs" (302). Such a diagnosis is plausible if misguided; more irresponsible is their suggestion that Macbeth manifests symptoms consistent with kuru, which is a disease localized almost entirely in Papua New Guinea. See Kotade, Patel, Hiremath, Somwanshi, and Dhamak, "Kuru: A Neurological Disorder."

62 *Macbeth*, in Shakespeare, *The Complete Works*, ed. Wells and Taylor, 1.3.147–51. Subsequent references to *Macbeth*, unless noted, will be made to this edition in the body of the text.

63 See *Macbeth*, ed. Kenneth Muir, 1.3.151 n.

64 See OED, s.v. "wrought, adj.," esp. I.1.a–c, 2, 3.b.

65 Both senses of *wrought* were in play during Shakespeare's time; see ibid. 3.b and 4.a.

66 Favila, "'Mortal Thoughts' and Magical Thinking in *Macbeth*," 1.

Chapter Three

1 Shakespeare, *King Lear*.

2 To cite but one important example, Mary Crane's seminal study of cognitive science and Shakespeare examines in detail male pregnancy in *Measure for Measure*; see Crane, *Shakespeare's Brain*, esp. 156–77. For other considerations of the conception metaphor and how it relates to (generally male) intellectual creativity, see Halio, "The Metaphor of Conception," 454–60; Maus, "A Womb of His Own," 266–88; Moncrief and McPherson, eds., *Performing Maternity in Early Modern England*; Pollard, "Conceiving Tragedy," 85–100; and Sacks, *Shakespeare's Imagery of Pregnancy*. My analysis here builds on these earlier fundamental works, but my concern is grounded first in historical biology and neuroscience, extending then to an examination of the dual nature of the conception metaphor.

3 Harvey, *Anatomical exercitations* and *Exercitationes de generatione animalium*. Harvey's translator George Ent uses *uterus*; here I shall use *uterus* and *womb* interchangeably.

4 By "their," Harvey is speaking of both the uterus/womb and the "future Egge or Conception" engendered in it, as well as the brain; see below for elaboration.

5 Goldberg, "A Dark Business," 425–30.

6 Eggert (*Disknowledge*) deploys the main part of Harvey's *Anatomical Exercitations*, focusing largely on Harvey's complete anatomical descriptions, in conjunction with Paracelsian texts, to propose an early modern strain of thought that "exclud[es] women from an active role in the generation of children and even from the realm of knowability." Eggert does briefly mention Harvey's passage on conception, arguing (and I agree) that "Harvey's analogy is this: the artist's brain is to his artistic work as the 'genitor's' (father's) idea is to the fetus. The uterus contains the fetus, but the uterus does not think" (159, 305n48). Eggert counters the oversimplified contention of Rumrich (*Milton Unbound*) that Harvey "makes the womb the equal of the brain" (104). Equivalency is, as we have seen, drastically different from metaphorical connection.

7 Note that Harvey's translator has used *slender threads* and *web* for Harvey's Latin *vel tenuia stamina* and *telis*; the root form *telis* in particular has, as we saw in chapter 1, etymological connections to the word *subtle* and thence to the *subtle net* or *rete mirabile*.

8 See Goldberg, "A Dark Business." See also van Diemerbroek, *The Anatomy of Human bodies*, and Ross, *Arcana microcosmi*, both of which quote Harvey's "De conceptione" for the purpose of dismantling the theory.

9 And, as Goldberg points out, the analogy itself dissolves when one takes into account that it is, for Harvey, the egg, and not the womb, that is the site

of actual generation: "Because Harvey is convinced that the egg is entirely responsible for the creation of the embryo, his analogy between the womb and brain falls apart, since even Harvey believes there are no structural similarities between egg and brain to support an inference to similar functions" (Goldberg, "A Dark Business," 424). At the same time, though, I would point out something of an elision, at least in the "De conceptione," between the eggs of oviparous animals (viz., Harvey's extensive embryological work with chicken eggs) and viviparous animals, particularly given Harvey's insistence that he can see nothing in the uterus after coition, and given Harvey's emphasis on the *structural* similarities between womb and brain as the underlying basis of his analogy. Goldberg does point out that the tenuousness of Harvey's theory caused its failure to be taken up in his own time; see 425–30.

10 Harvey also lists birdsong and spider webs as examples of novel beings generated by never-before-encountered and non-material essences.

11 See, for example, Sakka, Coll, and Chazal, "Anatomy and Physiology of Cerebrospinal Fluid."

12 See Smith et al., *The Animal Spirit Doctrine*, 37. See also Galen, *De placitis Hippocratis et Platonis*, 442.

13 Smith et al., *The Animal Spirit Doctrine*, 37–8.

14 Walter Pagel ("Medieval and Renaissance Contributions," 99) points out that for Nemesius, "the soul itself could not be localized, but the functions of the mind could … [T]he brain substance seemed to be less suitable as an intermediary between body and non-corporeal soul than the 'pneumatic' spaces of the ventricles."

15 Nemesius of Emesa, *On the Nature of Man*, 101–2. Nemesius also held that the nerves themselves were extensions of the brain, or even identifiable with the brain; he reasoned that, when we stub a toe, the nerves in our foot bring the sensation of pain to the brain, but the fact that we feel the pain in our foot means that the nerves of the foot carry a capacity for sensation equal to that of the brain (111–12).

16 See Hebermann et al., eds., *The Catholic Encyclopedia*, V, 749; and Lorenz, "Ancient Theories of the Soul." It should also be noted that Aristotelian authors generally did not subscribe to ventricular localization theory, as the principle underlying this philosophy located the soul and its functions in the heart, rather than in the brain. See, for example, Pagel, "Medieval and Renaissance Contributions," 103.

17 See Smith et al., *The Animal Spirit Doctrine*, 72–3.

18 Ibid., 76–7.

19 See Clarke and Dewhurst, eds., *An Illustrated History of Brain Function*. Clarke and Dewhurst point out that this "Cell Doctrine," a religiously

oriented version of the ventricular localization theory, was "virtually universal in the medieval West and East": "the images created by the sensations ('sensus com[m]unis') in the first cell were manipulated in cell two (reasoning) and whatever was left over was stored in cell three (memory)" (10).

20 See, for example, D.H.M. Woollam ("Concepts of the Brain"), who points out that "it was the fresh unpreserved brain with which alone [Galen and his contemporaries] were familiar, and the brain preserved in formalin, on which so much of our modern anatomical ideas are based, bears very little resemblance to the brain that Galen … knew" (5). Woollam also points out that the brain being dissected would likely have been an ox brain; see chapter 1 for a discussion of the importance of the ungulate brain in the problem of the *rete mirabile*.

21 Vesalius, *Fabrica*, Book VII, ch. 1, 623; translation in Clarke and O'Malley, *The Human Brain and Spinal Cord*, 468.

22 Vesalius remarks that the ventricles are "cavities and spaces in which the inhaled air, added to the vital spirit from the heart, is, by power of the peculiar substance of the brain, transformed into animal spirit." See Singer, *Vesalius on the Human Brain*, 39.

23 As Vesalius puts it, "And yet we clearly see in dissecting that men do not excel those animals by [possessing] any special cavity [in the brain]. Not only is the number [of ventricles] the same, but also all the other things [in the brain] are similar." Translation in Singer, *Vesalius on the Human Brain*, 40.

24 Geminus/Udall, *Compendiosa*, sig. B_2^r.

25 OED, s.v. "venter 1," esp. 2a, 3a–6.

26 See chapter 1.

27 OED, s.v. "chorion, n.," esp. 1, and "choroid, adj. & n.," esp. 2–3.

28 Aristotle [pseud.], *Aristotle's compleat and experienc'd midwife*, 120.

29 Sharpe, *The Midwives Book*, 69.

30 It is possible for the uterus to be partially divided (bicornuate or septate uterus) or wholly divided (didelphic uterus).

31 Crooke, *Mikrokosmographia*, sig. Rr_5^v–Rr_6^r. "Yard" is another word for penis; see, for example, Stevenson, *The Yard of Wit*, 25–92. See also OED, s.v. "yard, n. 2," 11.a.

32 Crooke, *Mikrokosmographia*, sig. Rr_6^v.

33 Ibid., sig. Oo_6^r.

34 Banister, *The Historie of Man*, sig. Ee_4^v.

35 Intriguingly, this terminology is not restricted to human anatomy. Snape, for example (*The Anatomy of an horse*, 114), observes of the pineal gland in the horse brain that "is also called the Yard or Prick of the Brain, from its being placed so near the Buttocks and Stones, as also because it resembleth a Man's Yard."

36 Rüff, *Expert Midwife*, 12.
37 Ibid., 33–4.
38 Crooke, *Mikrokosmographia*, sig. Pp_6^{r-v}.
39 Hesiod, *Theogeny*, ll. 929a–929t.
40 Homer, *The Homeric Hymns*, ll. 28.5–16. Susan Deacy (*Athena*, 28) elaborates on the significance of the names involved: "Zeus's epithet Metieta, [is] sometimes translated as 'wise' but more specifically denoting something like 'Metis-ized.' When Zeus gives birth to Athena, he does so as one who possesses Metis."
41 Le Grand, *An entire body of philosophy*, Preface, Section II, sig. A_1^v.
42 Philipot, *A brief historical discourse*, 117.
43 Brome, *The Love-Sick Court*, 127 (Act 3, scene 2).
44 Willis, *Dr. Willis's practice of physick*, 64.
45 Alsop, "The Author to his Book," sig. A_6^r–A_7^v.
46 T.B., "To the Religious, Vertuous, and Discreet Lady," sig. A_3^r. This is a good example of the difficulty of authorial provenance that attends the prefatory material of books in this period: authors, translators, and colleagues could all supply epistles or honorary poems, and often the author of a given piece will not sign the work or will sign only in initials, as here. In this case, given the language of the dedication, it seems likely that T.B. is an alias of Alius Medicus.
47 Arnot, *An essay on nothing …*, iii.
48 Vines, "To The truly Religious and Virtuous," sig. A_3^{r-v}.
49 See, for example, Tristram Shandy's disastrous birth at the hands of "man-midwife" Dr Slop, in Laurence Sterne's *The Life and Opinions of Tristram Shandy, Gentleman* – especially Walter Shandy's reasons for employing Dr Slop over the more traditional female midwives preferred by Elizabeth Shandy (45), Dr Slop's utilization of the then-novel invention of the forceps (168–70), and the nasal consequences of the instrument (193 et passim).
50 Ent, "To the Venerable," sig. A_8^v.
51 Anonymous, *The Court and City Vagaries*, sig. A_1^r.
52 Blunt, "To the Reader Gentile or Gentle," sig. A_2^r–A_3^v.

Chapter Four

1 Cavendish, "It is hard to believe, that there are other Worlds in this World," in *Poems and Fancies*, 44. Subsequent references will be made by page in the body of the text as PF.
2 Cavendish, *Blazing World*, 224; and Letter CXLIII, *CCXI Sociable Letters*, 296. Subsequent references will be made in the body of the text as BW and SL respectively. See also the introduction to *Paper Bodies*, where Bowerbank and

Mendelson explain that Cavendish's use of the phrase *paper bodies* "suggests that for Cavendish, her writings serve as surrogate bodies that will keep her identity and ideas alive. This way of thinking raises certain questions. What sorts of embodiments are texts? What is the relationship between a writer's 'paper bodies' and the actual lived context of the past?" (11).

3 Pepys, *Diary*, 30 March 1667. The actual author of the play was Cavendish's husband, William.

4 Ibid., 11 April 1667.

5 Osborne, *Letters*, 79.

6 Mary Evelyn, Letter from Mary Evelyn to Ralph Bohum, in John Evelyn, *Diary and Correspondence* vol. 4, 8–9.

7 Whitaker, *Mad Madge*, 293. See also Battigelli, *Margaret Cavendish and the Exiles of the Mind*; Rees, *Margaret Cavendish: Gender, Genre, Exile*; and Stimson, *Scientists and Amateurs*. Modern critics have tended to have mixed views about Cavendish's work (see in particular Virginia Woolf's withering dismissal), when they have dealt with her at all: she seems to have been easy to overlook until rather recently.

8 Mary Evelyn, Letter from Mary Evelyn to Ralph Bohum, in John Evelyn, *Diary and Correspondence* vol. 4, 8–9.

9 Lilley, "Note on This Edition," xxxiii.

10 Cavendish, *The World's Olio*, 94. Subsequent references will be made in the text as WO.

11 Siegfried, "Anecdotal and Cabalistic Forms in *Observations upon Experimental Philosophy*," 59.

12 See Cunning, "Cavendish on the Intelligibility of the Prospect of Thinking Matter," 120.

13 Cavendish's natural philosophy certainly isn't always internally consistent, as critics have noted. See, for example, Boyle ("Margaret Cavendish's Nonfeminist Natural Philosophy"), who argues that, particularly on the subject of atomism, Cavendish tends toward fluidity, both in the laying-out of her philosophy and in her responses to critics. See, too, Keller ("Producing Petty Gods"), who points out that Cavendish is less interested in absolute, monolithic resolution (of the kind the Royal Society generally hoped to achieve) and more invested in the act of argument itself. Whitaker makes a similar point in *Mad Madge* (esp. 158).

14 Cavendish, *Philosophical and Physical Opinions*, 21. Subsequent references will be made in the text as PPO.

15 Cavendish, *Philosophical Letters*, 12. Subsequent references will be made in the text as PL.

16 Clucas, "The Atomism of the Cavendish Circle," 261–2.

17 Shaw, *Certaine Helpes*, title page.

18 See Whitaker, *Mad Madge*, 133–59 for dating of the two texts. I sidestep here Cavendish's exhaustively revised theories on atomism; though these are related to her vitalism, they are in far more flux in Cavendish's writings. A proper consideration of her atomism is thus beyond the scope of this chapter – and it has been carefully addressed elsewhere. See, for example: Clucas, "The Atomism of the Cavendish Circle"; Clucas, "Poetic Atomism"; Sarasohn, "A Science Turned Upside Down"; and Stevenson, "The Mechanist-Vitalist Soul of Margaret Cavendish."

19 Cavendish, *Philosophicall Fancies*, 1–2. Subsequent references will be made in the text as *PF*. Her argument concerning motion derives in part from her anti-mechanist stance; mechanists held that it was possible for a resting body to lack motion entirely. Cavendish's response demonstrates the degree to which she considers motion inherent in matter: "For how can motion, being no substance but only a mode, quit one body and pass into another? One body may either occasion or imitate another's motions, but it can neither give nor take away what belongs to its own or another's substance, no more than matter can quit its nature from being matter" (*PL* 98). See, too, James, "The Philosophical Innovations," esp. 222–4.

20 There are for Cavendish three degrees of matter: inanimate, sensitive, and rational. Thus the spirits animating each type of matter vary by degree of movement, as well. James ("The Philosophical Innovations," 225) sums up the interaction: "Although inanimate or gross matter is inert, it is always intermixed with, and carried along by, self-moving sensitive matter, which in its turn intermingles with self-moving rational matter."

21 Clucas ("The Atomism of the Cavendish Circle," 265–6) points out that such terminological borrowing was endemic in natural philosophy broadly, and in texts dealing with atomism specifically: "It is interesting to note that, when faced with the most abstruse level of physical process, [members of the Cavendish circle] moved to borrow terminology current in contemporary chemistry and medicine: 'digestion,' 'fermentation,' 'resolution.' All of them are, in effect, driven outside of traditional, mechanical atomic explanations into something which to them appeared more conceptually appropriate."

22 Thought, and, by extension, certain elements of sensation and perception; this essential distinction would drive animal experimentation for centuries.

23 Stevenson ("The Mechanist-Vitalist Soul of Maragaret Cavendish," 532) sums up Cavendish's general philosophical viewpoint in relation to her contemporary vitalists and mechanists: "Like other mechanists, most notably Hobbes, she believes the universe comprises divisible, independent parts. Unlike other mechanists, Cavendish believes that these parts are animated,

or 'self-moving,' and that therefore they cannot be studied empirically to yield consistent, objectively reliable principles. Like other vitalists, such as Spinoza and Ann Conway, Cavendish believes that matter is informed with rational and sensitive power. Unlike other vitalists, she believes this power does not inhere in a single hierarchical order, but is fragmented and often oppositional."

24 James, "The Philosophical Innovations," 232.

25 See also *PL* 20, *PPO* 294–304. Patterning is distinct from imprinting; instead it could be best figured as a kind of sensory copying or mimicking. See *PL* 539–40. She recants slightly in *OEP* (170), stating that "I cannot certainly affirm, that all [perceptions] are made by imitation or patterning" (albeit she still believes this pretty strongly to be the case).

26 See, for example, Boyle ("Margaret Cavendish's Nonfeminist Natural Philosophy," 196, 227), who finds "very little evidence for attributing such protofeminism to Cavendish"; while she does "display an awareness of the ways in which power can be used to limit knowledge," her "attitude toward such uses of power is one that contemporary feminists would abhor." See, too, Lewis ("The Legacy of Margaret Cavendish," 344, 346), who posits that it is tempting to link Cavendish's "rationalist methodology, vitalistic metaphysics, and lamentations about women's plight in society" so as to form "a portrait of an early pioneer in feminism"; but he suggests that she may play up the lack of education of women in general to excuse her own unschooled status, and her "vision of women's role in society proves difficult to expose." Hutton ("Anne Conway," 231) also notes that "[i]n spite of everything Margaret Cavendish said about the educational and social disadvantages suffered by women … [she was] undoubtedly blessed by opportunities which very few of [her] contemporaries enjoyed."

27 From *Elements of Philosophy*, quoted in Stephen, *Hobbes*, 93.

28 See also Sarasohn, "*Leviathan* and the Lady: Cavendish's Critique of Hobbes in the *Philosophical Letters*," esp. 42–9, for further discussion of the relationship between Hobbes and Cavendish, and for further exploration of Cavendish's connections to mechanistic philosophies.

29 Cavendish, *Wits Cabal*, 286.

30 See, for example, Montaigne, "Of a monstrous child," in *The Complete Essays*, 539.

31 Paré, *Workes*, 978. Whereas one might be tempted to find a disturbing undercurrent of fears of miscegenation in the first example, for Paré the incident is less about race and more about monstrosity, insofar as he construes monstrosity as anything against the ordinary course of nature. Thus there are no changes in register as he relates the story of the white woman who gives birth

to a black child, the black woman who gives birth to a white child (a story from Heliodorus about the queen of Ethiopia), and the (presumably non-hirsute) woman who gives birth to a hairy child. This is not to suggest that such fears did not exist in the early modern period – quite the opposite – but they were not prominent for Paré in this particular instance.

32 Critics have observed in Paré's explanations a tendency to categorize imagination as causing physical, rather than supernatural, monstrosity. Pallister ("Introduction," xx), for example, points out that his focus "is on the architecture – the infinitely varied shapes and sizes – of nature, its engineering, especially as regards mechanics and anatomy, its order … and its accidental anomalies and disorders." Shildrick (*Embodying the Monster*, 35) notes that, while Paré "did not attempt to explain how such a corporeal process might actually work in the body," he does move against an "existing tradition" in that he "places maternal imagination among the mechanical rather than supernatural causes of monstrosity." Another especial point to consider is the interpretation of monsters, which may be read as marvels and prodigies. Hence, as Crawford (*Marvelous Protestantism*, 64) points out, some accounts of monsters "draw correspondences between monstrosity and specific women's behaviors, particularly as they pertain to controversial post-Reformation debates over the legitimate forms of marriage and reproduction." That is, in some cases monsters are read as punishments for women who deviated from prescribed societal norms; in such cases cause and interpretation depend intimately on one another.

33 Aristotle [pseud.], *Aristotle's compleat and experienc'd midwife*, 30.

34 Rüff, *The Expert Midwife*, 154–5.

35 Aristotle [pseud.], *Aristotle's compleat and experience'd midwife*, 122.

36 Anonymous, *The English Midwife Enlarged…*, 191.

37 Paré, *Workes*, 978.

38 Ibid., 979.

39 Sharpe, *The Midwives Book*, 118–19.

40 Even in her pioneering account of the problems of maternal imagination, Huet (*Monstrous Imagination*, 6) focuses more on the *effects* of the imagination than the *causes*. She argues for the prominence of maternal imagination, while recognizing that the actual relationship between maternal mind and womb remained unclear: "Although the mother's imagination was never considered to be the *only* possible cause of monstrosity, and did not receive exclusive medical attention at any time in the history of thought on the process of generation, it nevertheless haunted centuries of medical research."

41 Hellwarth, *The Reproductive Unconscious*, 2.

42 O'Neill, "Introduction," xxxiii. The passage in OEP is on p. 140.

43 O'Neill, "Introduction," xxxiii–xxxiv.

44 Cavendish is responding to van Helmont's *Oriatrike*, in particular ch. 35, "The Image of the Minde," 262–9.

45 Cavendish's sympathy with some of Thomas Hobbes's theories about the nature of animate matter has been observed by various critics. See, for example, Lobis, *The Virtue of Sympathy*, esp. 69–109. See also Battigelli, *Margaret Cavendish and the Exiles of the Mind*, esp. 62–84; and Hutton, "In Dialogue with Thomas Hobbes."

46 Michealian, "Margaret Cavendish's Epistemology," 39.

47 Sidney, *An apologie for poetrie*, sig. G_4^v. Though it was published under this title, the work is now generally known as *The Defence of Poesy*.

48 Ibid, sig. F_1^r.

Chapter Five

1 See Massey, "The Alchemical Womb," 224–5. The translation is Massey's.

2 Ibid., 224. Massey suggests that the flower's power to cure gout combines "medicine with salvation … [C]olchium stands for the alliance between medicine and alchemy, and associates the resulting productive knowledge with the male body and nobility" (225).

3 See the introduction.

4 There is precedent for modelling anatomical men after Vesalius, as de Lint ("Fugitive Anatomical Sheets," 79) points out: "Evidently Schoenborn [D. Bart Schoenbornio] tried to imitate Vesalius' figures and to show this as clearly as possible he represented his figure of a man with the face of Vesalius!" Buckley suggests ("Johann Remmelin's *Catoptrum*"), on the other hand, that the man here is modelled after Remmelin himself, author of the original text.

5 As we saw in chapter 3, Vesalius did appear to subscribe to the theory of ventricular localization, which posits that each of the brain's ventricles (sometimes in the period counted as three, sometimes as four) houses some aspect of the soul; typically (and in Vesalius's version) the anterior ventricle received impressions from the five senses and integrated them as commonsense, the second ventricle (connected to the first by a channel) received the commonsense and allowed the soul to cogitate, and the posterior ventricle created memories. See Vesalius's *Fabrica*, Book VII, ch. I, 623; translation in Clarke and O'Malley, *The Human Brain and Spinal Cord*, 468. See also Smith et al., *The Animal Spirit*, 85–7.

6 See Carlino, *Books of the Body*; Sawday, *The Body Emblazoned*; and Roberts and Tomlinson, *The Fabric of the Body*.

7 Carlino explains that the term "fugitive sheet" derives from the German *fliegende Blätter.* See Carlino, *Paper Bodies*, 1.

8 In addition to Carlino's *Paper Bodies*, see de Lint, "Fugitive Anatomical Sheets"; Faust, "Anatomical Flap Illustration"; Wells, "A Remarkable Pair of Anatomical Fugitive Sheets"; Wells, "The 'Sabio' and 'Sylvester' Families"; and Wells, "Anatomical Fugitive Sheets."

9 See Buckley, "Johann Remmelin's *Catoptrum Microcosmicum*"; Carlino, *Paper Bodies*; Massey, "The Alchemical Womb"; Russell, *A Bibliography of Johann Remmelin.*

10 See Cazort et al., "Catalogue of Works," esp. 121.

11 See Schmidt and Nichols, *Altered and Adorned*, 85.

12 See Buckley, "Johann Remmelin's *Catoptrum Microcosmicum*"; Carlino; *Paper Bodies*; Massey, "The Alchemical Womb."

13 Schmidt and Nichols observe that the extant copies of the *Catoptrum* and its descendants contain "no printed notations or instructional marks indicating where to cut or form a tab and in what order to assemble the parts, thereby making it a time-consuming process that only specially trained apprentices could have carried out" (*Altered and Adorned*, 103).

14 See Massey, "The Alchemical Womb," 209.

15 Buckley, "Johann Remmelin's *Catoptrum*," 24.

16 Schmidt and Nichols, *Altered and Adorned*, 91.

17 Cazort et al., *The Ingenious Machine of Nature*, 37.

18 Ibid., 82.

19 Von Calcar is thought to have been an associate of Titian.

20 Though Choulant grumbles that, as Vesalian derivatives, Remmelin's illustrations "represent the clumsiest study of anatomy," Crummer finds them "the most carefully planned and executed" images of their kind. See Choulant, *History and Bibliography*, 232; and Crummer, "A Check List," 136. See also Longo and Reynolds, *Wombs with a View*, 56.

21 Buckley and Massey both discuss this cloud in more historical and artistic detail; I am interested here in laying out a clearer sense of the original *Catoptrum* from which Moxon worked.

22 Benjamin, *The Origin of German Tragic Drama*, 176.

23 Mitchell's reading of the echoes between the *Catoptrum microcosmicum* and Phineas Fletcher's *The Purple Island* notes that the crushing of the serpent's head by the crucifix is in fact keyed in the 1639 edition to a Latin poem, "*Mediatoris Christi Solatum*" [The solace of Christ the mediator] (*The Purple Island and Anatomy*, 134.)

24 These are the pop-up brain in Bartisch's *Opthalmodouleia* (1583), which is contained in a textbook and thus cannot be counted a flap anatomy for lay readers;

the *Four Seasons of Humanity*, commonly called the Robert Fludd prints (c. 1630), which are indebted to Remmelin but only tangentially describe the brain, and that in the last season in a body about to die; and the odd pop-up pineal gland in Schuyl's 1662 edition of Descartes's *De Homine*, which attempts to present not an anatomy but a philosophical and physical argument. See also Cavalcanti et al., "Anatomy, Technology, Art, and Culture," 33–4.

25 Benjamin, *The Origin of German Tragic Drama*, 184–5.

26 Webster, *The Duchess of Malfi*.

27 The publishing history of the *Catoptrum* is a convoluted one. There was for centuries confusion about its authorship. Remmelin's friend Stephan Michelspacher, an occultist whose philosophy and 1616 *Cabala* (English title: *Cabala, mirror of Art and Nature: in alchemy*; dedicated to Remmelin) influenced the imagery of Remmelin's *Catoptrum*, and with whom Remmelin collaborated on the *Pinax*, was often mistakenly ascribed as the author of the *Catoptrum*. (See for example de Rola, *The Golden Game*, 52.) The earliest apparent iteration of the book, printed by Lucas Kilian, appears in Augsburg in 1613 [sometimes given as *Catoptri microcosmici*], and it consists only of flap-images and lacks any accompanying explanatory text; Michelspacher is listed as the author/printer (see Carlino, *Paper Bodies*, 284; Massey, "The Alchemical Womb," 209n4; and Schmidt, "Printed Bodies"). In 1615, Michelspacher also produced (again as author/printer, again in Augsburg) a pair of works bound together, comprising the *Pinax microcosmographicus* (1615), which comprises the *visios* and the *Elucidarius* (1614), an accompanying explanatory table (see Massey, "The Alchemical Womb," 209n4; McDaniel, "The Affair," 62 et passim; and Schmidt, "Printed Bodies," n5). As Massey, McDaniel, and Russell (*A Bibliography of Johann Remmelin*) all point out, these texts encouraged scholars through the nineteenth century to attribute Remmelin's work to Michelspacher (see, for example, Brunnhofer, *Fach-Katalog*, 872; and Power, "Dr. Walter Bayley," 451). McDaniel in particular suggests the eighteenth-century English anatomist James Douglas as a "plausible, though perhaps unintentional, originator" of the claim of Michelspacher's authorship (McDaniel, "The Affair," 61; see also Douglas, *Bibliographiae anatomicae*). Michelspacher, in a dedication to Remmelin in this 1615 edition, claims that, while Remmelin had wanted his name withheld, he had nevertheless granted permission for the work to be published (McDaniel, "The Affair," 62). Remmelin, in his 1619 edition of the *Catoptrum microcosmicum*, claims that friends of his had pressured him to put into print an anatomy he had been working on for his own edification; these same (fairweather) friends, he claims, also secretly printed the *Catoptrum* in 1613 (see McDaniel, "The Affair," 63, for a translation of the relevant passage). Remmelin himself corrected Kilian's plates and issued a corrected edition of the *Catoptrum* in

1619, the *Catoptrum microcosmicum*, which this time bore his name. A set of the plates was printed in Ulm, and one in Amsterdam; the Dutch plates are the ones that find their way to England as the 1670 *Survey*, as evidenced by the Dutch "visio" designations on each page. See also Cazort, *The Ingenious Machine of Nature*; Massey, "Alchemical Womb"; and Schmidt and Nichols, *Altered and Adorned*. See, too, Longo and Reynolds, *Wombs with a View*, for a useful summary of Remmelin's work (56). For catalogue information, see Russell, *British Anatomy*, 164–6; and Eimas, *Heirs of Hippocrates*, 168–70.

28 I am not alone in my dissatisfaction with neural representations in this period; Clarke and Dewhurst, for example, complain of Vesalian images – ostensibly the most accurate representations of the time – that "the artist seems to have gone out of his way to make the cerebral convolutions look like small intestines" (*An Illustrated History of Brain Function*, 62).

29 Wollam, "Concepts of the Brain," 5.

30 One of the images from the *De anima brutorum* is shown above; I shall return to the other later in the chapter. The figure in the *Survey* seems to be an approximation drawn from the two Willis brains; most of the larger structures (i.e., the corpora striata, the fornix, the cerebellum, and the medulla oblongata) are rendered more to the scale of the Willis image above, though the shading in the ventricles corresponds to the other Willis image, as does some of the lettering.

31 As Bryden wryly observes: "Joseph Moxon's record of the activities of the smith, joiner, carpenter, turner and printer [i.e., the *Mechanick Exercises*; see discussion below] are achievements that stand in stark contrast to the rhetoric of the Royal Society, whose widely trumpeted concerns to create a 'history of trades' produced remarkably little" ("Capital in the London Publishing Trade," 311.) See also: Hunter, *Establishing the New Science*; and Ochs, "The Royal Society of London's History of Trades Programme."

32 See, for example, Jacopo Berengario da Carpi's *Isagogae brevis* (Bologna, 1523).

33 The most striking example of this is van Spiegel and Casseri's *De formato foetu liber singularis* (Padua, 1626). See chapter 3 for more detailed discussion of the connection between brain and womb.

34 See Choulant, *History and bibliography of anatomic illustration*, 205–8; and Cushing, *A Bio-Bibliography of Andreas Vesalius*, 146–8.

35 It isn't unusual that these do not seem all to be the same man (note the facial hair or lack thereof); given the practical necessities of dissection and the relatively slower hand needed for illustration, it is likely that different cadavers had to be employed for different sections.

36 "Ornamental and landscape elements, architectural structures and classical remains, the attributes and gestures of figures, inscriptions and cartouches … form the anatomical iconography in such a way that each figure can in itself

tell a story which is no longer merely scientific or purely descriptive … These are not only decorat[iv]e or appended elements … but iconographical devices used to bolster an enjoyment of the anatomy, a mode of transmission of knowledge and meaning" (Carlino, *Paper Bodies*, 23, 26). See also Singer and Rabin, *A Prelude to Modern Science*, xix.

37 A series of seven women, for example, reveal successive layers of the womb and internal reproductive system. These women in particular, reclining on pillows or beds, or genially spreading their legs for the viewer's better instruction, bring to the fore the bibliographic history of these particular engravings. One can, looking closely, make out where sections of the original figures have been excised and anatomy replaced; likewise anatomical instruments and numbering/lettering systems have been superimposed on what with very little by way of imagination we can presume were originally rather titillating pornographic images. See Carlino, *Paper Bodies*, 22–7.

38 Then, too, Vesalius's dissected head is allowed to keep its scalp.

39 Entry for *Anatomy of the principal parts of the human body*, J. Blanchin after J. Dumoulin, 1675, Wellcome Library Catalogue.

40 Curiously, in the *Epitome* version of Estienne's anatomy, the explanatory text for this image runs along the side of the page, leaving the cartouche floating empty above the man. Thus the ink from the engraving on the recto side of this page tends to bleed through, so that the faint afterimages of clouds are visible in this cartouche. This is a point that will be useful to recall when we turn to a discussion of Remmelin's *Catoptrum*.

41 I am grateful to Cristelle Baskins for pointing out this connection.

42 Derrida, "The Purveyor of Truth," 34. Alternate translation: *The Postcard*, 415. See also Simons, "The Visual Dynamics of (Un)veiling."

43 Though an edition of the *Survey* is published in 1691 bearing his name, a fact mentioned neither by Bliss nor by the DNB.

44 Other items included: Camden's *Britannia* (variously bound); Vignola's *Architecture*; *Tutor to Astrology*; Moxon's *Use of the Globes*; *Passage to the North Pole*; books on the use of various playing cards; *Mechanick Exercises*; mathematical dictionaries; *Castelmain's Globe*; *Mechanick Dialling*; *Napier's Bones*; *Military geography*; various maps of the world, the North and South Hemispheres, England, Ireland, and Scotland; and various globes. The most expensive items were "Copernican" spheres, which went for £8 apiece (today roughly £887). At this lottery the *Survey* was listed at £2 for a bound and coloured copy, £1 10*s*. for a coloured copy, and 15*s*. for a plain copy; adjusting for time, they would be valued at roughly £265, £200, and £100, respectively, today. See Anonymous, "Denominations," which has reasonable conversion charts for the mid-eighteenth century. See also Bryden, "Capital in the London Publishing Trade," 300–3.

45 As discussed below and in Bryden's piece, in 1698 James Moxon disposed of a good deal of his father's stock in a lottery. Midwinter and Leigh may have acquired the *Survey* at that point and prepared it for publication four years later. Comparing the 1704 catalogue of Midwinter and Leigh's stock with the 1698 disposal stock list is instructive; Midwinter and Leigh published a number of the Moxon house's items.

46 In this sense, Tauvry's book accords well with other philosophical titles in the list, including Malebranche's *Search after Truth* and an anonymous set of *Remarks on Cartesius's Philosophy*.

47 Half a dozen such books of medical observations are listed, and these appear to be aimed at home remedies, as most are printed in octavo. One such title is *Every Man His Own Doctor* by John Archer (1671), the "Chymical Physitian in Ordinary" to King Charles II. The *Yearbook of Pharmacy*'s issue on *A Century of Old Books Relating to Pharmacy and Kindred Subjects* remarks drolly and witheringly that "the title page is the book, and with it the interest ends. The treatise never rises out of platitude, nor has it the merit of being quaint." See Tilden et al., *The Yearbook of Pharmacy*, 26–7. (Author Joseph Ince.)

48 See Bliss, *Some Aspects of Seventeenth Century English Printing*; Jagger, "Joseph Moxon, F.R.S., and the Royal Society"; and Johns, *The Nature of the Book*, 79–108. However, Bryden does mention it as part of the catalogue of stock James Moxon sold at lottery in 1698 (see below); see also Russell (*A Bibliography of Johann Remmelin*).

49 Bryden mentions that Midwinter and Leigh had acquired the plates from James Moxon, after the 1698 stock disposal; if we examine their list of publications in 1704, the *Survey* is apparently not the only stock the pair received from the younger Moxon. See Bryden, "Capital," 310. But it is disappointing that such elaborate readings of Remmelin's work as those proffered by Anderson, Buckley, and Massey devote little if any attention to the *Catoptrum*'s second life in England. The present piece here thus serves as corrective in that it joins together these two historical strands.

50 The DNB lists *A Book of Drawing, Limning, Washing or Colouring of Mapps and Prints* (1647), for the map seller Thomas Jenner, as the notable exception.

51 Bliss, *Some Aspects of Seventeenth Century English Printing*, 15.

52 Ibid., 16.

53 Public Record Office SP 29/49 f. 66 recto. A "Waggoner" is a term coined after the Anglicized spelling of Lucas Janszoon Waghenaer, whose *Mariner's Mirror* (Holland, 1583) was long held to be a useful volume of navigation. See Hall, *The Scientific Renaissance*, 209.

54 Johns, *The Nature of the Book*, 81–2.

55 Public Record Office SP 29/49, f. 66 verso.

56 Jagger, "Joseph Moxon, F.R.S., and the Royal Society," 196–7, includes a copy and transcription of the petition.

57 Ibid., 198.

58 Bliss, *Some Aspects of Seventeenth Century English Printing*, 16–17.

59 See Bryden, "Capital," 332; van der Krogt, "Globes Made Portable," 8; and Wynter and Turner, *Scientific Instruments*, 173.

60 Bryden, "Capital," 332.

61 Taylor, "Pocket-Sized Globes."

62 *Advertisement. There is invented by the Right Honourable the Earl of Castlemain, a new kind of globe* (London, n.d.); quoted by Bryden, "Capital," 335, who notes that the globes in question were likely originally introduced by Joseph, rather than James, Moxon.

63 Indeed, several entries in Robert Hooke's diary make it clear that Moxon was very well acquainted with members of the Royal Society even prior to his election into it; see Jagger, "Joseph Moxon, F.R.S., and the Royal Society," 199. Hooke purchased cards and the first volume of the *Mechanick Exercises*, and also turned his nose up at a copy of *Napier's Bones* which Moxon showed him.

64 Quoted by Jagger, "Joseph Moxon, F.R.S., and the Royal Society," 198, who references Birch, *The History of the Royal Society of London*, vol. 1, 469.

65 Moxon, *Tutor to Astronomy*, 200.

66 Puttenham, *The Arte of Englishe Poesie*, 4–6. See chapter 1 and the Introduction, this volume, for more discussion of this quote. Note, too, that a later edition of Puttenham's *Art of English Poetry* is included in Midwinter and Leigh's 1704 stock catalogue.

67 Moxon, *Mechanick Exercises*, sig. A24^{v}.

68 Bliss, *Some Aspects of Seventeenth Century Publishing*, 19–20.

69 Jagger, "Joseph Moxon, F.R.S., and the Royal Society," 199.

70 Ibid.

71 Public Record Office SP 29/405, p. 339.

72 DNB entry on Moxon. As Jagger notes: "The first and indeed only occurrence of negative voting occurred at the Anniversary meeting held on 30 November 1678 at which Moxon was elected with 27 votes for, and four against. At the same meeting a further seven candidates were elected, the only other instance of negative voting being the one cast against Dr. John Mayow, an Oxford physician and natural philosopher" ("Joseph Moxon, F.R.S., and the Royal Society," 200). See also Birch, *History*, vol. 3, 442.

73 Diary of Robert Hooke, 1672–1680, edited by Robinson and Adams (London, 1935), 287 (see Bliss, *Some Aspects of Seventeenth Century Publishing*, 17–18). Moxon would, however, be expelled from the Royal Society in 1682 for not paying his dues; Jagger hypothesizes, pragmatically, that Moxon's lapse was due

to his resentment at not having been elected as the Printer of the Royal Society. See DNB and Jagger, "Joseph Moxon, F.R.S., and the Royal Society," 201–2.

74 Russell (*A Bibliography of Johann Remmelin*) hypothesizes that Ireton is Moxon himself.

75 See *OED*, s.v. "survey, n.," 1a.

76 *OED*, s.v. "delineate, v.," 1.

77 Dobson, "Pioneers," 706.

78 Havers's publisher for the *Osteologia Nova*, Samuel Smith, also published work by Robert Boyle, as well as the monthly *Philosophical Transactions* of the Royal Society. It seems possible that Havers was put in touch with the Moxons through Smith, given that Smith operated out of the Princes Arms in St Paul's Church-Yard and at that time James Moxon had moved to Warwick Lane, selling still under his father's sign of Atlas. Warwick Lane is one street over from St Paul's Church-Yard, where Midwinter and Leigh would also sell under the Rose and Crown.

79 Dobson, "Pioneers," 702.

80 Russell, *A Bibliography of Johann Remmelin*, 93n18.

81 Dobson, "Pioneers," 704; though he notes that the French claimed his discoveries had been previously reported by others. The DNB claims only two editions in Latin.

82 Dobson ("Pioneers," 704) believes Havers may have been the first to hold this position.

83 Butler, *Sermon*, sig. C_3^v, D_1^r.

84 Russell, *A Bibliography of Johann Remmelin*, 93n18.

85 Ibid., 9.

86 Ibid.

87 de Graaf, *Epistola*, translation by Catchpole. See Catchpole, "Regnier de Graaf," 1265. Sylvius was, as Lindeboom points out, "one of the first physicians on the continent of Europe to accept Harvey's theory of the circulation of the blood"; see Lindeboom, *Dutch Medical Biography*, col. 1940.

88 Lindeboom, *Dutch Medical Biography*, col. 705.

89 Ibid., col. 152–3.

90 See Tubbs et al., "Richard Lower"; and Fastag et al., "Richard Lower."

91 See Fastag et al., "Richard Lower"; Boyle, "Tryals Proposed by Mr. Boyle"; and Coga, "An Account of the Experiment of Transfusion."

92 Franklin, "The Work of Richard Lower." See also Feindel, "Thomas Willis"; Molnár, "Thomas Willis"; O'Connor, "Thomas Willis"; and Symonds, "The Circle of Willis."

93 Willis, *The Anatomy of the Brain*. See also Franklin, "The Work of Richard Lower," esp. 114; and Felts, "Richard Lower."

Coda

1 I would like to thank one of the members of McGill-Queen's University Press's editorial board for bringing this convergence to my attention; see Barker, *The Tremulous Private Body.*

2 See Mitchell, "*The Anatomy Lesson*"; see also Binstock, "Seeing Representations: or, The Hidden Master in Rembrandt's *Syndics*," esp. 3, 7–9.

3 See Sawday, *The Body Emblazoned*, 155. See also IJpma, Middelkoop, and van Gulik, "Rembrandt's *Anatomy Lesson*," 381–5. This latter piece is of especial interest, given that the authors compare Deijman's anatomy to a contemporary brain dissection they have carried out, in order to comment on and indeed reinforce the accuracy of Rembrandt's depiction.

4 See Sawday, *The Body Emblazoned*; Bordin and D'Ambrosio, *Medicine in Art*, 277; Choulant, *History and Bibliography of Anatomic Illustration*, 391; Garrison and Streeter, "Sculpture and Painting as Modes of Anatomical Illustration," esp. 321; Mitchell, "Rembrandt's *The Anatomy Lesson of Dr. Tulp*," 155n13.

5 The assistant is Gysbrect Matthijsz Calcoen, of the Amsterdam guild of surgeons. See Haas, "The Anatomy Lesson of Dr. Joan Deyman, by Rembrandt (1606–69)," 908.

6 It has also been suggested that pre-1800s brain dissections may have been performed on severed heads rather than on bodies in context, as it were; see Lakke, "Autopsy Practices for Brain Dissections and Rembrandt's Anatomy Lesson of Dr. Deyman," 101–7.

7 See also IJpma, Middelkoop, and van Gulik, "Rembrandt's *Anatomy Lesson*," who note that in such a pose, "[n]ormally, the brain would have barely been visible from the viewpoint of the soles of the feet. Rembrandt combined long views with close-up views by depicting the body in a foreshortened way … It is evident that Rembrandt had done everything in his power to focus our attention on the act of the praelector," i.e., Deijman himself (384).

8 Binstock, "Seeing Representations," 7–8.

9 Ibid., 9. Binstock is referencing the critical semiotic notion of the *suture* as described by Kaja Silverman, in which the viewer recognizes himself or herself as a participant of the artistic text. See Silverman, *The Subject of Semiotics*, 205.

10 Sawday, *The Body Emblazoned*, 155.

11 Ibid., 156.

12 IJpma, Middelkoop, and Gulik, "Rembrandt's *Anatomy Lesson*," 383. They do note, however, that Rembrandt's dissection depicts the falx in an axial rotation that they could not replicate in their own real-world dissection, so, they conclude, Deijman's *Lesson* "did not represent an exact anatomic dissection of the falx cerebri" (384).

13 See, for example, Burnell, *The Augustinian Person*, 41.
14 Burke, "The Hands of Mr. Ottermole," 496. This short story also served as the basis for an episode of *Alfred Hitchcock Presents*, season 2, episode 32, originally airing in 1957.
15 Burke, "The Hands of Mr. Ottermole," 496.

Bibliography

Adler, Hans, and Sabine Gross. "Adjusting the Frame: Comments on Cognitivism and Literature." *Poetics Today* 23, no. 2 (2002): 195–200.

Alsop, George. "The Author to his Book." In *A character of the province of Maryland*. London, 1666.

Anderson, John, Lynne Reder, and Christian Lebiere. "Working Memory: Activation Limitations on Retrieval." *Cognitive Psychology* 30 (1996): 221–56.

Annese, Jacopo, Natalie M. Schenker-Ahmed, Hauke Bartsch, Paul Maechler, Colleen Sheh, Natasha Thomas, Junya Kayano, Alexander Ghatan, Noah Bresler, Matthew P. Frosch, Ruth Klaming, and Suzanne Corkin. "Postmortem Examination of Patient H.M.'s Brain Based on Histological Sectioning and Digital 3D Reconstruction." *Nature Communications* 5, no. 3122 (2014): 2–9.

Anonymous. *The Court and City Vagaries, or Intrigues, of Both Sexes. Written by one of the fair Sex*. London, 1710[?].

Anonymous. "Denominations." http://www.umich.edu/~ece/student_projects/money/denom.html (accessed 1 June 2016).

Anonymous. *The English Midwife Enlarged*… London, 1682.

Arceus, Franciscus. *A most excellent and Compendious Method of curing woundes in the head, and in other partes of the body, with other precepts of the same Arte, practiced and written by that famous man Franciscus Arceus, Doctor in Phisicke & Chirurgery: and translated into English by John Read, Chirurgion*. Translated by John Read. London, 1588.

Arikha, Noga. *Passions and Tempers: A History of the Humours*. New York: Ecco, 2007.

Ariosto, Ludovico. *Orlando Furioso*. Edited by E. Sanguinetie and M. Turchi. Milan, Italy: Garzanti, 1964.

– *Orlando Furioso*. Translated by Guido Waldman. Oxford, UK: Oxford University Press, 1974.

– *Orlando Furioso in English Heroical Verse*. Translated by Sir John Harrington. London, 1607.

Aristotle [pseud.]. *Aristotle's compleat and experienc'd midwife*, 7th ed. Translated by W[illiam] S[almon]. London, n.d. [1740?].

Armstrong, Paul B. *How Literature Plays with the Brain: The Neuroscience of Reading and Art*. Baltimore, MD: The Johns Hopkins University Press, 2013.

Arnot, Hugo. *An essay on nothing* … London, 1776.

Ayers, P.K. "Reading, Writing, and *Hamlet*." *Shakespeare Quarterly* 44, no. 4 (1993): 423–39.

Aziz-Zadeh, L., S.M. Wilson, G. Rizzolatti, and M. Iacoboni. "Congruent Embodied Representations for Visually Presented Actions and Linguistic Phrases Describing Actions." *Current Biology* 16 (2006): 1818–23.

B., T. "To the Religious, Vertuous, and Discreet Lady, Mrs. Elizabeth Moore of Spargrave." In Alius Medicus, *Animadversions on the medicinal observations of the Heidelberg, Palatinate, Dorchester practitioner of Physick, Mr. Frederick Loss*. London, 1674.

Baig, Mirza, Faheem Chishty, Phillip Immesoete, and Chris Karas. "The Eastern Heart and Galen's Ventricle: A Historical Review of the Purpose of the Brain." *Neurosurgical Focus* 23 (2007): E3–9.

Banister, John. *The Historie of Man, Sucked from the Sappe of the Most Approved Anathomistes*. London, 1578.

Barker, Francis. *The Tremulous Private Body: Essays on Subjection*. Ann Arbor, MI: University of Michigan Press, 1995.

Barsalou, L.W. "Perceptual Symbol Systems." *Behavioral & Brain Sciences* 22 (1999): 577–660.

Bartholin, Thomas, and Caspar Bartholin. *Bartholinus Anatomy*. London, 1663.

Bartisch, Georg. *Opthalmodouleia*. Dresden, 1583.

Battigelli, Anna. *Margaret Cavendish and the Exiles of the Mind*. Lexington, KY: University Press of Kentucky, 1998.

Benjamin, Walter. *The Origin of German Tragic Drama*. London: Verso Books, 1998.

Biagioli, Mario. "Stress in the Book of Nature: The Supplemental Logic of Galileo's Realism." *MLN* 118, no. 3 (2003): 557–85.

Binstock, Benjamin. "Seeing Representations: Or, the Hidden Master in Rembrandt's *Syndics*." *Representations* 83, no. 1 (2003): 1–37.

Birch, T. *The History of the Royal Society of London*. 4 vols. London, 1756–1757.

Blake, Liza. "The Grounds of Literature and Science: Margaret Cavendish's Creature Manifesto." In *The Palgrave Handbook of Early Modern Literature and Science*, edited by Howard Marchitello and Evelyn Tribble, 3–26. London: Palgrave Macmillan, 2017.

Bliss, Carey S. *Some Aspects of Seventeenth Century English Printed with Special Reference to Joseph Moxon*. Los Angeles, CA: William Andrews Clark Memorial Library, University of California, Los Angeles, 1965.

Bloom, Harold. *John Donne and the Metaphysical Poets*. New York: Bloom's Literary Criticism, 2010.

Blumenberg, Hans. *Paradigms for a Metaphorology*. Translated by Robert Savage. Ithaca, NY: Cornell University Press, 2010.

Blunt, Edward. "To the Reader Gentile or Gentle." In John Earle, *Micro-cosmographie. Or, A peece of the world discovered*. London, 1628.

Böhme, Jakob. *Forty Questions of the Soul Concerning Its Original, Essence, Substance, Nature or Quality and Property*. Translated by John Sparrow. London, 1665.

Bono, James J. "Science, Discourse, and Literature: The Role/Rule of Metaphor in Science." In *Literature and Science: Theory and Practice*, edited by Stuart Peterfreund, 59–89. Boston: Northeastern University Press, 1990.

Bordin, Giorgio, and Laura Polo D'Ambrosio. *Medicine in Art*. Translated by Jay Hyams. Los Angeles, CA: The J. Paul Getty Museum, 2010.

Bossy, Michel-André. "Medieval Debates of Body and Soul." *Comparative Literature* 28, no. 2 (1987): 144–63.

Bowerbank, Sylvia, and Sara Mendelson, eds. *Paper Bodies: A Margaret Cavendish Reader*. Orchard Park, NY: Broadview Literary Texts, 1999.

Boyle, Deborah. "Margaret Cavendish's Nonfeminist Natural Philosophy." *Configurations* 12, no. 2 (2004): 195–227.

Boyle, Robert. "Tryals Proposed by Mr. Boyle to Dr. Lower, to be Made by Him, for the Improvement of Transfusing Blood out of One Life Animal into Another." *Philosophical Transactions of the Royal Society* 1 (1665–66): 385–8.

Bray, Patrick M. "Forgetting the Madeleine: Proust and the Neurosciences." In *Literature, Neurology, and Neuroscience: Historical and Literary Connections*, edited by Anne Stiles, Stanley Finger, and Francois Boller, 41–54. Oxford, UK: Elsevier, 2013.

Brazier, Mary A.B. *A History of Neurophysiology in The Seventeenth and Eighteenth Centuries: From Concept to Experiment*. New York: Raven Press, 1984.

Brome, Richard. *The Love-Sick Court*. In *Five New Playes*. London, 1659.

Brooks, Douglas A. "'Within the Book and Volume of My Brain': *Hamlet* and the Crisis of Memory – Past and Present." *Foreign Literature Studies / Wai Guo Wen Xue Yan Jiu* 28, no. 117 (2006): 24–31.

Brown, Harcourt. "The Renaissance and Historians of Science." *Studies in the Renaissance* 7 (1960): 27–42.

Brunnhofer, Hermann. *Fach-Katalog der Aargauischen Kantonsbibliothek*. Aarau, Switzerland, 1887.

Bryden, D.U. "Capital in the London Publishing Trade: James Moxon's Stock Disposal of 1698, a 'Mathematical Lottery.'" *The Library* series 6, 19.4 (1997): 293–350.

Buckley, Cali. "Johann Remmelin's *Catoptrum Microcosmicum* and the End of an Era." *Bodleian Library Report* (August 2013): 18–33.

Burke, Thomas. "The Hands of Mr. Ottermole." In *Murderous Schemes: An Anthology of Classic Detective Stories*, edited by Donald E. Westlake and J. Madison Davis, 480–96. Oxford, UK: Oxford University Press, 1996.

Burnell, Peter. *The Augustinian Person*. Washington, DC: The Catholic University of America Press, 2005.

Butler, Lilly. *A sermon preach'd at the funeral of Clopton Havers, M.D. April 29. 1702 at Willingale-Doe in Essex*. London, 1702.

Carlino, Andrea. *Books of the Body: Anatomical Ritual and Renaissance Learning*. Translated by John and Anne Tedeschi. Chicago, IL: The University of Chicago Press, 1999.

– *Paper Bodies: A Catalogue of Anatomical Fugitive Sheets 1587–1687*. Translated by Noga Arikha. *Medical History*, Supplement No. 19, p. 222–91. London: Wellcome Institute for the History of Medicine, 1999.

Carruthers, Mary J. *The Book of Memory. A Study of Memory in Medieval Culture*. Cambridge: Cambridge University Press, 1990.

Cassell's Latin and English Dictionary, 1st ed. Edited by D.P. Simpson. New York: Macmillan Publishing Company, 1987.

Catchpole, Hubert R. "Regnier de Graaf, 1641–1673." *Bulletin of the History of Medicine* 8, no. 9 (1940): 1261–1300.

Cavalcanti, Daniel D., William Feindel, James T. Goodrich, T. Forcht Dagi, Charles J. Prestigiacomo, and Mark C. Preul. "Anatomy, Technology, Art, and Culture: Toward a Realistic Perspective of the Brain." *Neurosurgical Focus* 27, no. 3 (2009): E2, p. 1–42.

Cavendish, Margaret. *CCXI Sociable Letters*. London, 1664.

– *The Description of a New World, Called The Blazing World*. In *The Blazing World and Other Writings*, edited by Kate Lilley. New York: Penguin Classics, 1994.

– *Observations upon Experimental Philosophy*. Edited by Eileen O'Neill. Cambridge: Cambridge University Press, 2001.

– *Philosophical Letters*. London, 1664.

– *The Philosophical and Physical Opinions*. London, 1655.

– *Philosophicall Fancies*. London, 1653.

– *Poems and Fancies*. London, 1653.

– *Wits Cabal*. In *Playes Written by the Thrice Noble, Illustrious and Excellent Princess, the Lady Marchioness of Newcastle*. London, 1662.

– *The World's Olio*. London, 1655.

Cazort, Mimi. *The Ingenious Machine of Nature: Four Centuries of Art and Anatomy*. Ottawa: National Gallery of Canada, 1996.

Cazort, Mimi, Monique Kornell, and Bruce Hugh Russell. "Catalogue of Works." In Mimi Cazort, *The Ingenious Machine of Nature: Four Centuries of Art and Anatomy*. Ottawa: National Gallery of Canada, 1996.

Chapman, Alexander. *Jesuitism described under the name of Babylons policy. In a sermon preached in the collegiate church at Westminster, March 25. 1610*. London, 1610.

Choulant, Ludwig. *History and bibliography of anatomic illustration*. Translated and annotated by Mortimer Frank. New York: Hafner, 1962.

Chu, Simon, and John J. Downes. "Proust Nose Best: Odors Are Better Cues of Autobiographical Memory." *Memory & Cognition* 30, no. 4 (2002): 511–18.

Clarke, Edwin, and Kenneth Dewhurst. *An Illustrated History of Brain Function: Imaging the Brain from Antiquity to the Present*. Oxford, UK: Sanford Publications, 1972.

Clarke, E., and C.D. O'Malley, eds. and trans. *The Human Brain and Spinal Cord. A Historical Study Illustrated by Writings from Antiquity to the Twentieth Century*. Berkeley, CA: University of California Press, 1968.

Clucas, Stephen. "The Atomism of the Cavendish Circle: A Reappraisal." *The Seventeenth Century* 9 (1994): 247–73.

– "Poetic Atomism in Seventeenth-Century England: Henry More, Thomas Traherne and 'Scientific Imagination.'" *Renaissance Studies* 5, no. 3 (1991): 327–40.

Coga, Arthur. "An Account of the Experiment of Transfusion, Practised upon a Man in London." *Philosophical Transactions of the Royal Society* 2 (1666–67): 557–9.

Cohen, Patricia. "Next Big Thing in English: Knowing They Know That You Know." *New York Times* (30 March 2010). http://www.nytimes.com/2010/04/01/books/01lit.html?pagewanted=all.

Cook, Amy. "Staging Nothing: *Hamlet* and Cognitive Science." *SubStance* 35 (2006): 83–99.

Crane, Mary Thomas. *Shakespeare's Brain: Reading with Cognitive Theory*. Princeton, NJ: Princeton University Press, 2001.

Crane, Mary, and Alan Richardson. "Literary Studies and Cognitive Science: Toward a New Interdisciplinarity." *Mosaic: A Journal for the Interdisciplinary Study of Literature* 32, no. 2 (1999): 123–40.

Crashaw, William. *The Complaint or Dialogue, Betwixt the Soule and the Body of a Damned Man*. London, 1632.

Crawford, Julie. *Marvelous Protestantism: Monstrous Births in Post-Reformation England*. Baltimore, MD: The Johns Hopkins University Press, 2005.

Crollius, O. *Traité des signatures*. Lyon, 1624.

Crooke, Helkiah. *Mikrokosmographia*. London, 1615.

Crummer, Leroy. "A Check List of Anatomical Books Illustrated with Cuts and Superimposed Flaps." *Bulletin of the Medical Library Association* 20 (1932): 131–9.

Cunning, David. "Cavendish on the Intelligibility of the Prospect of Thinking Matter." *History of Philosophy Quarterly* 23, no. 2 (2006): 117–36.

Cunningham, Andrew. *The Anatomical Renaissance: The Resurrection of the Anatomical Projects of the Ancients*. Aldershot, UK: Scholar Press, 1997.

Curtis, Clayton E., and Mark D'Esposito. "The Effects of Prefrontal Lesions on Working Memory Performance and Theory." *Cognitive, Affective, & Behavioral Neuroscience* 4, no. 4 (2004): 528–39.

Cushing, Harvey. *A Bio-Bibliography of Andreas Vesalius*. New York: Schuman's, 1943.

da Carpi, Jacopo. *Isagogae brevis*. Bologna, 1523.

Daston, Lorraine. "Galilean Analogies: Imagination at the Bounds of Sense." *Isis* 75, no. 2 (1984): 302–10.

Daston, Lorraine, and Katherine Park. *Wonders and the Order of Nature 1150–1750*. New York: Zone Books, 2001.

Davies, John. *Nosce teipsum*. London, 1622.

Deacy, Susan. *Athena*. New York: Routledge, 2008.

de Graaf, Renier. *Epistola de nonnullis circa partes genitals novis inventis. Ad Franciscum de le Boe Sylvium*. Leiden, 1668.

de Lint, Jan Gerard. "Fugitive Anatomical Sheets." *Janus* 28 (1924): 78–91.

de Rola, Stanislas Klossowski. *The Golden Game: Alchemical Engravings of the Seventeenth Century*. London: Thames and Hudson, 1988.

Derrida, Jacques. "Fors." Translated by Barbara Johnson. *The Georgia Review* 31, no. 1 (Spring 1997): 64–116.

– *The Postcard*. Translated by Alan Bass. Chicago, IL: University of Chicago Press, 1987.

– "The Purveyor of Truth." *Yale French Studies* 52 (1975): 31–113.

Descartes, René. *De Homine, Renatus Des Cartes de homine*. Edited by Florent Schuyl. Leiden, Netherlands, 1662.

Dittrich, Luke. *Patient H.M.: A Story of Memory, Madness, and Family Secrets*. New York: Random House, 2016.

Dobson, Jessie. "Pioneers of Osteogeny: Clopton Havers." *The Journal of Bone and Joint Surgery, British Volume* 34-B(4) (1952): 702–7.

Donne, John. *The Poems of John Donne*. Edited by H.J.C. Grierson. London: Oxford University Press, 1912.

– *The Sermons of John Donne*. Edited by George R. Potter and Evelyn M. Simpson. Berkeley, CA: University of California Press, 1953–62.

– *The Songs and Sonnets of John Donne*, 2nd ed. Edited by Theodore Redpath. New York: St Martin's Press, 1983.

Donoghue, Denis. *Metaphor*. Cambridge, MA: Harvard University Press, 2014.

Douglas, James. *Bibliographiae anatomicae: Specimen sive catalogus*, 2nd ed. Leiden, 1734.

Draaisma, Douwe. *Metaphors of Memory: A History of Ideas about the Mind*. Translated by Paul Vincent. Cambridge: Cambridge University Press, 2000.

Drake, Stillman. "Introduction: Fourth Part." In Galileo, *Discoveries and Opinions of Galileo*, translated by Stillman Drake, 217–27. New York: Anchor Books, 1957.

Du Boisgueheneuc, Foucaud, Richard Levy, Emmanuelle Volle, Magali Seassau, Hughes Duffau, Serge Kinkingnehun, Yves Samson, Sandy Zhang, and Bruno Dubois. "Functions of the Left Superior Frontal Gyrus in Humans: A Lesion Study." *Brain* 129 (December 2006): 3315–28.

Dunn, P.M. "Galen (AD 129–200) of Pergamun: Anatomist and Experimental Physiologist." *Archives of Disease in Childhood: Fetal and Neonatal* E 88 (2003): F441–2.

Eggert, Katherine. *Disknowledge: Literature, Alchemy, and the End of Humanism in Renaissance England*. Philadelphia, PA: University of Pennsylvania Press, 2015.

Eimas, Richard, compiler and ed. *Heirs of Hippocrates. The Development of Medicine in a Catalogue of Historic Books in the Hardin Library for the Health Sciences, The University of Iowa*, 3rd ed. Iowa City, IA: University of Iowa Press, 1990.

Ent, George. "To the Venerable, & highly Accomplish't, The President & Fellows of the College of Physitians in London." In William Harvey, *Anatomical Exercitations, Concerning the Generation Of Living Creatures*, translated by George Ent. London, 1653.

Evelyn, John. *Diary and Correspondence of John Evelyn, F.R.S.*, vol. 4. Edited by William Bray. London: George Bell and Sons, 1889.

Fastag, Eduardo, Joseph Varon, and George Sternbach. "Richard Lower: The Origins of Blood Transfusion." *The Journal of Emergency Medicine* 44, no. 6 (2013): 1146–50.

Faust, Fred W. "Anatomical Flap Illustration, XVI to XX Century." *Marquette Medical Review* no. 21 (1956): 151–8.

Favila, Marina. "'Mortal Thoughts' and Magical Thinking in *Macbeth*." *Modern Philology* 99, no. 1 (2001): 1–25.

Feindel, William. "Thomas Willis (1621–1675) – The Founder of Neurology." *Canadian Medical Association Journal* 87 (1962): 289–96.

Felts, John H. "Richard Lower: Anatomist and Physiologist." *Annals of Internal Medicine* 132, no. 5 (2000): 420–3.

Finger, Stanley. *Origins of Neuroscience: A History of Explorations into Brain Function*. Oxford, UK: Oxford University Press, 2001.

Fleming, Juliet. *Graffiti and the Writing Arts of Early Modern England*. Philadelphia, PA: University of Philadelphia Press, 2001.

Fletcher, Joseph. *The Historie of the Perfect-Cursed-Blessed Man: Setting forth Mans {Excellency / Miserie / Felicitie} by his {Generation, / Degeneration, / Regeneration}*. London, 1629.

Fludd, Robert (attributed to). *The Four Seasons of Humanity*. Flemish or German, c. 1630.

Foakes, R.A. *Hamlet Versus Lear: Cultural Politics and Shakespeare's Art*. Cambridge: Cambridge University Press, 1993.

Fornaźzari, Luis, Ann Mansur, Tom A. Schweizer, and Corinne Fisher. "Proust and Madeleine: Together in the Thalamus?" *Neurology: Clinical Practice* 5, no. 4 (2015): 355–6.

Forrester, John M. "The Marvellous Network and the History of Enquiry into Its Function." *Journal of the History of Medicine and Allied Sciences* 57 (2002): 198–217.

Foucault, Michel. *The Order of Things: An Archaeology of the Human Sciences*. New York: Vintage Books, 1994.

Franklin, K.J. "The Work of Richard Lower (1631–1691)." *Proceedings of the Royal Society of Medicine* 25 (1931): 113–18.

Freud, Sigmund. "A Note Upon the 'Mystic-Pad.'" In *Collected Papers*, volume 5, edited by James Strachey, 175–80. New York: Basic Books, 1959.

Funahashi, Shintaro, Matthew Chafee, and Patricia Goldman-Rakic. "Prefrontal Neuronal Activity in Rhesus Monkeys Performing a Delayed Anti-saccade Task." *Nature* 365 (1993): 753–6.

Galen. *Claudii Galeni opera omnia*. Edited by C.G. Kühn. Lipsiae, 1821–33.

– *De placitis Hippocratis et Platonis*. In *On the Doctrines of Hippocrates and Plato*, edited and translated by P. De Lacy (3 vols.). Berlin: Akademie-Verlag, 2005.

Galileo. "The Assayer." In *Discoveries and Opinions of Galileo*, translated by Stillman Drake, 229–80. New York: Anchor Books, 1957.

Garrison, Fielding H., and Edward C. Streeter. "Sculpture and Painting as Modes of Anatomical Illustration." Edited by Francis R. Packard. *Annals of Medical History* 2, no. 4 (1919): 305–29.

Geminus, Thomas. *Compendiosa totius anatomie delineatio, aere exarata*. Translated by Nicholas Udall. London, 1553.

Geraci, Alessandra, Luca Surian, Marco Ferraro, and Anna Cantagallo. "Theory of Mind in Patients with Ventromedial or Dorsolateral Prefrontal Lesions Following Traumatic Brain Injury." *Brain Injury* 24, no. 7–8 (2010): 978–87.

Glenberg, Arthur M., and V. Gallese. "Action-Based Language: A Theory of Language Acquisition, Comprehension, and Production." *Cortex* 48 (2012): 905–22.

Glenberg, Arthur M., Jessica K. Witt, and Janet Metcalfe. "From the Revolution to Embodiment: 25 Years of Cognitive Psychology." *Perspectives on Psychological Science* 8, no. 5 (2013): 573–85.

Glickstein, Mitchell. *Neuroscience: A Historical Introduction*. Cambridge, MA: The MIT Press, 2014.

Goldberg, Benjamin. "A Dark Business, Full of Shadows: Analogy and Theology in William Harvey." *Studies in History and Philosophy of Biological and Biomedical Sciences* 44 (2013): 419–32.

Goldberg, Jonathan. *Writing Matter: From the Hands of the English Renaissance*. Stanford, CA: Stanford University Press, 1990.

Graziani, René. "John Donne's 'The Extasie' and Ecstasy." *Review of English Studies*, New Series 19 (1968): 121–36.

Greenblatt, Stephen. *Hamlet in Purgatory*. Princeton, NJ: Princeton University Press, 2001.

Gross, Charles G. "Galen and the Squealing Pig." *The Neuroscientist* 4 (1998): 216–21.

– *A Hole in the Head: More Tales in the History of Neuroscience*. Cambridge, MA: The MIT Press, 2009.

Guevara, Perry. "'His fine pate full of fine dirt' (*Hamlet* 5.1.100): Deep Brains and Transhistorical Depressions." Tempe, AZ: Paper presented to the Society for Literature, Science and the Arts Conference, November 2017. Personal communication.

Haarmann, Henk, and Marius Usher. "Maintenance of Semantic Information in Capacity-Limited Short-Term Memory." *Psychonomic Bulletin & Review* 8 (2001): 568–78.

Haas, L.F. "The Anatomy Lesson of Dr. Joan Deyman, by Rembrandt (1606–69)." *The Journal of Neurology, Neurosurgery, and Psychiatry* 55, no. 10 (1992): 908.

Halio, Jay. "The Metaphor of Conception and Elizabethan Theories of Imagination." *Neophilologus* 50 (1966): 454–60.

Hall, Mary Boas. *The Scientific Renaissance 1450–1630*. Mineola, NY: Dover Publications, 1994. Originally published as volume II of *The Rise of Modern Science* series, edited by A. Rupert Hall. New York: Harper & Brothers, 1962.

Harcourt, Glenn. "Andreas Vesalius and the Anatomy of Antique Sculpture." *Representations* 17 (1987): 28–61.

Harvey, William. *Anatomical exercitations, concerning the generation of living creatures*. Translated by George Ent. London, 1653.

– *Exercitationes de generatione animalium*. London, 1651.

Haworth, Samuel. *Anthropologia*. London, 1680.

Hayles, N. Katherine. *Chaos Bound: Orderly Disorder in Contemporary Literature and Science*. Ithaca, NY: Cornell University Press, 1990.

– *The Cosmic Web: Scientific Field Models and Literary Strategies in the Twentieth Century*. Ithaca, NY: Cornell University Press, 1984.

Hebermann, Charles, Edward Pack, Conde Pallen, Thomas Shahan, and John Wynne, eds. *The Catholic Encyclopedia: An International Work of Reference on the Constitution of Doctrine, Discipline, and History of the Catholic Church* (15 vols.). New York: Robert Appleton Company, 1907.

Helgerson, Richard. "What Hamlet Remembers." *Shakespeare Studies* 10 (1977): 67–97.

Hellwarth, Jennifer Wynne. *The Reproductive Unconscious in Medieval and Early Modern England*. New York: Routledge, 2002.

Herculano-Houzel, Suzana. "The Human Brain in Numbers: A Linearly Scaled-Up Primate Brain." *Frontiers in Human Neuroscience* 3 (November 2009): 1–11.

Heseler, Baldasar. *Andreas Vesalius' First Public Anatomy at Bologna, 1540. An eyewitness report by Baldasar Heseler, Medicinae Scolaris, together with his notes on Matthaeus Curtis' Lectures on Anatomia Mundini*. Translated by Ruben Eriksson. Uppsala and Stockholm, Sweden: Almquist & Wiksells Botryckeri Aktiebolag, 1959.

Hesiod. *The Theogeny of Hesiod*. Translated by Hugh G. Evelyn-White. Cambridge, MA: Loeb Classical Library, 1914.

Heywood, Thomas. *A Woman Killed with Kindness*. In *A Woman Killed with Kindness and Other Domestic Plays*, edited by Martin Wiggins, 69–127. Oxford, UK: Oxford University Press, 2008.

Hirsch, David A. Hedrich. "Donne's Atomies and Anatomies: Deconstructed Bodies and the Resurrection of Atomic Theory." *Studies in English Literature* 31 (1991): 69–94.

Hodgson, Edward S. "Long-Range Perspectives on Neurobiology and Behavior." *American Zoologist* 30, no. 3 (1990): 403–505.

Holleran, James V. "Maimed Funeral Rites in *Hamlet*." *English Literary History* 19 (1989): 65–93.

Homer. *The Homeric Hymns*. Translated by Apostolos N. Athanassikis. Baltimore, MD: The Johns Hopkins University Press, 2004.

Huet, Marie-Hélène. *Monstrous Imagination*. Cambridge, MA: Harvard University Press, 1993.

Hunt, James. "On the Localisation of the Functions of the Brain, with Special Reference to the Faculty of Language." *Anthropological Review* 6 (1868): 329–45.

Hunter, M. *Establishing the New Science: The Experience of the Early Royal Society*. New York: Boydell & Brewer, 1989.

Hutton, Sarah. "Anne Conway, Margaret Cavendish and Seventeenth-Century Scientific Thought." In *Women, Science and Medicine 1500–1700*, edited by

Lynette Hunter and Sarah Hutton, 218–34. Phoenix Mill, UK: Sutton Publishing Limited, 1997.
– "In Dialogue with Thomas Hobbes: Margaret Cavendish's Natural Philosophy." *Women's Writing* 4, no. 3 (1997): 421–32.
Huxley, Aldous. *Literature and Science*. London: Chatto & Windus, 1963.
IJpma, Frank F.J., Norbert E. Middelkoop, and Thomas M. van Gulik. "Rembrandt's *Anatomy Lesson of Dr. Deijman* of 1656 Dissected." *Neurosurgery* 73, no. 3 (2013): 381–5.
Inge, Denise. "Introduction." In Thomas Traherne, *Thomas Traherne, Poetry and Prose, The Golden Age of Spiritual Writing*, edited by Denise Inge, xiii–xxviii. London: Society for Promoting Christian Knowledge, 2002.
Jagger, Graham. "Joseph Moxon, F.R.S., and the Royal Society." *Notes and Records of the Royal Society of London* 49, no. 2 (1995): 193–208.
Jakobson, Roman. "Two Aspects of Language and Two Types of Aphasic Disturbances." In Jakobson, *On Language*, edited by Linda R. Waugh and Monique Monville-Burston, 115–33. Cambridge, MA: Harvard University Press, 1995.
James, Susan. "The Philosophical Innovations of Margaret Cavendish." *British Journal for the History of Philosophy* 7, no. 2 (1999): 219–44.
Johns, Adrian. *The Nature of the Book: Print and Knowledge in the Making*. Chicago, IL: University of Chicago Press, 2000.
Johnson, Laurie. *The Tain of Hamlet*. Newcastle-upon-Tyne, UK: Cambridge Scholars Publishing, 2013.
Jordan, Richard Douglas. "Thomas Traherne and the Art of Meditation." *Journal of the History of Ideas* 46, no. 3 (1985): 381–403.
Keller, Eve. "Producing Petty Gods: Margaret Cavendish's Critique of Experimental Science." *English Literary History* 642 (1997): 447–71.
Kerrigan, John. "Hieronimo, Hamlet and Remembrance." *Essays in Criticism* 31, no. 2 (1981): 105–26.
Kotade, K.B., S.I. Patel, S.M. Hiremath, S.B. Somwanshi, and K.B. Dhamak. "Kuru: A Neurological Disorder." *International Journal of Current Trends in Pharmaceutical Research* 2, no. 2 (2014): 405–16.
Knowles, Ronald. "*Hamlet* and Counter-Humanism." *Renaissance Quarterly* 52, no. 4 (1999): 1046–69.
Lakke, J.P. "Autopsy Practices for Brain Dissections and Rembrandt's Anatomy Lesson of Dr. Deyman." *Journal of the History of Neuroscience* 7, no. 2 (1998): 101–7.
Lakoff, George, and Mark Johnson. *Metaphors We Live By*. Chicago, IL: The University of Chicago Press, 1980.
Lange, Marjory E. "Humourous Grief: Donne and Burton Read Melancholy." In *Speaking Grief in English Literary Culture: Shakespeare to Milton*, edited by

Margo Swiss and David Kent, 69–97. Pittsburgh, PA: Duquesne University Press, 2002.

Latour, Bruno. *We Have Never Been Modern*. Translated by Catherine Porter. Cambridge, MA: Harvard University Press, 1993.

Laurentius, M. Andreas. *A Discourse of the Preservation of the Sight: of Melancholike diseases; of Rheumes, and of Old age*. Translated by Richard Surphlet. London, 1599.

Le Grand, Antoine. *An entire body of philosophy according to the principles of the famous Renate Des Cartes*. London, 1694.

Lehrer, Jonah. *Proust Was a Neuroscientist*. Boston: Houghton Mifflin Company, 2007.

Lewalski, Barbara Kiefer. *Protestant Poetics and the Seventeenth-Century Religious Lyric*. Princeton, NJ: Princeton University Press, 1979.

Lewis, Eric. "The Legacy of Margaret Cavendish." *Perspectives on Science* 9, no. 3 (2001): 341–65.

Lewis, Rhodri. "Hamlet, Metaphor, and Memory." *Studies in Philology* 109, no. 5 (2012): 609–41.

Lilley, Kate. "Note on This Edition." In Margaret Cavendish, *The Blazing World and Other Writings*, edited by Kate Lilley, xxxiii. New York: Penguin Books, 1994.

Lind, R.L. *Jacopo Berengario Da Carpi. A Short Introduction to Anatomy (Isagogae Breves)*. Chicago, IL: The University of Chicago Press, 1959.

Lindeboom, G.A. *Dutch Medical Biography. A Biographical Dictionary of Dutch Physicians and Surgeons 1475–1975*. Amsterdam: Rodopi Press, 1984.

Lindemann, Mary. *Medicine and Society in Early Modern Europe*. Cambridge: Cambridge University Press, 1999.

Littlefield, Melissa M., and Jenell M. Johnson. "Theorizing the Neuroscientific Turn – Critical Perspectives on a Translational Discipline." In *The Neuroscientific Turn: Transdisciplinarity in the Age of the Brain*, edited by Melissa M. Littlefield and Jenell M. Johnson, 1–25. Ann Arbor, MI: The University of Michigan Press, 2012.

Lobis, Seth. *The Virtue of Sympathy: Magic, Philosophy, and Literature in Seventeenth-Century England*. New Haven, CT: Yale University Press, 2015.

Longo, Lawrence D., and Lawrence P. Reynolds. *Wombs with a View: Illustrations of the Gravid Uterus from the Renaissance through the Nineteenth Century*. New York: Springer International Publishing, 2016.

Lorenz, Hendrick. "Ancient Theories of the Soul." In *The Stanford Encyclopedia of Philosophy*, Summer 2009 edition, edited by Edward N. Zalta. https://plato.stanford.edu/archives/sum2009/entries/ancient-soul/.

Lower, Richard. *Tractatus de corde*. Amsterdam, 1669.

MacLean, P.D. "Some Psychiatric Implications of Physiological Studies on Frontotemporal Portion of Limbic System (Visceral Brain)." *Electroencephalography and Clinical Neurophysiology* 4, no. 4 (1952): 407–18.

Maier, Michael. *Atalanta fugiens*. Oppenheim, 1617.

Manns, J.R., R.O. Hopkins, and L.R. Squire. "Semantic Memory and the Human Hippocampus." *Neuron* 38, no. 1 (2003): 127–33.

Manzoni, T. "The Cerebral Ventricles, the Animal Spirits and the Dawn of Brain Localization of Function." *Archives Italiennes de Biologie* 136, no. 2 (1998): 103–52.

Marchitello, Howard. "Artifactual Knowledge in *Hamlet*." In *Knowing Shakespeare: Senses, Embodiment, and Cognition*, edited by Lowell Gallagher and Shankar Raman, 137–53. London: Palgrave Macmillan, 2010.

Margoliouth, H.M. "Introduction." In Thomas Traherne, *Thomas Traherne: Centuries, Poems, and Thanksgivings, volume 1: Introduction and Centuries*, edited by H.M. Margiliouth, xxviii–xxxi. Oxford, UK: Oxford University Press, 1958.

Marvell, Andrew. *The Complete Poems*. Edited by Elizabeth Story Donno. New York: Penguin Books, 1996.

Massey, Irving. *Neural Imagination: Aesthetic and Neuroscientific Approaches to the Arts*. Austin: The University of Texas Press, 2009.

Massey, Lyle. "The Alchemical Womb: Johann Remmelin's *Catoptrum microcosmicum*." In *Visual Cultures of Secrecy in Early Modern Europe*, edited by Timothy McCall, Sean Roberts, and Giancarlo Fiorenza, 208–28. Kirksville, MI: Truman State University Press, 2013.

Maus, Katharine Eisaman. "A Womb of His Own: Male Renaissance Poets in the Female Body." In *Sexuality and Gender in Early Modern Europe*, edited by James Grantham Turner, 266–88. Cambridge: Cambridge University Press, 1993.

McDaniel, Walton Brooks. "The Affair of the '1613' Printing of Johannes Rümelin's *Catoptron*." *Transactions and Studies of the College of Physicians of Philadelphia* 6 (1938): 60–72.

Michealian, Kourken. "Margaret Cavendish's Epistemology." *British Journal for the History of Philosophy* 17, no. 1 (2009): 31–53.

Michelspacher, Stephan. *Cabala, Speculum artis et naturae in alchymia*. Dresden, 1616.

Miller, Earl, Cynthia Erickson, and Robert Desimone. "Neural Mechanisms of Visual Working Memory in Prefrontal Cortex of the Macaque." *The Journal of Neuroscience* 16 (1996): 5154–67.

Mitchell, Charles. "Donne's 'The Extasie': Love's Sublime Knot." *Studies in English Literature* 8 (1968): 91–101.

Mitchell, Dolores. "Rembrandt's *The Anatomy Lesson of Dr. Tulp*: A Sinner among the Righteous." *Artibus et Historiae* 15, no. 30 (1993): 145–56.

Mitchell, Peter. *The Purple Island and Anatomy in Early Seventeenth-Century Literature*. Cranbury, NJ: Associated University Presses, 2007.

Molnár, Zoltán. "Thomas Willis (1621–1675), the Founder of Clinical Neuroscience." *Nature Reviews Neuroscience* 5 (2004): 329–35.

Moncrief, Kathryn M., and Kathryn R. McPherson, eds. *Performing Maternity in Early Modern England*. Aldershot, UK: Ashgate, 2007.

Montaigne, Michel de. *The Complete Essays of Montaigne*. Translated by Donald Frame. Stanford, CA: Stanford University Press, 1958.

More, Henry. *An antidote against atheism or, an appeal to the naturall faculties of the minde of man, whether there be not a God*. London, 1655.

– *The immortality of the soul, So farre forth as it is demonstrable from the Knowledge of Nature and the Light of Reason*. London, 1659.

Morris, Harry. "'Hamlet' as 'Memento Mori' Poem." PMLA 85, no. 5 (1970): 1035–40.

Moxon, Joseph. *A Tutor to Astronomy and Geography*. London, 1659.

– *Mathematicks Made Easie* … London, 1692.

– *Mechanick Exercises: Or the Doctrine of Handy-Works* … London, 1703 [later edition].

Murison, Justine S. "'The Paradise of Non-Experts': The Neuroscientific Turn of the 1840s United States." In *The Neuroscientific Turn: Transdisciplinarity in the Age of the Brain*, edited by Melissa M. Littlefield and Jenell M. Johnson, 29–48. Ann Arbor, MI: The University of Michigan Press, 2012.

Nemesius of Emesa. *On the Nature of Man* (*De natura hominis*). Translated by R.W. Sharples and P. van der Eijk. Liverpool, UK: Liverpool University Press, 2008.

Norton, Scott A., Robert M. Paris, and Kimberly J. Wonderlich. "'Strange things I have in head': Evidence of Prion Disease in Shakespeare's *Macbeth*." *Clinical Infectious Diseases* 42, no. 2 (2006): 229–302.

Ochs, K.H. "The Royal Society of London's History of Trades Programme: An Early Episode in Applied Science." *Notes and Records of the Royal Society of London* 39 (1985): 129–58.

O'Connor, James P.B. "Thomas Willis and the Background to *Cerebri Anatome*." *Journal of the Royal Society of Medicine* 96 (2003): 139–43.

O'Neill, Eileen. "Introduction." In Margaret Cavendish, *Observations upon Experimental Philosophy*, edited by Eileen O'Neill, x–xxxvi. Cambridge: Cambridge University Press, 2001.

Osborne, Dorothy. *Letters from Dorothy Osborne to Sir William Temple*. Edited by Edward Abbott Parry. New York: E.B. Dutton & Co., 1888.

Pagel, Walter. "Medieval and Renaissance Contributions to Knowledge of the Brain and Its Functions." In *The History and Philosophy of Knowledge of the*

Brain and Its Functions, edited by Frederick Poynter, 95–114. Oxford, UK: Blackwell Scientific Publications, Ltd., 1958.

Pallister, Janis. "Introduction." In Ambroise Paré, *On Monsters and Marvels*, translated by Janis Pallister, xv–xxxii. Chicago, IL: The University of Chicago Press, 1982.

Papez, J.W. "A Proposed Mechanism of Emotion." *Journal of Neuropsychiatry and Clinical Neuroscience* 7, no. 1 (1995): 103–12.

Paré, Ambroise. *The Workes of that famous Chirurgion Ambrose Pare …* Translated by Thomas Johnson. London, 1634.

Parker, Amanda, Henny Ngu, and Helen J. Cassaday. "Odour and Proustian Memory: Reduction of Context-Dependent Forgetting and Multiple Forms of Memory." *Applied Cognitive Psychology* 15 (2001): 159–71.

Parkin, Alan J. "H.M.: The Medial Temporal Lobes and Memory." In *Classic Cases in Neuropsychology*, edited by Chris Code, Claus-W. Wallesch, Yves Joanette, and André Lecours, 337–48. East Sussex, UK: Psychology Press, 1996.

Paster, Gail Kern. "The Pith and Marrow of Our Attribute: Dialogue of Skin and Skull in *Hamlet* and Holbein's *The Ambassadors*." *Textual Practice* 23, no. 2 (2009): 247–65.

Paul, Jean. "Vorschule der Ästhetik. [Introductory course in aesthetics.]" In *Sämtliche Werke*, edited by Norbert Miller, 7–456. Frankfurt, Germany: Zweitausendeins, 1996 [1813].

Pepys, Samuel. *The Diary of Samuel Pepys M.A. F.R.S.* Transcribed from the shorthand manuscript in the Pepysian Library, Magdalene College, Cambridge, by the Rev. Mynors Bright, M.A., Late Fellow and President of the College. Edited by Henry B. Wheatley. London: George Bell & Sons, 1893.

Persaud, T.V.N. *The Early History of Human Anatomy: From Antiquity to the Beginning of the Modern Era*. Springfield, IL: Thomas, 1989.

Philipot, Thomas. *A brief historical discourse of the original and growth of heraldry*. London, 1672.

Pinker, Steven. *The Blank Slate: The Modern Denial of Human Nature*. New York: Penguin, 2003.

Plato. *Phaedrus*. Translated by Alexander Nehamas and Paul Woodruff. Cambridge, MA: Hackett Classics, 1995.

Pollard, Tanya. "Conceiving Tragedy." In *Shakespearean Sensations: Experiencing Literature in Early Modern England*, edited by Katharine A. Craik and Tanya Pollard, 85–100. Cambridge: Cambridge University Press, 2013.

Pontius, Anneliese A. "Overwhelming Remembrance of Things Past: Proust Portrays Limbic Kindling by External Stimulus – Literary Genius Can Presage Neurobiological Patterns of a Puzzling Behavior." *Psychological Reports* 73 (1993): 615–21.

Popham, Edward. *A Looking-glasse for the Soule, and a Definition thereof.* London, 1619.

Porter, Roy. *Flesh in the Age of Reason: The Modern Foundations of Body and Soul.* New York: W.W. Norton & Co., 2004.

Power, D'Arcy. "Dr. Walter Bayle and His Works, 1529–1592." *Medico-Chirurgical Transactions* 90 (1907): 415–54.

Poynter, Frederick, ed. *The History and Philosophy of the Brain and Its Functions. An Anglo-American Symposium. London, July 15–17, 1957.* Oxford, UK: Blackwell Scientific Publications Ltd., 1958.

Preston, Claire. *The Poetics of Scientific Investigation in Seventeenth-Century England.* Oxford, UK: Oxford University Press, 2015.

Puttenham, George. *The Arte of English Poesie. Contrived into three Bookes: The first of Poets and Poesie, the second of Proportion, the third of Ornament.* London, 1589.

Read, Alexander. *Manuall of the anatomy of the body of man.* London, 1650.

– *A Treatise of the First Part of Chirurgery …: The part which teacheth the re-unition of the parts of the body disjoyned. Containing The Methodicall Doctrine of Wounds: Delivered in Lectures in the Barber-Chirurgeons-Hall, upon Tuesdaies, appointed for these Exercises, and the keeping of their Courts.* London, 1650.

Rees, Emma L.E. *Margaret Cavendish: Gender, Genre, Exile.* Manchester, UK: Manchester University Press, 2003.

Remmelin, Johann. *Catoptrum microcosmicum* [sometimes given as *Catoptri microcosmici*]. [Author given as Stephan Michelspacher.] Augsburg, Ulm [?], 1613.

– *Catoptrum microcosmicum.* Ulm, 1619.

– *Elucidarius, Tabulis Synopticis, Microcosmici Laminis Incisi Aeneis, Admirandam Partium Hominis Creaturarum Divinarum Praestantissimi Universarum Fabricam repraesentantis.* [Author given as Stephan Michelspacher.] Augsburg, 1614/1615.

– *An exact survey of the Microcosmus, or Little World …* Translated by John Ireton. London: Joseph Moxon, 1670, 1675, 1691.

– *An exact survey of the Microcosmus, or Little World …* Translated and edited by Clopton Havers. London: J. Moxon, 1695, 1702.

– *Pinax microcosmographicus, hoc est, Admirandae Partium Hominis Creaturarum Divinarum praestantissimi Universarum Fabricae.* [Author given as Stephan Michelspacher.] Augsburg, 1615.

Richardson, Alan, and Francis F. Steen. "Literature and the Cognitive Revolution: An Introduction." *Poetics Today* 23, no. 1 (2002): 1–8.

Roberts, K.B., and J.D.W. Tomlinson, *The Fabric of the Body: European Traditions of Anatomical Illustration.* Oxford, UK: Clarendon Press, 1992.

Rocca, Julius. *Galen on the Brain: Anatomical Knowledge and Physiological Speculation in the Second Century A.D.* Leiden, Netherlands: Brill, N.H.E.J., N.V. Koninklijke, Boekhandel en Drukkerij, 2003.

Rorden, Chris, and Hans-Otto Karanth. "Using Human Brain Lesions to Infer Function: A Relic from a Past Era in the fMRI Age?" *Neuroscience* 5 (2004): 813–19.

Rose, Clifford, ed. *Neurology of the Arts. Painting, Music, Literature.* London: Imperial College Press, 2004.

Ross, Alexander. *Arcana microcosmi, or, The hid secrets of a man's body discovered.* London, 1652.

Rousseau, George S. *Nervous Acts: Essays on Literature, Culture and Sensibility.* New York: Palgrave Macmillan, 2004.

Rüff, Jakob [James Rueff]. *The Expert Midwife, Or, An Excellent and most necessary Treatise of the generation and birth of Man.* Translated [?] by E. G[riffin]. London, 1637.

Rumrich, John. *Milton Unbound: Controversy and Reinterpretation.* Cambridge: Cambridge University Press, 1996.

Russell, Kenneth F. *British Anatomy 1525–1800: A Bibliography of Works Published in Britain, America and on the Continent*, 2nd ed. Winchester, UK: St. Paul's Bibliographies, 1987.

– *A Bibliography of Johann Remmelin the Anatomist.* East St. Kilda, VIC: Heirs & Assigns of the Late K.F. Russell, 1991.

Sacks, Elizabeth. *Shakespeare's Imagery of Pregnancy.* London: Macmillan, 1980.

Saive, Anne-Lise, Nadine Ravel, Marc Thevent, Jean-Pierre Royet, and Jane Plailly. "A Novel Experimental Approach to Episodic Memory in Humans Based on the Privileged Access of Odors to Memories." *Journal of Neuroscience Methods* 213, no. 1 (2015): 22–31.

Sakka, L., G. Coll, and J. Chazal. "Anatomy and Physiology of Cerebrospinal Fluid." *European Annals of Otorhinolaryngology, Head and Neck Diseases* 128, no. 6 (2011): 309–16.

Sarasohn, Lisa. "*Leviathan* and the Lady: Cavendish's Critique of Hobbes in the *Philosophical Letters.*" In *Authorial Conquests: Essays on Genre in the Writings of Margaret Cavendish*, edited by Line Cottegnies and Nancy Weitz, 40–58. Madison: Fairleigh Dickinson University Press, 2003.

– "A Science Turned Upside Down: Feminism and the Natural Philosophy of Margaret Cavendish." *Huntington Library Quarterly* 47, no. 4 (1984): 289–307.

Sawday, Jonathan. *The Body Emblazoned: Dissection and the Human Body in Renaissance Culture.* New York: Routledge, 1995.

– "Towards the Renaissance Computer." In *The Renaissance Computer: Knowledge Technology in the First Age of Print*, edited by Neil Rhodes and Jonathan Sawday, 27–41. London: Routledge, 2000.

Scarborough, John. "Galen and the Gladiators." *Episteme* 5 (1971): 98–111.

– "Galen Redivivus: An Essay Review." *Journal of the History of Medicine and Allied Sciences* 43 (1988): 313–21.

Schmidt, Suzanne Karr. "Printed Bodies and the Materiality of Early Modern Prints." *Art in Print* 1, no. 1 (2011). http://artinprint.org/article/printed-bodies-and-the-materiality-of-early-modern-prints/ (accessed 1 June 2016).

Schmidt, Suzanne Karr, and Kimberly Nichols. *Altered and Adorned: Using Renaissance Prints in Daily Life*. New Haven, CT: Yale University Press, 2011.

Schmitt, C.B., Q. Skinner, E. Kessler, and J. Kraye, eds. *The Cambridge History of Renaissance Philosophy*. Cambridge: Cambridge University Press, 1991.

Shakespeare, William. *The Complete Works*. Edited by Stanley Wells and Gary Taylor. Oxford, UK: Clarendon Press, 2005.

– *Hamlet*. Edited by Harold Jenkins. The Arden Shakespeare 2nd series. Surrey, UK: Thomas Nelson & Sons Ltd., 1997.

– *King Lear*. Edited by R.A. Foakes. New York: Arden Editions (3rd series), 1997.

– *Macbeth*. Edited by Kenneth Muir. Arden Shakespeare. London: Methuen & Co. Ltd, 1962.

– *The Plays and Poems of William Shakespeare, With the Corrections and Illustrations of Various Commentators: Comprehending A Life of the Poet, and An Enlarged History of the Stage, by The Late Edward Malone, With a New Glossarial Index*. Volume VII. London, 1821.

– *The Tragical Historie of Hamlet Prince of Denmarke*. London, 1603.

Shapin, Steven, and Simon Schaffer. *Leviathan and the Air-Pump: Hobbes, Boyle, and the Experimental Life*. Princeton, NJ: Princeton University Press, 1985.

Sharpe, Jane. *The Midwives Book. Or the whole Art of Midwifery Discovered*. London, 1671.

Shaw, John. *Certaine Helpes and Remedies Under God to Prevent Dearth and Scarcity*. London, 1638.

Shepherd, Gordon M., and Kirsten Shepherd-Barr. "Proust Effect." In *Encyclopedia of Neurocience*, edited by Marc D. Binder, Nobutaka Hirokawa, and Uwe Windhorst, 3333–5. New York: Springer, 2009.

Shildrick, Margrit. *Embodying the Monster: Encounters with the Vulnerable Self*. London: Sage Publications Ltd., 2002.

Sidney, Philip. *An apologie for poetrie*. London, 1595.

– *Sir P.S. His Astrophel and Stella*. London, 1591.

Siegfried, Brandie R. "Anecdotal and Cabalistic Forms in *Observations upon Experimental Philosophy*." In *Authorial Conquests: Essays on Genre in the*

Writings of Margaret Cavendish, edited by Line Cottegnies and Nancy Weitz, 59–79. Madison: Farleigh Dickinson University Press, 2003.

Silverman, Kaja. *The Subject of Semiotics*. New York: Oxford University Press, 1983.

Simons, Patricia. "The Visual Dynamics of (Un)veiling in Early Modern Culture." In *Visual Cultures of Secrecy in Early Modern Europe*, edited by Timothy McCall, Sean Roberts, and Giancarlo Fiorenza, 24–53. Kirksville, MI: Truman State University Press, 2013.

Singer, Charles. *Vesalius on the Human Brain*. Edited by E. Ashworth Underwood. London: Oxford University Press, 1952.

Singer, C., and C. Rabin. *A Prelude to Modern Science, Being a Discussion of the History, Sources and Circumstances of the "Tabulae anatomicae sex" of Vesalius*. Cambridge: Cambridge University Press, 1946.

Siraisi, Nancy G. *Medieval and Early Renaissance Medicine: An Introduction to Knowledge and Practice*. Chicago, IL: University of Chicago Press, 1990.

Skinner, Quentin, Eckhard Kessler, and Jill Kraye, eds. *The Cambridge History of Renaissance Philosophy*. Cambridge: Cambridge University Press, 1988.

Slotine, J.J.-E., and W. Lohmiller. "Modularity, Evolution, and the Binding Problem: A View from Stability Theory." *Neural Networks* 14, no. 2 (2001): 137–45.

Smith, A.J. "Preface." In John Donne, *The Complete English Poems*, 13–16. New York: Penguin Books, 1996.

Smith, C.U.M., Eugenio Frixione, Stanley Finger, and William Clower. *The Animal Spirit Doctrine and the Origins of Neurophysiology*. New York: Oxford University Press, 2012.

Snape, Andrew. *The Anatomy of an horse*. London, 1683.

Snow, C.P. *The Two Cultures*. Cambridge: Cambridge University Press, 1993.

Spiller, Elizabeth. *Science, Reading, and Renaissance Culture: The Art of Making Knowledge, 1580–1670*. Cambridge: Cambridge University Press, 2004.

Squire, Larry R., and Stuart M. Zola. "Episodic Memory, Semantic Memory, and Amnesia." *Hippocampus* 8, no. 3 (1998): 205–11.

Stallybrass, Peter, Roger Chartier, John Mowery, and Heather Wolfe. "Hamlet's Tables and the Technologies of Writing in Renaissance England." *Shakespeare Quarterly* 55, no. 4 (2004): 379–419.

Sterne, Laurence. *The Life and Opinions of Tristram Shandy, Gentleman*. Edited by Melvin New and Joan New. New York: Penguin Classics, 2003.

Steven, Leslie. *Hobbes*. London: The Macmillan Company, 1904.

Stevens, Scott Manning. "Sacred Heart and Secular Brain." In *The Body in Parts: Fantasies of Corporeality in Early Modern Europe*, edited by David Hillman and Carla Mazzio, 263–84. New York: Routledge, 1997.

Stevenson, Jay. "The Mechanist-Vitalist Soul of Margaret Cavendish." *Studies in English Literature* 36, no. 3 (1996): 527–43.

Stevenson, Raymond. *The Yard of Wit: Male Creativity and Sexuality, 1650–1750*. Philadelphia, PA: University of Pennsylvania Press, 2004.

Stimson, Dorothy. *Scientists and Amateurs: A History of the Royal Society*. New York: Henry Schuman, 1948.

Sudhoff, Walter. "Die Lehre von den Hirnventrikeln in textlicher und graphischer Tradition des Altertums und Mittelalters." *Archiv für Geschichte der Medizin* Bd. VII, H. 3 (1913): 149–205.

Sugg, Richard. "Donne and the Uses of Anatomy." *Literature Compass* 1 (2003–04): 1–13.

– *John Donne: Critical Issues*. New York: Palgrave, 2007.

– *Murder after Death: Literature and Anatomy in Early Modern England*. Ithaca, NY: Cornell University Press, 2007.

Sullivan, Garret A. Jr., *Memory and Forgetting in English Renaissance Drama. Shakespeare, Marlowe, Webster*. Cambridge: Cambridge University Press, 2005.

Sutton, John. *Philosophy and Memory Traces. Descartes to Connectionism*. Cambridge: Cambridge University Press, 1998.

Symonds, Charles. "The Circle of Willis." *British Medical Journal* (15 January 1955): 119–24.

Targoff, Ramie. *John Donne, Body and Soul*. Chicago, IL: The University of Chicago Press, 2008.

Taylor, Katie. "Pocket-Sized Globes." Explore Whipple Collections, Whipple Museum of the History of Science, University of Cambridge, 2009. http://www.hps.cam.ac.uk/whipple/explore/globes/pocketsizedglobes/ (accessed 11 February 2015).

Thompson, Ann, and John O. Thompson. *Shakespeare: Meaning & Metaphor*. Iowa City, IA: University of Iowa Press, 1987.

Tilden, W.A., James Collins, Daniel Hanbury, A.F. Haselden, F.T. Marzials, John Moss, Richard Reynolds, and Joseph Ince, eds. *The Yearbook of Pharmacy*. Liverpool, England, 1870.

Todd, Rebecca M., and Adam K. Anderson. "The Neurogenetics of Remembering Emotions Past." *Proceedings of the National Academy of Sciences of the United States of America* 106, no. 45 (2009): 18881–2.

Tomey, Matthew, Ricardo Komotar, and J. Mocco. "Herophilus, Erasistratus, Aretaetus, and Galen: Ancient Roots of the Bell-Magendie Law." *Neurosurgical Focus* 23 (2007): E12–14.

Traherne, Thomas. *Thomas Traherne: Centuries, Poems, and Thanksgivings, volume 1: Introduction and Centuries*. Edited by H.M. Margiliouth. Oxford, UK: Oxford University Press, 1958.

– *Thomas Traherne, Poetry and Prose, The Golden Age of Spiritual Writing*, edited by Denise Inge. London: Society for Promoting Christian Knowledge, 2002.

Triplet, Roger. *Writing Tables with a Kalender for xxiii. yeeres, with sundry necessarie rules*. London, 1604.
Tubbs, R. Shane, Marios Loukas, Mohammadli M. Shoja, Mohammad R. Ardalan, and W. Jerry Oakes. "Richard Lower (1631–1691) and His Early Contributions to Cardiology." *International Journal of Cardiology* 128 (2008): 17–21.
van der Krogt, Peter. "Globes Made Portable for the Pocket." *Bulletin of The Scientific Instrument Society* 7 (1985): 8–15.
van Diemerbroek, Ysbrand. *The Anatomy of Human bodies*. London, 1694.
van Helmont, Jan Baptiste. *Oriatrike or, Physick Refined*. London, 1662.
van Spiegel, Adriann, and Giulio Casseri. *De format foetu liber singularis*. Padua, 1626.
Vargha-Khadem, F., D.G. Gadian, K.E. Watkins, A. Connelly, W. Van Paesschen, and M. Mishkin. "Differential Effects of Early Hippocampal Pathology on Episodic and Semantic Memory." *Science* 277, no. 5324 (1997): 376–80.
Vesalius, Andreas. *De humani corporis fabrica*. Basel, 1543.
– *Tabulae anatomicae sex*. Venice, 1538.
Vicary, Thomas. *A Profitable Treatise of the Anatomy of Mans Bodie*. London, 1577.
Vines, Richard. "To The truly Religious and Virtuous, Mr. Joan Purefoy …" In George Abbot, *Brief Notes Upon the whole Book of Psalms*. London, 1651.
Waddington, Raymond. "'All in All': Shakespeare, Milton, Donne, and the Soul-in-Body Topos." *English Literary Renaissance* 20 (1990): 40–68.
Webster, John. *The Duchess of Malfi*. Edited by John Russell Brown. Revels Student Editions. Manchester, UK: Manchester University Press, 1997.
Wells, L.H. "Anatomical Fugitive Sheets with Superimposed Flaps, 1538–1540." *Texts and Documents* (1968): 403–7.
– "The 'Sabio' and 'Sylvester' Families of Anatomical Fugitive Sheets: Note on a Pair of Sheets in the National Library of Medicine." *Bulletin of the History of Medicine* 40, no. 5 (1966): 467–75.
– "A Remarkable Pair of Anatomical Fugitive Sheets …" *Bulletin of the History of Medicine* (1964): 470–6.
Werstine, Paul. "The Textual Mystery of *Hamlet*." *Shakespeare Quarterly* 39, no. 1 (1988): 1–26.
Whitaker, Katie. *Mad Madge: The Extraordinary Life of Margaret Cavendish, Duchess of Newcastle, the First Woman to Live by Her Pen*. New York: Basic Books, 2002.
Willard, Thomas. "Donne's Anatomy Lesson: Vesalian or Paracelsian?" *John Donne Journal* 3 (1984): 35–61.
Williams, George Walton. "Hamlet's Tables." *Hamlet Studies* 1, no. 2 (October 1979): 141.

Williams, William Proctor. "Hamlet's Pockets: Problems with Stage Directions." In *Inside Shakespeare: Essays on the Blackfriars Stage*, edited by Paul Menzer, 192–9. Selinsgrove, PA: Susquehanna University Press, 2006.

Willis, John. *Mnemonica; Or, The Art of Memory, Drained out of the pure Fountains of Art & Nature* ... London, 1661.

– *Mnemonica; sive Ars Reminiscendi: e puris artis naturaeque fontibus hauta* ... London, 1618.

Willis, Thomas. *The Anatomy of the Brain*. Tuckahoe, NY: USV Pharmaceutical Corp., 1971.

– *The Anatomy of the Brain and Nerves*. Edited by William Feindel. Montreal, QC: McGill University Press, 1965.

– *Cerebri anatome*. London, 1664; Amsterdam, 1672.

– *Dr. Willis's practice of physick*. Translated by Samuel Pordage. London, 1684.

– *Five Treatises*. London, 1681.

Woollam, D.H.M. "Concepts of the Brain and Its Functions in Classical Antiquity." In *The History and Philosophy of Knowledge of the Brain and Its Functions*, edited by Frederick Noël Lawrence Poynter, 5–18. Oxford, UK: Blackwell Scientific Publications Ltd., 1958.

Woolnor, H[enry?]. *The true originall of the soule. Proving both by divine and naturall reason, that the production of mans Soule is neither by creation nor propagation, but a certain meane way between both*. London, 1641.

Wynter, Harriet, and Anthony J. Turner. *Scientific Instruments*. London: Charles Scribner's Sons, 1975.

Yates, Frances. *The Art of Memory*. London: Routledge and Kegan Paul, 1966.

Index

Page numbers in italics refer to figures.

Abbas, Haly, 56, 224n63
Alsop, George, 114–15
Anatomia Magistri Nicolai Physici, 100
anatomical illustration, 32–4, 41, 48, 57, 60, 153–82, 199–204, 241n13; of the brain, 155, *170–2*, *174–6*, *178–83*, 182–9, 200–4, 243n28; and choice to represent Adam and Eve, 164–71; and Christian iconography, 170–1, 173–4, 201, 207, 208–9; and the flap anatomy, 32–4; similarity of to map-making, 198–9, 203; and symbolism of the colchium flower, 153–5
anatomy, study and teaching of, 4, 7, 12, 41, 71–7, 99; based on physical and metaphorical correspondences, 96, 105–6, 109, 175–6, 234n35; based on reasoned principles rather than observed facts, 96–7; Cartesian, 210, 213; to discover the nature of God, 73; etymology of terms in, 194; and humoural theory, 13, 79–80, 138–9; legal restrictions on, 159; and *pneuma*, 14, 21, 99–100; and theory of spirits (vegetative, vital, animal), 41–2, 43, 44, 49, 54, 57–8, 80, 96, 99–100, 102–4, 105, 127, 156, 221n20, 222n38, 228–9n22; and theory of "venters," 13–14; using animals, 45, 101, 139, 201, 234n20; using dissection, 41, 45, 46, *47*, 48, 77, 101, *108*, 109, 159, 160, 164, 173, *179*, *180*, *181*, *183*, 184, *197*, 201, 204, 220n17; and ventricular localization theory, 96, 99–105, *101*, 121, 233n16. *See also* brain, anatomy of
anatomy texts, 4, 34, 42, 63, 153–204; and flap anatomy, 32–3, 155, 159–62, 164, 166, 190, 191, 203, 211, 241n13; and fugitive sheets, 32, 159–60, 164, 191; printing and publishing of, 33, 34, 155, 160–2, 164, 166, 172–3, 180–2, 184, 190–1, 194, 202–4, 210, 211–12. *See also* anatomical illustration; Avicenna: *De generatione embryonis*; *Catoptrum microcosmicum*; Crooke, Helkiah: *Mikrokosmographia*; *Exact Survey of the Microcosmus*; flap anatomy; Geminus, Thomas: *Compendiosa totius anatomie delineatio*;

Vesalius, Andreas: *De humani corporis fabrica*; Vesalius, Andreas: *Tabulae anatomicae sex*; Willis, Thomas: *Cerebri anatome*
Anglicus, Bartholomaeus, 42, 43
Arceus, Francis, 71–3, 74–7; *A most excellent and Compendious Method of curing woundes in the head*, 71–2; and the *rete mirabile*, 72
Ariosto, Ludovico, 8, 31; *Orlando Furioso*, 8–9, 31, 32, 216n17
Aristotle, 13, 15, 31, 32, 41, 79, 94, 100, 104, 138, 221n18, 221n20, 233n16
Arnot, Hugo, 116
Augustine, Saint, 212
Austen, Jane, 19
Avicenna, 100, *101*, 220–1n18; *De generatione embryonis*, 100

Bacon, Francis, 112, 145; *The New Atlantis*, 145; *Sylva Sylvarum*, 145
Banister, John, 56, 57, 106
Barker, Francis: *The Tremulous Private Body*, 205
Barsalou, Lawrence, 19
Benjamin, Walter, 169, 173
Berengario, Jacopo, 41, 220n18
Binstock, Benjamin, 209, 248n9
Blankaart, Steven, 201–2
Bliss, Carey S., 191, 194, 196
Blumenberg, Hans, 26–7; and theory of metaphorology, 26–7
Blunt, Edward, 118
Borghi, Anna M., 19
Boswell, James (the younger), 81
Boyle, Robert, 5
brain: damage to, 20, 21, 38, 68, 69, 71–7, 79, 82–5, 86–90, 227–8n6; described/defined, 8–15; male and female differentiated, 129–31; and memory, 4, 21, 69, 70, 73, 77–9, 83–4, 86–7, 100, 102, 103, 104; relationship of to body, 5, 212–13; relationship of to mind, 10, 20, 21, 22, 27, 29, 71, 74, 77, 86, 90, 123–7, 129; relationship of to soul, 4, 5, 6–7, 21–2, 38, 43, 46, 57–63, 65–7, 96, 99–105, *101*, 121, 129, 224n68, 225n77; relationship of to thought, 7–8, 19, 28–30, 38, 77, 96–7, 122, 123–35, 138, 140, 155, 224n68. *See also* brain, anatomy of; brain, metaphors for
brain, anatomy of, 15, 34, 38–48, 49, 56, 57–63, 99–109, *107*, *108*, 174–82, 205–13, 227n3; amygdala, 73; choroid plexus, 43–4, 102, 104, 156, 175; circle of Willis, *62*, 63, 222n38; correspondence of to womb, 93, 95–6, 98, 99, 103–4, 105, *107*, *108*, 109, 136–9, 233n9; cortex, 48, *107*, 167, 184, 188; and dissection, 45, *47*, 48, 77, *108*, 109, 173, *179*, *180*, *181*, *183*, 184, 187–9, 211, 234n20, 248n3, 248n6; *dura mater*, *46*, 109, *171*, 175, *181*, 184, 188; gendering of, 105, 106; hippocampus, 4, 70, 73; and imaging (CAT/PET scan, fMRI, MRI), 34, 74, 162, 204, 210, 211–12, 217n38, 227n3; and lesion study, 21, 30, 73, 84, 227n3; limbic system, 4, 70; and mapping of cognitive function, 6–7, 21, 69, 73–4, 217n38, 227n4; *pia mater*, 175, *181*, 188; pineal gland, 105, 157, 175, 210, 222n37, 224n67; ventricles, 6, 43, 44, 48, 52, 65, 73, 74, 83–4, 85, 95–6, 99–105, 106, 121, 156, 175, 228–9n22, 234n22, 240n5
brain, metaphors for, 4, 5, 7, 8, 22–30; book, 7, 26, 30, 32–3, 77, 153, 155,

162–4, 172–3, 180, 182–9, 204, 205, 208, 211; cut/lesion, 7, 20–1, 26, 30; handmaiden to the soul, 5, 99; house/seat of the soul, 5, 6, 12–13, 96, 99–105, 121, 129, 211; microcosm, 26, 39–40, 57, 67; palace, 4, 6, 10; prison of the soul, 5, *11*, 12, 15, 53, 67; and story of birth of Athena, 110–13, 114; used by Margaret Cavendish, 124–9, 130–1; used in modern neuroscience, 7, 30; web/net/knot, 7, 14–15, 26, 37, 38–48, 49, 51–2, 53, 55–63, *60*, 65–7, 155, 208, 226n85; womb, 3, 7, 26, 28–9, 30, 92–8, 99, 114–18, 119, 122, 132–4, 136–9, 141; writing machine, 7, 29, 122

Bright, Timothy, 80

Brome, Richard: *The Love-Sick Court*, 112–13

Brooks, Douglas, 82

Bryden, D.U., 193, 243n31, 245n49

Buckley, Cali, 160

Burke, Thomas: "The Hands of Mr. Ottermole," 212–13

Butler, Lilly, 200

Carlino, 184

Catoptrum microcosmicum, 32, 33–4, 155, 160, *165*, 166–82, *168*, 190, 191, 193, 198, 199, 201, 202, 203, 241n13, 245n45; brain represented in, 34, 155, 170–3, 174–7, 202; as a flap anatomy, 33–4, 160–2, 203, 241n13; publishing history of, 33–4, 155, 169, 180, 190–1, 196–9, 201, 203–4, 242–3n27, 245n49. See also *Exact Survey of the Microcosmus, or Little World, An*

Cavendish, Charles, 130

Cavendish, Margaret, 16, 20, 28–9, 30, 31, 119–35, 138, 140–52, 155, 196, 208; and "disease of wit," 119–23; as a feminist, 129–30, 238n26; and integrity of mind and brain, 123–7, 129, 131–2, 140, 144; and male and female brains differentiated, 129–31; and metaphor of "paper bodies," 119, 141–2, 236n2; and metaphor of patterning/figuring, 122, 123, 128, 138, 140, 142, 143–4, 238n25; and metaphors for the brain, 124–6, 127–9, 130–1, 132–4, 140, 144; and relationship between act of writing and printing, 140–5; response of to critics, 150–1; and role of reason, 147–50; and subversion of male intellectual conception, 130–1

Cavendish, Margaret, philosophy of: and atomism, 129, 236n13, 237n18; of creation, 151–2; interdependence of with her poetry/fiction, 128–9, 145–50, 237n23; materialist, 123–8, 129, 133–5, 142, 155, 237n20; mechanistic, 122, 127, 237n19, 237–8n23; natural, 122, 126–8, 129, 236n13; and rejection of Cartesian dualism, 29, 124, 140, 143, 155; and vitalism, 30, 129, 237n18, 237–8n23

Cavendish, Margaret, works of: *The Animall Parliament*, 131; *The Blazing World*, 20, 123, 145–52; "The Circle of the Brain cannot be Squar'd," 131–2; *Observations upon Experimental Philosophy*, 121, 145–7, 149, 152; *Philosophical and Physical Opinions*, 124; *Philosophical Letters*, 124, 133, 142; *Philosophicall Fancies*, 126; *Poems and Fancies*, 120, 126, 130–1,

144; "Preface to the Reader" (*The World's Olio*), 130; *The World's Olio*, 121, 124–6, 128–9, 140–1, 144
Cazort, Mimi, 164–6
Chapman, Alexander, 58–9
Chartier, Roger, 80, 81
childbirth, as metaphor for thought, 208; and role of metaphorical midwife, 117–18, 141; and story of birth of Athena, 110–13
Clucas, Stephen, 124
cognition, 14, 73, 84; as aspect of soul, 96, 97; embodied, 19, 29, 212, 213; extended, 7–8, 19, 29; grounded, 7–8, 19, 30, 212; modern theories of, 7–8, 19, 29, 30
conception, as metaphor, 91–8, 99, 116, 138, 141, 232n2; and female body perceived as dangerous creative space, 92; and male writers co-opting female space and experience, 93, 98, 113, 118, 123, 133, 141, 232n2
Crane, Mary, 10, 19
Crooke, Helkiah, 4, 5, 6, 10, 12–13, 16, 26, 47, *47*, 48, 57–8, *60*, 60–1, 63, 67, 73, 74, 83–4, 105–6, *107*, *108*, 109, 173, 184, 190, 206, 224nn67–8, 224n73; *Mikrokosmographia*, 47, 57, 58, *60*, *107*, *108*
Cunning, David, 123

da Carpi, Berengario, 43–4; *Commentary on Mundinus*, 44; and description of the *rete mirabile*, 43
Damasio, Antonio, 16
Daston, Lorraine, 5, 32
Davies, John, 53, 54, 55, 56; *Nosce teipsum*, 53, 55
Dawkins, Richard, 64
de Graaf, Reinier, 201; *Collectanea*, 202; *De mulierum organis*, 201; *De virorum organis generationi inservientibus*, 201; *Tractatus novus de circulatione sanguinis per fibras*, 202
Derrida, Jacques, 82, 85, 149, 189; "Fors," 85
de Saussure, Ferdinand, 24
Descartes, 21, 26, 29, 95, 105, 122, 127, 140, 222n37, 224n67, 225n76; and Cartesian anatomy, 210, 213; and Cartesian dualism, 21, 29, 95, 122, 124, 140; and Cartesian mechanism, 112, 124, 127
Dobson, Jessie, 247n81–2
Donne, John, 13, 15, 16, 20, 35, 38–9, 41, 48–53, 54–6, 59–60, 63, 66–7, 208; "The Ecstasy," 15, 35, 38–9, 48, 49–53, 54–6, 67, 222n45, 223n46, 223n49; *First Anniversary*, 66–7; and "interinanimation," 50–1; "The Progress of the Soul," 59; sermons of, 13, 20, 49; *Songs and Sonnets*, 51
Donoghue, Denis, 23
Dumoulin, I., *182*
Dürer, Albrecht: *Adam and Eve*, *161*, 166, *167*, 170

Earle, John: *Micro-cosmographie*, 118
Eggert, Katherine, 37–8, 232n6; and "disknowledge," 37
Einstein, Albert, 24
Ent, George, 117
Estienne, Charles, *180–1*, *183*, 184–9, 211; anatomical drawings by, 184–9, 244n37; *De dissectione partium corporis*, *180–1*, *183*, 184–9
Evelyn, John, 120, 196
Evelyn, Mary, 120–1, 122, 196

Exact Survey of the Microcosmus, or Little World, An (English edition of Remmelin's *Catoptrum microcosmicum*), 33–4, *154*, 155, 160, 167, *170–2*, 172, 174, 176, *176–7*, 177–80, 182, *186*, 190–1, *192*, 193, 194, *195*, 196–204, *197*, 244n44, 245n45, 245n49; 1695 edition of, 33–4, 176, 177, 180, 190, 191, *195*, 199–204; 1702 edition of, *174–7*, 176, 177, *197*; and "Adam," *154*, *170–2*, *174–5*; alterations and revisions to, 167, 177, 191, 198, 199–204; brain represented in, 200–4; and "Eve," *176–7*, 177. See also *Catoptrum microcosmicum*

Fabrica. See Vesalius, Andreas: *De humani corporis fabrica*
Favila, Marina, 90
flap anatomy, 155, 159–82, 190, 191, 203, 211, 241n13; general absence of brain from, 172–3; as *memento mori*, 164, 171, 173, 198, 203; and vanitas emblems, 173
Fleming, Juliet, 83
Fletcher, Joseph, 36
Fonteyn, Joris, 207, 213
Forrester, John, 60
Foucault, Michel, 24–6, 136; and analogy, 25–6; and doctrine of signatures, 24–5, 136; *The Order of Things*, 24
Franklin, K.J., 202

Galen, 13, 14–15, 20–1, 41–6, 48, 53–60, *60*, 63, 74, 77, 84, 99–100, 104, 127, 156, 174, 175, 221n18, 221n20, 223–4n61, 224n66; and animal experimentation, 74, 228n20; and model of anatomy, 21, 54–5, 100, 228n22; and the *rete mirabile*, 42–4, 46, 48, 60, 63; and role of the brain's ventricles, 74, 84, 99; and theory of spirits (vegetative, vital, animal), 41–4, 49, 54, 57–8, 80, 96, 99–100, 102–4, 105, 127, 156, 221n20, 222n38, 228–9n22; and treatment of wounded gladiators, 20–1, 74, 77, 84
Galileo, 25, 30–2, 53, 218n41, 218n59; *The Assayer* (*Il Saggiatore*), 30–2; and "book of nature" metaphor, 30–2
Geminus, Thomas, 4–5, 47, 56, 57, 102–3, 155, *163*, 166, 173, 222n36; *Compendiosa totius anatomie delineatio*, 47, *163*, 166, 221–2n35; and "strangeness" of the brain, 4–5, 155; and ventricular localization, 102–3
Gibbs, R.W., 19
Glenberg, Arthur M., 19
Goldberg, Benjamin, 94, 96, 232–3n9
Goldberg, Jonathan, 83
Greenblatt, Stephen, 80
Greene, Brian, 64
Guevara, Perry, 80, 84

Hamlet, 21, 68–71, 77–90, 140, 208, 211; concern with memory and brain damage in, 69–71, 77–90, 208, 228n8, 229n27, 230n44, 230n48; and table-books, 79, 81–2, 230n43; variations in editions of, 77–9, 229n27
Harrington, John, 9
Harvey, William, 16, 28, 93–8, 99, 103, 112, 113, 114, 117, 123, 133–5, 139, 140, 202, 208, 222n38, 232n6, 232–3n9; on conception and the principle of generation, 28, 93–8, 99, 103, 114, 133, 208, 232–3n9; "De conceptione," 28, 93, 99, 103, 133, 233n9; *Exercitationes de generatione*

animalium (*De generatione*, or *Anatomical exercitations, concerning the generation of living creatures*), 28, 93, 117, 139, 232n6; and "plastick generative power," 93–8, 112, 113, 135
Havers, Clopton, 180, *195*, 199–203, 247n78; and Haversian canals, 200; *Osteologia Nova, or some New Observations of the Bones*, 200, 201; and research on human skeleton, 200
Hayles, N. Katherine, 24
heart, 5, 10, 13, 73, 78–9; in Christian iconography, 10; dissection of, *197*; as possible locus of memory, 78–9; as possible locus of soul, 6, 10, 233n16; and theory of spirits (vegetative, vital, animal), 41, 44, 80; and theory of "venters," 13–14
Herbert, George, 54
Herophilus, 174, 221n18
Heseler, Baldasar, 44
Hesiod, 111
Heywood, Thomas: *A Woman Killed with Kindness*, 78
Hippocrates, 13, 53, 135
Hobbes, Thomas, 26, 132, 219n5
Homer, 194; *The Iliad*, 31, 32
Hooke, Robert, 194, 196
human body, metaphors for: book, 153, 162, 164; microcosm, 203
humoural theory, 13, 79–80, 138–9; and the brain, 79–80

IJpma, Frank, 210–11, 248n3
imagination, 14, 31–2, 45, 90, 91, 92, 123, 129, 143–4, 152, 211; as faculty of the soul, 6, 93, 96, 97, 100, 121; as generative power, 97; located in the brain, 100, 121; and memory, 68, 80, 85, 87; role of in conception, 93, 99. *See also* imagination, maternal
imagination, maternal, 28–9, 92, 123, 135–40, 145, 238n31, 239n32, 239n40; and fear of woman's power, 139–40; and monstrosity, 28, 29, 92, 123, 135, 140, 145, 238n31, 239n32, 239n40
Ireton, John, *154*, 155, *170–2*, *192*, 198, 199, 203

Jagger, Graham, 192–3, 246n72, 246–7n73
James, Susan, 128
Johns, Adrian, 192
Johnson, Jenell, 64–5; *The Neuroscientific Turn*, 65
Johnson, Laurie, 19
Johnson, Mark: *Metaphors We Live By*, 23

King Lear, 91, 210; and metaphor of conception, 91
Kuhn, Thomas, 63

Lakoff, George: *Metaphors We Live By*, 23
Latour, Bruno, 16
Laurentius, 79–80
Le Grand, Antoine, 112
Leigh, Thomas, *186*, 190, 245n45, 245n49, 247n78
Lewalski, Barbara, 54
Lewis, Rhodri, 80, 230n44
Lilley, Kate, 121
Lind, R.L., 41
Lindeboom, G.A., 202
literature, relationship of to scientific study, 5–6, 7, 16, 17, 18–20, 39,

64–5; in early modern period, 16, 17; fiction, outside the true/false opposition, 31–2, 148–9
Littlefield, Melissa, 64–5; *The Neuroscientific Turn*, 65
liver: as possible locus of soul, 6; and theory of "venters," 13–14
Lower, Richard, *197*, 201, 202; *Tractatus de corde*, *197*, 202

Macbeth, 21, 71, 86–90, 231n61; and the damaged brain, 21, 88–9; and memory, 89; and prion disease, 87, 231n61
Magnus, Albertus, 100
Maier, Michael, 110, *110*, 111; *Atalanta fugiens*, *110*
Mantegna: *Cristo in scurto*, 207
Marvell, Andrew, 12, 53, 54, 60, 67; "A Dialogue between the Soul and the Body," 12, 53
Massey, Lyle, 153
medical science, 12, 14–15, 20, 21, 35, 40, 49, 63, 70, 71–7, 160, 191, 203
melancholy, 79–80, 84; effect of on memory, 79–80
memory, 21, 42, 68–9, 70, 73, 77–85, 91, 97, 100, 104; and crypt metaphor, 85; and damage/violence to brain, 21, 38, 69, 73, 82, 84, 86–7; and effacement/erasure, 21, 68–9, 70, 77, 79, 83, 84, 85, 90, 230n44; effect of melancholy on, 79–80; as faculty of the soul, 6, 12, 14, 96, 97, 100; forgetting and remembering as violent acts, 83, 85, 86, 87–9; in *Hamlet*, 21, 69–71, 77–86, 88; and imagination, 68, 80, 85, 87; impairment of, 38, 69, 70, 73, 82; located in the brain, 4, 21, 69, 70, 73, 77, 78, 83–4, 86, 87, 100, 102, 103, 104; located in the heart, 78–9, 90; in *Macbeth*, 86–90; wax metaphor for, 79, 80–1; and writing metaphor, 21, 77, 79, 81, 82, 83, 86, 87
Menenius, 86
mesmerism, 65
metaphor, 4, 7–10, 12–13, 14–15, 20, 22–31, 69, 213; dangers of, 220n13; faded, 30; as philosophical tool, 26; as rhetorical device, 7, 26–7. *See also* brain, metaphors for; soul, metaphors for
Metcalfe, Janet, 19
Michaelian, Kourken, 143
Michelspacher, *192*, *195*, 242n27
Middelkoop, Norbert, 210–11, 248n3
midwifery, 136–40, 235n49
Midwinter, Daniel, *186*, 190, 245n45, 245n49, 247n78
mind, 7, 15, 38, 41, 86, 123–9, 143–4, 155, 208, 211; and maternal imagination, 135–40; relationship of to body, 8, 29, 42, 71, 140, 213 (*see also* Descartes: Cartesian dualism); relationship of to brain, 10, 20, 21, 22, 27, 29, 71, 74, 77, 86, 90, 123–6, 129; restraining role of, 213; and story of birth of Athena, 110–13
Mitchell, Charles, 48
Mitchell, Dolores, 205
Molaison, Henry Gustav (Patient H.M.), 21, 73, 228n18
monstrosity, 123, 135, 145, 239n32; caused by maternal imagination, 28–9, 92, 123, 135–40, 238n31, 239n32, 239n40
Montaigne, Michel de, 135
More, Henry, 36–7, 193, 219n5, 222n37
Mowery, John, 80, 81

Moxon, James, 180, 189, 190, 191, *195*, 199, 201, 203, 245n45, 245n49, 246n62, 247n78
Moxon, James Sr (father to Joseph), 191
Moxon, Joseph, 33, 155, 167, 172, 182, *185*, 189–99, 201, 203, 246n62–3, 246n72–3, 247n78; as Hydrographer to the King, 191–3; as maker of maps and globes, 190, 191–2, 193–4; *Mathematicks Made Easie...*, *185*, 189, 191; *Mechanick Exercises*, 194–6; publishing house of, 33, 34, 155, 180–2, *186*, 189–99, 203; *Tutor to Astronomy and Geography*, 194
Mundinus, 41, 104, 220n18
Munk, William, 201
Murison, Justine: "'The Paradise of Non-Experts,'" 65

Nemesius, 99–100, 233n14–15; on role of ventricles, 99–100
net/knot metaphor for brain/soul, 3, 7, 14, 15, 26, 37, 38–48, 49, 51–2, 53, 55–63, *60*, 65–7, 155, 208, 226n85; as emblem of imprisonment, 67; and surrounding metaphors of knitting/chaining, 53, 66, 67. See also *rete mirabile*
neuroanatomy. *See* brain, anatomy of
neuroscience, 15–16, 21–2, 29, 38–40, 57–8, 64–6, 69, 77, 99; birth and development of, 3–5, 17, 20, 21–2, 34, 38, 207; connection of to literary studies, 16, 18–20, 38–40, 64–5, 77, 216–17n34, 217n36, 226n80; interdisciplinary nature of, 15–16, 20, 38, 57–8; modern, and metaphors for the brain, 7; and neuroforensics, 20–1, 71; and the "neuroscientific turn," 64–5. *See also* brain; brain, anatomy of
Newton, Isaac, 65, 193

O'Neill, Eileen: and "sympathetic affinity," 142
Osborne, Dorothy, 120

Palmer, Roger: *Essay on the Use of the Celestial and Terrestrial Globe*, 193
Paré, Ambroise, 135, 137–8, 140, 238n31, 239n32; and printing/imprinting in maternal imagination, 138, 140, 238n31, 239n32
Park, Katharine, 5
Paster, Gail Kern, 19
Patient H.M. *See* Molaison, Henry Gustav
Paul, Jean, 30
penis cerebri. *See* brain, anatomy of: pineal gland
Pepys, Samuel, 120, 167, 193, 200; and commentary on "The Humorous Lovers," 120
Philipot, Thomas, 112–13
philosophy, 20, 26, 31, 97, 112; materialist, 123–9, 133–5; mechanistic, 122, 127, 190; natural, 5, 6, 12, 20, 26, 40, 122, 126–8, 129, 237n21
Pinker, Steven, 16, 64, 216–17n34
Plato, 13, 15, 31, 41, 53, 82, 100
poetry, 66–7, 151, 194; and birth of knowledge, 112, 113, 114, 144, 145; of John Donne, 48–56; on the soul, 36, 38–40
Popham, Edward, 35–6, 38, 219n2; *A Looking-glasse for the Soule*, 35–6, 219n2
Preston, Claire, 27, 218n41

Proust, Marcel, 18, 217n36; and memory, 18; and modern neuroscience, 18, 217n36, 226n80; *Swann's Way*, 18
Puttenham, George, 5–6, 22–3, 39, 194, 220n13; *The Arte of English Poesie*, 22–3

Rabin, C., 184
Ramachandran, V.S., 16, 227–8n6
Read, Alexander, 72, 75
Rembrandt, 205–13, *206–7*, 248n3, 248n7; *Anatomy Lesson of Dr. Deijman*, 205–13, *206–7*, 248n3, 248n12; *The Anatomy Lesson of Dr. Tulp*, 205, 208, 209
Remmelin, Johann, 32, *154*, 155, 159, 160, 162, 164, *165*, *168*, 169–82, *170–2*, *174–7*, *186*, 188, 190, 191, *192*, *195*, *197*, 198, 199, 202, 203, 208, 210, 211, 213, 240n4, 242–3n27, 245n49; *Catoptrum microcosmicum*, 32, 33–4, 155, 160–1, *165*, 166–82, *168*, 190–1, 193, 196–9, 201, 202, 203–4, 241n13, 242n27, 245n45, 245n49. See also *Catoptrum microcosmicum*; *Exact survey of the Microcosmus, or Little World, An*
rete mirabile, 14–15, 29–30, 35–67, *60*, *61*, 72, 102, 127, 156, 210, 223–4n61, 224n66, 225n74, 225n76, 226n85; arguments against existence of, 43–4, 49, 60, 222n38, 225n76; and choroid plexus, 43–4; and circle of Willis, *62*, 63, 222n38; as container for the soul, 14–15, 37–48, 58, 62; as net/knot, 14, 29–30, 37–48, 49, 51–2, 55–6, 57–9
retiform plexus. See *rete mirabile*
Richardson, Alan, 9, 19
Roberts, K.B., 45
Rocca, Julius, 42, 43
Rousseau, George, 42, 63, 65, 225–6n77
Royal Society, the, 20, 34, 180–2, 194, 196, 200, 201, 202, 236n13, 246n63, 246–7n73
Rüff, Jakob, *107*, *108*, 109, 136; *The Expert Midwife*, *107*, *108*
Russell, Kenneth, 201

Sarsi, 31, 32
Sawday, Jonathan, 67, 206, 208, 210, 226n85
Schaffer, Simon, 18
Schmidt, Suzanne Karr, 160
science, relationship of to literature, 5–6, 7, 16, 17, 18–20, 39, 64–5; in early modern period, 5–6, 16, 17
Shakespeare, William, 16, 17, 21, 91, 231n61, 232n2; *Hamlet*, 21, 68–71, 77–85, 86–90, 140, 208, 211, 228n8, 230n48; *King Lear*, 91, 210; *Macbeth*, 21, 71, 86–90, 231n61; tragedies, concern with brain in, 86, 89, 90. See also *Hamlet*; *Macbeth*
Shapin, Steven, 18
Sharpe, Jane, 104–5, 138–9; and anatomy of the womb, 104–5; and maternal imagination, 138–9
Sidney, Philip, 31, 114, 116, 148, 151; *Astrophil and Stella*, 114; *Defense of Poetry*, 148; and philosophy's dependence on poetry, 148
Siegfried, Brandie, 122
Singer, C., 184
Smith, Samuel, 201, 247n78
Snow, C.P., 16, 17–18, 64, 226n79; and the "two cultures" problem, 16, 17–19, 64, 226n79; and "third

culture," 16–18, 19, 64, 226n79; and "fourth culture," 18–19
soul: compared to the Holy Trinity, 36; described, 10–12, 35–48; faculties of, 6–7, 12, 13, 14, 36, 37, 38, 93, 96, 97, 100, 102–3, 121, 124, 132, 147–9; as immaterial substance, 5, 6, 14–15, 23, 25, 36–8, 46, 53–5, 66, 95, 224n68; locatability of, 4, 7, 8, 14, 37–48, 59, 66, 70, 163–4, 210; and the *pneuma psychikon*, 99–100, 102; relationship of to body, 6–7, 12, 14–15, 21, 35–63, 65–7, 70, 132, 164, 204, 212, 219n6, 219–20n10, 222n37, 222–3n45; relationship of to brain, 4, 5, 6–7, 14–15, 21–2, 23–4, 36–8, 43, 46, 54–5, 57–63, 65–7, 95, 96, 99–105, *101*, 121, 129, 224n68, 225–6n77; restraining role of, 213; and the *rete mirabile*, 14–15, 35–67; and theory of spirits (vegetative, vital, animal), 6, 13–15, 41–2, 49, 55, 59, 80, 96, 99, 102–3, 105, 127, 228–9n22; and ventricular localization theory, 96, 99–105
soul, located in the brain (neurocentric), 6, 102, 121, 129; and ventricular localization theory, 96, 99–105, 233n16, 233–4n19, 240n5
soul, metaphors for, 7, 12–13, 23, 55; net/knot, 14–15, 37–48, 49, 51–2, 53, 55–63, *60*, 65–7, 226n85; palace, 6, 10
Spiller, Elizabeth, 5
spirits, theory of, 6, 13–15, 41–2, 43, 44, 49, 54, 55, 57–9, 80, 96, 99–100, 102, 103, 104, 105, 127, 156; animal, 41–2, 44, 57, 58, 59, 80, 96, 99–100, 102, 103, 104, 105, 127, 156; and *pneuma*, 14, 21, 99–100, 102; vegetative, 41, 104, 127; vital, 41, 42, 44, 57, 80, 104, 127, 156
Stallybrass, Peter, 80, 81
Steen, Francis F., 9
Stevens, Scott Manning, 10
Sugg, Richard, 39, 49, 55, 57, 221n19, 222n37
Sullivan, Garrett, 83
Sutton, John, 19, 80, 230n44
Sylvius, Franciscus, 201

table-books, 81–2, 230n43
Targoff, Ramie, 50–1
Tauvry, Daniel, 190, 245n46; *A New Rational Anatomy*, 190
Thompson, Ann, 78
Thompson, John, 78
Tomlinson, J.D.W., 45
Traherne, Thomas, 53–4; "Thanksgivings for the Soul," 54
Tribble, Evelyn, 19
"two cultures" problem, 16, 17–19, 64, 226n79; and "third culture," 16–18, 19, 64, 226n79; and "fourth culture," 18–19

Udall, Nicholas, 102, *163*

Valverde, Juan, *178*, 184; *Historia de la composicion del cuerpo humano*, *178*
van Gulik, Thomas, 210–11, 248n3, 248n7, 248n12
Van Helmont, John Baptista, 142
ventricular localization theory, 96, 99–105, *101*, 121
Vesalius, Andreas, 15, 21, 33, 40–1, 43, 44–8, *46–7*, 56, 57, 58, 60, *61*, 66, 101–2, 103, 104, 155–6, *156*, *162–3*, 166, 167, 173, 184, 187, 188, 190, 191,

194, 201, 206, 208, 210, 211, 221n18, 234nn22–3, 240nn4–5, 243n28; "Adam and Eve," *162–3*, 166, 173; and the choroid plexus, 102, 104, 156; *De humani corporis fabrica (On the Fabric of the Human Body)*, 15, 33, 40–1, 44, 45, 46, *46*, 47–8, 60, 101–2, 155–6, *156*, 157, 159, *162*, 166, 173, 187, 222n35, 224n73; and rejection of *rete mirabile*, 60, 156, 224n66, 225n76; *Tabulae anatomicae sex*, 44, 46, 60–1, *61*, 156, 224n73
Vines, Richard, 116–17
Virgil, 31
Vogtherr, Heinrich, *158*, 160, 191; *Anathomia oder abconterfettung eines Weibs leib*, *158*; "Eve" (flap anatomy), *158*, 160
von Calcar, Stephen, 166
von Gerstorff, Hans, *179*, 184; *Feldbuch der Wundartzney*, *179*

Webster, John: *The Duchess of Malfi*, 173
Whitaker, Katie, 120
Williams, George, 78
Williams, William Proctor, 83
Williamson, Joseph, 196
Willis, John, 68, 80, 85; *Mnemonica; Or, The Art of Memory*, 68
Willis, Thomas, 32, 34, 38, 63, *76*, 76–7, 113, *174–7*, 176–7, 190, 194, 200, 202, 226n78; *Cerebri anatome*, 32, 34, 38, 63, *175–7*, 176, 177, 194, 202; and circle of Willis, *62*, 63, 222n38, 226n78; *De anima brutorum*, *174*, 176; *Five Treatises*, *76*; on intellectual conception, 113; *Receipts for the Cure of All Distempers*, 190
Witt, Jessica K., 19
Wolfe, Heather, 80, 81
womb: and anatomical connection with brain, 93, 95–6, 99, 105, *107–8*, 109; and linguistic/etymological connection with brain, 98, 103–4; and metaphorical connection with brain, 3, 7, 26, 28–9, 30, 92, 93–8, 99, 103–4, 109–13, 114–18, 119, 133, 136–9, 141; and the process of writing, 92, 95, 141
womb, anatomy of, 104–5, *107–8*, 109, 157, 234n30; representations of, 93, 109, 157, 184, 189
Woollam, D.H.M., 174, 234n20
Woolnor, Henry, 36
Wren, Christopher, 34, 38, *62*, 194, 196, 202; *Cerebri anatome*, 34, 38, *62*, 194